Exploratory Data Analysis with MATLAB®

Third Edition

Chapman & Hall/CRC
Computer Science and Data Analysis Series

The interface between the computer and statistical sciences is increasing, as each discipline seeks to harness the power and resources of the other. This series aims to foster the integration between the computer sciences and statistical, numerical, and probabilistic methods by publishing a broad range of reference works, textbooks, and handbooks.

SERIES EDITORS
David Blei, Princeton University
David Madigan, Rutgers University
Marina Meila, University of Washington
Fionn Murtagh, Royal Holloway, University of London

Proposals for the series should be sent directly to one of the series editors above, or submitted to:

Chapman & Hall/CRC
Taylor and Francis Group
3 Park Square, Milton Park
Abingdon, OX14 4RN, UK

Published Titles

Semisupervised Learning for Computational Linguistics
Steven Abney

Visualization and Verbalization of Data
Jörg Blasius and Michael Greenacre

Design and Modeling for Computer Experiments
Kai-Tai Fang, Runze Li, and Agus Sudjianto

Microarray Image Analysis: An Algorithmic Approach
Karl Fraser, Zidong Wang, and Xiaohui Liu

R Programming for Bioinformatics
Robert Gentleman

Exploratory Multivariate Analysis by Example Using R
François Husson, Sébastien Lê, and Jérôme Pagès

Bayesian Artificial Intelligence, Second Edition
Kevin B. Korb and Ann E. Nicholson

Published Titles cont.

Computational Statistics Handbook with MATLAB®, Third Edition
Wendy L. Martinez and Angel R. Martinez

Exploratory Data Analysis with MATLAB®, Third Edition
Wendy L. Martinez, Angel R. Martinez, and Jeffrey L. Solka

Statistics in MATLAB®: A Primer
Wendy L. Martinez and MoonJung Cho

Clustering for Data Mining: A Data Recovery Approach, Second Edition
Boris Mirkin

Introduction to Machine Learning and Bioinformatics
Sushmita Mitra, Sujay Datta, Theodore Perkins, and George Michailidis

Introduction to Data Technologies
Paul Murrell

R Graphics
Paul Murrell

Correspondence Analysis and Data Coding with Java and R
Fionn Murtagh

Pattern Recognition Algorithms for Data Mining
Sankar K. Pal and Pabitra Mitra

Statistical Computing with R
Maria L. Rizzo

Statistical Learning and Data Science
Mireille Gettler Summa, Léon Bottou, Bernard Goldfarb, Fionn Murtagh, Catherine Pardoux, and Myriam Touati

Music Data Analysis: Foundations and Applications
Claus Weihs, Dietmar Jannach, Igor Vatolkin, and Günter Rudolph

Foundations of Statistical Algorithms: With References to R Packages
Claus Weihs, Olaf Mersmann, and Uwe Ligges

Chapman & Hall/CRC
Computer Science and Data Analysis Series

Exploratory Data Analysis with MATLAB®

Third Edition

Wendy L. Martinez
Angel R. Martinez
Jeffrey L. Solka

CRC Press
Taylor & Francis Group
Boca Raton London New York

CRC Press is an imprint of the
Taylor & Francis Group, an **informa** business
A CHAPMAN & HALL BOOK

CRC Press
Taylor & Francis Group
6000 Broken Sound Parkway NW, Suite 300
Boca Raton, FL 33487-2742

First issued in paperback 2021

© 2017 by Taylor & Francis Group, LLC
CRC Press is an imprint of Taylor & Francis Group, an Informa business

No claim to original U.S. Government works

ISBN-13: 978-1-4987-7606-6 (hbk)
ISBN-13: 978-1-03-217905-6 (pbk)
DOI: 10.1201/9781315366968

Publisher's Note

The publisher has gone to great lengths to ensure the quality of this reprint but points out that some imperfections in the original copies may be apparent.

Visit the Taylor & Francis Web site at
http://www.taylorandfrancis.com

and the CRC Press Web site at
http://www.crcpress.com

Angel and Wendy dedicate this book to their children:

Deborah,
Jeffrey,
Robbi (the middle child),
and Lisa (Principessa)

Jeffrey dedicates this book to his wife Beth,
and sons Stephen and Rob

Table of Contents

Part I
Introduction to Exploratory Data Analysis

Chapter 1
Introduction to Exploratory Data Analysis

Part II
EDA as Pattern Discovery

Chapter 2
Dimensionality Reduction — Linear Methods

Chapter 3
Dimensionality Reduction — Nonlinear Methods

Chapter 4
Data Tours

Chapter 5
Finding Clusters

Chapter 6
Model-Based Clustering

Chapter 7
Smoothing Scatterplots

Part III
Graphical Methods for EDA

Chapter 8
Visualizing Clusters

Chapter 9
Distribution Shapes

Chapter 10
Multivariate Visualization

Chapter 11
Visualizing Categorical Data

Appendix A
Proximity Measures

Appendix B
Software Resources for EDA

Appendix C
Description of Data Sets

Appendix D
MATLAB® Basics

Preface to the Third Edition

As usual, we did not have enough time to describe all of the techniques we wanted to include in our previous two editions. Furthermore, statisticians and data scientists continue to advance the state-of-the-art in visualization and exploratory data analysis. So, we are grateful to have this chance to update this book with recent advancements and missing topics.

We list below some of the major changes and additions in the third edition.

- Chapter 2 has new content on random projections and estimating the local intrinsic dimensionality.
- Chapter 3 includes a description of deep learning, autoencoders, and stochastic neighbor embedding (t–SNE).
- Chapter 5 contains a clustering approach that has been around for many years and is based on the minimum spanning tree. It also has a discussion of several cluster validity indices that have been added to the MATLAB® Statistics and Machine Learning Toolbox.
- Chapter 9 now has a section on kernel density estimation, which is useful for understanding how are data distributed. We also added violin plots, beanplots, and new variants of boxplots.
- Chapter 11 on visualizing categorical data is a new addition to the book. We think this is an important area, which is often neglected in exploratory data analysis. This chapter includes methods to visualize the distribution shapes of univariate categorical data and tabular data.

The makers of MATLAB have extended the capabilities of their product, especially in the area of statistics and data analysis. In fact, their relevant toolbox is now called the MATLAB Statistics and Machine Learning Toolbox. For ease of exposition, we will continue to use the previous name of the Statistics Toolbox in this text. However, readers should understand that we are referring to the latest MATLAB toolbox.

The MATLAB code for the examples, the EDA Toolbox, and the data sets used in the book are available for download. They can be downloaded from the CRC Press website for this book found at

`https://www.crcpress.com/9781498776066`

As we mentioned previously, there seems never to be enough time to accomplish all we wanted to do when writing this book. In fact, we had hoped to write an R Shiny app implementing many of the methods we describe. This is our next task after this book gets published. So, please keep an eye out for the R Shiny app; we will post it to the CRC Press website for this book. Our hope is that this will make the techniques in the book available to a wider audience. For more on the R statistical computing environment, see

https://cran.r-project.org/

For MATLAB product information, please contact:

The MathWorks, Inc.
3 Apple Hill Drive
Natick, MA, 01760-2098 USA
Tel: 508-647-7000
Fax: 508-647-7001
E-mail: info@mathworks.com
Web: www.mathworks.com

We would like to acknowledge the invaluable help of those researchers who wrote MATLAB code for methods described in this book and made it available for free of charge. We are grateful to MoonJung Cho and Eungchun Cho for reviewing portions of the book. We greatly appreciate the help and patience of those at CRC press, especially David Grubbs. As always, we are indebted to The MathWorks, Inc. for their special assistance with MATLAB.

Disclaimers

1. Any MATLAB programs and data sets that are included with the book are provided in good faith. The authors, publishers, or distributors do not guarantee their accuracy and are not responsible for the consequences of their use.
2. Some of the MATLAB functions provided with the EDA Toolbox were written by other researchers, and they retain the copyright. References are given in Appendix B and in the **help** section of each function. Unless otherwise specified, the EDA Toolbox is provided under the GNU license specifications:

   http://www.gnu.org/copyleft/gpl.html

3. The views expressed in this book are those of the authors and do not represent the views of the U.S. Government or its components.

Wendy L. Martinez, Angel R. Martinez, and Jeffrey L. Solka
February 2017

Preface to the Second Edition

In the past several years, many advancements have been made in the area of exploratory data analysis, and it soon became apparent that it was time to update this text. In particular, many innovative approaches have been developed for dimensionality reduction, clustering, and visualization.

We list below some of the major changes and additions in the second edition.

- We added significant content to the chapter on linear dimensionality reduction. The new methods we discuss are nonnegative matrix factorization and linear discriminant analysis. We also expanded the set of methods that are available for estimating the intrinsic dimensionality of a data set.

- Curvilinear component analysis is a nonlinear dimensionality reduction method that is now described in Chapter 3. Curvilinear component analysis was developed as an improvement to self-organizing maps.

- A description of independent component analysis has been added to the chapter on data tours.

- Several new clustering methods are now included in the text. These include nonnegative matrix factorization, probabilistic latent semantic analysis, and spectral-based clustering.

- We included a discussion of smoothing splines, along with a fast spline method that works with uniformly spaced data.

- Several visualization methods have been added to the text. These are a rangefinder boxplot for bivariate data, scatterplots with marginal histograms, biplots, and a new method called Andrews' images.

In a spirit similar to the first edition, this text is *not* focused on the theoretical aspects of the methods. Rather, the main focus of this book is on the *use* of the EDA methods. So, we do not dwell so much on implementation and algorithmic details. Instead, we show students and practitioners how the methods can be used for exploratory data analysis by providing examples and applications.

The MATLAB® code for the examples, the toolboxes, the data sets, and color versions of all figures are available for download. They can be downloaded from the Carnegie Mellon StatLib site found here:

http://lib.stat.cmu.edu

or from the book's website:

http://pi-sigma.info

Please review the **readme** file for installation instructions and information on any changes.

For MATLAB product information, please contact:

> The MathWorks, Inc.
> 3 Apple Hill Drive
> Natick, MA, 01760-2098 USA
> Tel: 508-647-7000
> Fax: 508-647-7001
> E-mail: info@mathworks.com
> Web: www.mathworks.com

We would like to acknowledge the invaluable help of those researchers who wrote MATLAB code for methods described in this book and also made it available for free. In particular, the authors would like to thank Michael Berry for helpful discussions regarding nonnegative matrix factorization and Ata Kaban for allowing us to use her PLSI code. We are also very grateful to Mia Hubert and Sabine Verboven for granting us permission to use their bagplot function and for their patience with our emails.

We thank the editors of the book series in Computer Science and Data Analysis for including this text. We greatly appreciate the help and patience of those at CRC press: David Grubbs, Bob Stern, and Michele Dimont. As always, we are indebted to Naomi Fernandes and Tom Lane at The MathWorks, Inc. for their special assistance with MATLAB.

Disclaimers

1. Any MATLAB programs and data sets that are included with the book are provided in good faith. The authors, publishers, or distributors do not guarantee their accuracy and are not responsible for the consequences of their use.

2. Some of the MATLAB functions provided with the EDA Toolboxes were written by other researchers, and they retain the copyright. References are given in Appendix B and in the **help** section of

each function. Unless otherwise specified, the EDA Toolboxes are provided under the GNU license specifications:

http://www.gnu.org/copyleft/gpl.html

3. The views expressed in this book are those of the authors and do not necessarily represent the views of the United States Department of Defense or its components.

Wendy L. Martinez, Angel R. Martinez, and Jeffrey L. Solka
October 2010

Preface to the First Edition

One of the goals of our first book, *Computational Statistics Handbook with MATLAB®* [2002], was to show some of the key concepts and methods of computational statistics and how they can be implemented in MATLAB.[1] A core component of computational statistics is the discipline known as exploratory data analysis or EDA. Thus, we see this book as a complement to the first one with similar goals: *to make exploratory data analysis techniques available to a wide range of users.*

Exploratory data analysis is an area of statistics and data analysis, where the idea is to first explore the data set, often using methods from descriptive statistics, scientific visualization, data tours, dimensionality reduction, and others. This exploration is done without any (hopefully!) pre-conceived notions or hypotheses. Indeed, the idea is to use the results of the exploration to guide and to develop the subsequent hypothesis tests, models, etc. It is closely related to the field of data mining, and many of the EDA tools discussed in this book are part of the toolkit for knowledge discovery and data mining.

This book is intended for a wide audience that includes scientists, statisticians, data miners, engineers, computer scientists, biostatisticians, social scientists, and any other discipline that must deal with the analysis of raw data. We also hope this book can be useful in a classroom setting at the senior undergraduate or graduate level. Exercises are included with each chapter, making it suitable as a textbook or supplemental text for a course in exploratory data analysis, data mining, computational statistics, machine learning, and others. Readers are encouraged to look over the exercises because new concepts are sometimes introduced in them. Exercises are computational and exploratory in nature, so there is often no unique answer!

As for the background required for this book, we assume that the reader has an understanding of basic linear algebra. For example, one should have a familiarity with the notation of linear algebra, array multiplication, a matrix inverse, determinants, an array transpose, etc. We also assume that the reader has had introductory probability and statistics courses. Here one should know about random variables, probability distributions and density functions, basic descriptive measures, regression, etc.

In a spirit similar to the first book, this text is *not* focused on the theoretical aspects of the methods. Rather, the main focus of this book is on the *use* of the

---

[1] MATLAB® and Handle Graphics® are registered trademarks of The MathWorks, Inc.

EDA methods. Implementation of the methods is secondary, but where feasible, we show students and practitioners the implementation through algorithms, procedures, and MATLAB code. Many of the methods are complicated, and the details of the MATLAB implementation are not important. In these instances, we show how to use the functions and techniques. The interested reader (or programmer) can consult the M-files for more information. Thus, readers who prefer to use some other programming language should be able to implement the algorithms on their own.

While we do not delve into the theory, we would like to emphasize that the methods described in the book have a theoretical basis. Therefore, at the end of each chapter, we provide additional references and resources; so those readers who would like to know more about the underlying theory will know where to find the information.

MATLAB code in the form of an Exploratory Data Analysis Toolbox is provided with the text. This includes the functions, GUIs, and data sets that are described in the book. This is available for download at

http://lib.stat.cmu.edu

Please review the **readme** file for installation instructions and information on any changes. M-files that contain the MATLAB commands for the exercises are also available for download.

We also make the disclaimer that our MATLAB code is not necessarily the most efficient way to accomplish the task. In many cases, we sacrificed efficiency for clarity. Please refer to the example M-files for alternative MATLAB code, courtesy of Tom Lane of The MathWorks, Inc.

We describe the EDA Toolbox in greater detail in Appendix B. We also provide website information for other tools that are available for download (at no cost). Some of these toolboxes and functions are used in the book and others are provided for informational purposes. Where possible and appropriate, we include some of this free MATLAB code with the EDA Toolbox to make it easier for the reader to follow along with the examples and exercises.

We assume that the reader has the Statistics Toolbox (Version 4 or higher) from The MathWorks, Inc. Where appropriate, we specify whether the function we are using is in the main MATLAB software package, Statistics Toolbox, or the EDA Toolbox. The development of the EDA Toolbox was mostly accomplished with MATLAB Version 6.5 (Statistics Toolbox, Version 4); so the code should work if this is what you have. However, a new release of MATLAB and the Statistics Toolbox was introduced in the middle of writing this book; so we also incorporate information about new functionality provided in these versions.

We would like to acknowledge the invaluable help of the reviewers: Chris Fraley, David Johannsen, Catherine Loader, Tom Lane, David Marchette, and Jeffrey Solka. Their many helpful comments and suggestions resulted in a better book. Any shortcomings are the sole responsibility of the authors. We owe a special thanks to Jeffrey Solka for programming assistance with finite mixtures and to Richard Johnson for allowing us to use his Data Visualization Toolbox and updating his functions. We would also like to acknowledge all of those researchers who wrote MATLAB code for methods described in this book and also made it available for free. We thank the editors of the book series in Computer Science and Data Analysis for including this text. We greatly appreciate the help and patience of those at CRC Press: Bob Stern, Rob Calver, Jessica Vakili, and Andrea Demby. Finally, we are indebted to Naomi Fernandes and Tom Lane at The MathWorks, Inc. for their special assistance with MATLAB.

Disclaimers

1. Any MATLAB programs and data sets that are included with the book are provided in good faith. The authors, publishers, or distributors do not guarantee their accuracy and are not responsible for the consequences of their use.

2. Some of the MATLAB functions provided with the EDA Toolbox were written by other researchers, and they retain the copyright. References are given in Appendix B and in the **help** section of each function. Unless otherwise specified, the EDA Toolbox is provided under the GNU license specifications:

http://www.gnu.org/copyleft/gpl.html

3. The views expressed in this book are those of the authors and do not necessarily represent the views of the United States Department of Defense or its components.

Wendy L. and Angel R. Martinez
October 2004

Part I

Introduction to Exploratory Data Analysis

Chapter 1

Introduction to Exploratory Data Analysis

We shall not cease from exploration
And the end of all our exploring
Will be to arrive where we started
And know the place for the first time.

T. S. Eliot, "Little Gidding" (the last of his *Four Quartets*)

The purpose of this chapter is to provide some introductory and background information. First, we cover the philosophy of exploratory data analysis and discuss how this fits in with other data analysis techniques and objectives. This is followed by an overview of the text, which includes the software that will be used and the background necessary to understand the methods. We then present several data sets that will be employed throughout the book to illustrate the concepts and ideas. Finally, we conclude the chapter with some information on data transforms, which will be important in some of the methods presented in the text.

1.1 What is Exploratory Data Analysis

John W. Tukey [1977] was one of the first statisticians to provide a detailed description of *exploratory data analysis* (EDA). He defined it as "detective work – numerical detective work – or counting detective work – or graphical detective work" [Tukey, 1977, page 1]. It is mostly a philosophy of data analysis where the researcher examines the data without any pre-conceived ideas in order to discover what the data can tell him or her about the phenomena being studied. Tukey contrasts this with *confirmatory data analysis* (CDA), an area of data analysis that is mostly concerned with statistical hypothesis testing, confidence intervals, estimation, etc. Tukey [1977] states that "Confirmatory data analysis is judicial or quasi-judicial in character." CDA methods typically involve the process of making inferences about or estimates of some population characteristic and then trying to evaluate the precision associated with the results. EDA and CDA should not be used separately from each other, but rather they should be used in a complementary way. The analyst explores the data looking for patterns and structure that leads to hypotheses and models.

Tukey's book on EDA was written at a time when computers were not widely available and the data sets tended to be somewhat small, especially by today's standards. So, Tukey developed methods that could be accomplished using pencil and paper, such as the familiar box-and-whisker plots (also known as boxplots) and the stem-and-leaf. He also included discussions of data transformation, smoothing, slicing, and others. Since our book is written at a time when computers are widely available, we go beyond what Tukey used in EDA and present computationally intensive methods for pattern discovery and statistical visualization. However, our philosophy of EDA is the same – that those engaged in it are *data detectives*.

Tukey [1980], expanding on his ideas of how exploratory and confirmatory data analysis fit together, presents a typical straight-line methodology for CDA; its steps follow:

1. State the question(s) to be investigated.
2. Design an experiment to address the questions.
3. Collect data according to the designed experiment.
4. Perform a statistical analysis of the data.
5. Produce an answer.

This procedure is the heart of the usual confirmatory process. To incorporate EDA, Tukey revises the first two steps as follows:

1. Start with some idea.
2. Iterate between asking a question and creating a design.

Forming the question involves issues such as: What can or should be asked? What designs are possible? How likely is it that a design will give a useful answer? The ideas and methods of EDA play a role in this process. In conclusion, Tukey states that EDA is an attitude, a flexibility, and some graph paper.

A small, easily read book on EDA written from a social science perspective is the one by Hartwig and Dearing [1979]. They describe the CDA mode as one that answers questions such as "Do the data confirm hypothesis XYZ?" Whereas, EDA tends to ask "What can the data tell me about relationship XYZ?" Hartwig and Dearing specify two principles for EDA: *skepticism* and *openness*. This might involve visualization of the data to look for anomalies or patterns, the use of resistant (or robust) statistics to summarize the data, openness to the transformation of the data to gain better insights, and the generation of models.

Some of the ideas of EDA and their importance to teaching statistics were discussed by Chatfield [1985]. He called the topic *initial data analysis* or IDA. While Chatfield agrees with the EDA emphasis on starting with the noninferential approach in data analysis, he also stresses the need for looking at how the data were collected, what are the objectives of the analysis, and the use of EDA/IDA as part of an integrated approach to statistical inference.

Hoaglin [1982] provides a summary of EDA in the *Encyclopedia of Statistical Sciences*. He describes EDA as the "flexible searching for clues and evidence" and confirmatory data analysis as "evaluating the available evidence." In his summary, he states that EDA encompasses four themes: resistance, residuals, re-expression, and display.

Resistant data analysis pertains to those methods where an arbitrary change in a data point or small subset of the data yields a small change in the result. A related idea is *robustness*, which has to do with how sensitive an analysis is to departures from the assumptions of an underlying probabilistic model.

Residuals are what we have left over after a summary or fitted model has been subtracted out. We can write this as

$$\text{residual} = \text{data} - \text{fit}.$$

The idea of examining residuals is common practice today. Residuals should be looked at carefully for lack of fit, heteroscedasticity (nonconstant variance), nonadditivity, and other interesting characteristics of the data.

Re-expression has to do with the transformation of the data to some other scale that might make the variance constant, might yield symmetric residuals, could linearize the data, or add some other effect. The goal of re-expression for EDA is to facilitate the search for structure, patterns, or other information.

Finally, we have the importance of *displays* or *visualization* techniques for EDA. As we described previously, the displays used most often by early practitioners of EDA included the stem-and-leaf plots and boxplots. The use of scientific and statistical visualization is fundamental to EDA, because often the only way to discover patterns, structure, or to generate hypotheses is by visual transformations of the data.

Given the increased capabilities of computing and data storage, where massive amounts of data are collected and stored simply because we can do so and not because of some designed experiment, questions are often generated *after* the data have been collected [Hand, Mannila, and Smyth, 2001; Wegman, 1988]. Perhaps there is an evolution of the concept of EDA in the making and the need for a new philosophy of data analysis.

1.2 Overview of the Text

This book is divided into two main sections: pattern discovery and graphical EDA. We first cover linear and nonlinear dimensionality reduction because sometimes structure is discovered or can only be discovered with fewer dimensions or features. We include some classical techniques such as principal component analysis, factor analysis, and multidimensional scaling, as well as some computationally intensive methods. For example, we discuss self-organizing maps, locally linear embedding, isometric feature mapping, generative topographic maps, curvilinear component analysis, and more.

Searching the data for insights and information is fundamental to EDA. So, we describe several methods that 'tour' the data looking for interesting structure (holes, outliers, clusters, etc.). These are variants of the grand tour and projection pursuit that try to look at the data set in many 2–D or 3–D views in the hope of discovering something interesting and informative.

Clustering or unsupervised learning is a standard tool in EDA and data mining. These methods look for groups or clusters, and some of the issues that must be addressed involve determining the number of clusters and the validity or strength of the clusters. Here we cover some classical methods such as hierarchical clustering and k-means, as well as nonnegative matrix factorization and minimum spanning trees. We also devote an entire chapter to a density-based technique called model-based clustering that includes a way to determine the number of clusters and to assess the resulting clusters.

Evaluating the relationship between variables is an important subject in data analysis. We do not cover the standard regression methodology; it is assumed that the reader already understands that subject. Instead, we include a chapter on scatterplot smoothing techniques such as loess and smoothing splines.

The second section of the book discusses many of the standard techniques of visualization for EDA. The reader will note, however, that graphical techniques, by necessity, are used throughout the book to illustrate ideas and concepts.

In this section, we provide some classic, as well as some novel ways of visualizing the results of the cluster process, such as dendrograms, silhouette plots, treemaps, rectangle plots, and ReClus. These visualization techniques can be used to assess the output from the various clustering algorithms that were covered in the first section of the book. Distribution shapes can tell us important things about the underlying phenomena that produced the data. We will look at ways to determine the shape of the distribution by using boxplots, bagplots, q-q plots, histograms, and others.

Finally, we present ways to visualize multivariate data, and a new chapter on visualizing categorical data is now included. We discuss parallel coordinate plots, scatterplot matrices, glyph plots, coplots, dot charts,

Andrews' curves, scatterplots, sieve plots, and mosaic plots. The ability to interact with the plot to uncover structure or patterns is important, and we present some of the standard methods such as linking and brushing. We also connect both sections by revisiting the idea of the grand tour and show how that can be implemented with Andrews' curves and parallel coordinate plots.

We realize that other topics can be considered part of EDA, such as descriptive statistics, outlier detection, robust data analysis, probability density estimation, and residual analysis. However, these topics are beyond the scope of this book. Descriptive statistics are covered in introductory statistics texts, and since we assume that readers are familiar with this subject matter, there is no need to provide explanations here. Similarly, we do not emphasize residual analysis as a stand-alone subject, mostly because this is widely discussed in other books on regression and multivariate analysis.

We do cover some density estimation, such as model-based clustering (Chapter 6) and histograms (Chapter 9). The reader is referred to Scott [2015] for an excellent treatment of the theory and methods of multivariate density estimation in general or Silverman [1986] for kernel density estimation. For more information on MATLAB implementations of density estimation the reader can refer to Martinez and Martinez [2015] and Martinez and Cho [2014]. Finally, we will likely encounter outlier detection as we go along in the text, but this topic, along with robust statistics, will not be covered as a stand-alone subject. There are several books on outlier detection and robust statistics. These include Hoaglin, Mosteller, and Tukey [1983], Huber [1981], and Rousseeuw and Leroy [1987]. A rather dated paper on the topic is Hogg [1974].

We use MATLAB® throughout the book to illustrate the ideas and to show how they can be implemented in software. Much of the code used in the examples and to create the figures is freely available, either as part of the downloadable toolbox included with the book or on other internet sites. This information will be discussed in more detail in Appendix B. For MATLAB product information, please contact:

> The MathWorks, Inc.
> 3 Apple Hill Drive
> Natick, MA, 01760-2098 USA
> Tel: 508-647-7000
> Fax: 508-647-7001
> E-mail: info@mathworks.com
> Web: www.mathworks.com

It is important for the reader to understand what versions of the software or what toolboxes are used with this text. The book was updated using MATLAB Version 9.0 (R2016a), and we made some use of the MATLAB Statistics Toolbox. Note that this toolbox is now called the Statistics and Machine Learning Toolbox. We continue to refer to it in this text with the shorter name for easier reading. We will refer to the Curve Fitting Toolbox in

Chapter 7, where we discuss smoothing. However, this particular toolbox is not needed to use the examples in the book.

To get the most out of this book, readers should have a basic understanding of matrix algebra. For example, one should be familiar with determinants, a matrix transpose, the trace of a matrix, etc. We recommend Strang [1988, 1993] for those who need to refresh their memories on the topic. We do not use any calculus in this book, but a solid understanding of algebra is always useful in any situation. We expect readers to have knowledge of the basic concepts in probability and statistics, such as random samples, probability distributions, hypothesis testing, and regression.

1.3 A Few Words about Notation

In this section, we explain our notation and font conventions. MATLAB code will be in Courier New bold font such as this: **function**. To make the book more readable, we will indent MATLAB code when we have several lines of code, and this can always be typed in as you see it in the book.

For the most part, we follow the convention that a vector is arranged as a column, so it has dimensions $p \times 1$. [1] In most situations, our data sets will be arranged in a matrix of dimension $n \times p$, which is denoted as **X**. Here n represents the number of observations we have in our sample, and p is the number of variables or dimensions. Thus, each row corresponds to a p-dimensional observation or data point. The ij-th element of **X** will be represented by x_{ij}. For the most part, the subscript i refers to a row in a matrix or an observation, and a subscript j references a column in a matrix or a variable. What is meant by this will be clear from the text.

In many cases, we might need to center our observations before we analyze them. To make the notation somewhat simpler later on, we will use the matrix $\mathbf{X}_c$ to represent our centered data matrix, where each row is now centered at the origin. We calculate this matrix by first finding the mean of each column of **X** and then subtracting it from each row. The following code will calculate this in MATLAB:

```
% Find the mean of each column.
[n,p] = size(X);
xbar = mean(X);
% Create a matrix where each row is the mean
% and subtract from X to center at origin.
Xc = X - repmat(xbar,n,1);
```

[1] The notation $m \times n$ is read "m by n," and it means that we have m rows and n columns in an array. It will be clear from the context whether this indicates matrix dimensions or multiplication.

1.4 Data Sets Used in the Book

In this section, we describe the main data sets that will be used throughout the text. Other data sets will be used in the exercises and in some of the examples. This section can be set aside and read as needed without any loss of continuity. Please see Appendix C for detailed information on all data sets included with the text.

1.4.1 Unstructured Text Documents

The ability to analyze free-form text documents (e.g., Internet documents, intelligence reports, news stories, etc.) is an important application in computational statistics. We must first encode the documents in some numeric form in order to apply computational methods. The usual way this is accomplished is via a term-document matrix, where each row of the matrix corresponds to a word in the lexicon, and each column represents a document. The elements of the term-document matrix contain the number of times the i-th word appears in j-th document [Manning and Schütze, 2000; Charniak, 1996]. One of the drawbacks to this type of encoding is that the order of the words is lost, resulting in a loss of information [Hand, Mannila, and Smyth, 2001].

We now present a novel method for encoding unstructured text documents where the order of the words is accounted for. The resulting structure is called the bigram proximity matrix (BPM).

Bigram Proximity Matrices

The **bigram proximity matrix** (BPM) is a nonsymmetric matrix that captures the number of times word pairs occur in a section of text [Martinez and Wegman, 2002a; 2002b]. The BPM is a square matrix whose column and row headings are the alphabetically ordered entries of the lexicon. Each element of the BPM is the number of times word i appears immediately before word j in the unit of text. The size of the BPM is determined by the size of the lexicon created by alphabetically listing the unique occurrences of the words in the corpus. In order to assess the usefulness of the BPM encoding we had to determine whether or not the representation preserves enough of the semantic content to make them separable from BPMs of other thematically unrelated collections of documents.

We must make some comments about the lexicon and the pre-processing of the documents before proceeding with more information on the BPM and the data provided with this book. All punctuation *within* a sentence, such as commas, semi-colons, colons, etc., were removed. All end-of-sentence punctuation, other than a period, such as question marks and exclamation points, were converted to a period. The period is used in the lexicon as a word, and it is placed at the beginning of the alphabetized lexicon.

Other pre-processing issues involve the removal of noise words and stemming. Many natural language processing applications use a shorter version of the lexicon by excluding words often used in the language [Kimbrell, 1988; Salton, Buckley, and Smith, 1990; Frakes and Baeza-Yates, 1992; Berry and Browne, 2005]. These words, usually called **stop words**, are said to have low informational content and are deleted from the document. However, not all researchers agree with this approach [Witten, Moffat, and Bell, 1994].

Taking the denoising idea one step further, one could also stem the words in the denoised text. The idea is to reduce words to their stem or root to increase the frequency of key words and thus enhance the discriminatory capability of the features. Stemming is routinely applied in the area of information retrieval (IR). In this application of text processing, stemming is used to enhance the performance of the IR system, as well as to reduce the total number of unique words and save on computational resources. The stemmer we used to pre-process the text documents is the Porter stemmer [Baeza-Yates and Ribero-Neto, 1999; Porter, 1980]. The Porter stemmer is simple; however, its performance is comparable with older established stemmers. Please see the following websites for different software implementations of the Porter stemmer:

`http://snowball.tartarus.org/`

and

`http://tartarus.org/~martin/PorterStemmer/`

We are now ready to give an example of the BPM. The BPM for the sentence or text stream,

"The wise young man sought his father in the crowd."

is shown in Table 1.1. We see that the matrix element located in the third row (*his*) and the fifth column (*father*) has a value of one. This means that the pair of words *his father* occurs once in this unit of text. It should be noted that in most cases, depending on the size of the lexicon and the size of the text stream, the BPM will be very sparse and very large.

By preserving the word ordering of the discourse stream, the BPM captures a substantial amount of information about meaning. Also, by obtaining the individual counts of word co-occurrences, the BPM captures the 'intensity' of the discourse's theme. Both features make the BPM a suitable tool for capturing meaning and performing computations to identify semantic similarities among text units of discourse (e.g., sentences, paragraphs, documents). Note that a BPM is created for each text unit.

TABLE 1.1

Example of a BPM

	.	crowd	his	in	father	man	sought	the	wise	young
.										
crowd	1									
his				1						
in								1		
father				1						
man							1			
sought			1							
the		1							1	
wise										1
young						1				

Note that the zeros are left out for ease of reading.

One of the data sets included in this book, which was obtained from text documents, came from the Topic Detection and Tracking (TDT) Pilot Corpus (Linguistic Data Consortium, Philadelphia, PA):

https://catalog.ldc.upenn.edu/LDC98T25.

The TDT corpus is comprised of close to 16,000 stories collected from July 1, 1994 to June 30, 1995 from the Reuters newswire service and CNN broadcast news transcripts. A set of 25 events are discussed in the complete TDT Pilot Corpus. These 25 topics were determined first, and then the stories were classified as either belonging to the topic, not belonging, or somewhat belonging (*Yes*, *No*, or *Brief*, respectively).

TABLE 1.2

List of 16 Topics

Topic Number	Topic Description	Number of Documents Used
4	Cessna on the White House	14
5	Clinic Murders (Salvi)	41
6	Comet into Jupiter	44
8	Death of N. Korean Leader	35
9	DNA in OJ Trial	29
11	Hall's Copter in N. Korea	74
12	Humble, TX Flooding	16
13	Justice-to-be Breyer	8
15	Kobe, Japan Quake	49
16	Lost in Iraq	30
17	NYC Subway Bombing	24
18	Oklahoma City Bombing	76
21	Serbians Down F-16	16
22	Serbs Violate Bihac	19
24	US Air 427 Crash	16
25	WTC Bombing Trial	12

In order to meet the computational requirements of available computing resources, a subset of the TDT corpus was used. A total of 503 stories were chosen that includes 16 of the 25 events. See Table 1.2 for a list of topics. The 503 stories chosen contain only the *Yes* or *No* classifications. This choice stems from the need to demonstrate that the BPM captures enough meaning to make a correct or incorrect topic classification choice.

There were 7,146 words in the lexicon after denoising and stemming. So each BPM has $7,146 \times 7,146 = 51,065,316$ elements, and each document (or observation) resides in a very high dimensional space. We can apply several EDA methods that require the interpoint distance matrix only and not the original data (i.e., BPMs). Thus, we only include the interpoint distance matrices for different measures of semantic distance: IRad, Ochiai, simple matching, and L_1. It should be noted that the match and Ochiai measures started out as similarities (large values mean the observations are similar), and were converted to distances for use in the text. See Appendix A for more information on these distances and Martinez [2002] for other choices, not included here. Table 1.3 gives a summary of the BPM data we will be using in subsequent chapters.

TABLE 1.3

Summary of the BPM Data

Distance	Name of File
IRad	`iradbpm`
Ochiai	`ochiaibpm`
Match	`matchbpm`
L_1 Norm	`L1bpm`

One of the EDA techniques we might want to apply to these data is dimensionality reduction so further processing can be accomplished, such as clustering or supervised learning. We could also be interested in visualizing the data in some manner to determine whether or not the observations exhibit some interesting structure. Finally, we might use these data with a clustering algorithm to see how many groups are found in the data or to find latent topics or subgroups.

1.4.2 Gene Expression Data

The Human Genome Project completed a map (in draft form) of the human genetic blueprint in 2001 (**www.nature.com/omics/index.html**), but much work remains to be done in understanding the functions of the genes and the role of proteins in a living system. The area of study called *functional genomics* addresses this problem, and one of its main tools is DNA *microarray* technology [Sebastiani et al., 2003]. This technology allows data to be collected on multiple experiments and provides a view of the genetic activity (for thousands of genes) for an organism.

We now provide a brief introduction to the terminology used in this area. The reader is referred to Sebastiani et al. [2003] or Griffiths et al. [2000] for more detail on the unique statistical challenges and the underlying biological and technical foundation of genetic analysis. As most of us are aware from introductory biology, organisms are made up of cells, and the nucleus of each cell contains DNA (deoxyribonucleic acid). DNA instructs the cells to produce proteins and how much protein to produce. Proteins participate in most of the functions living things perform. Segments of DNA are called *genes*. The *genome* is the complete DNA for an organism, and it contains the genetic code needed to create a unique life. The process of gene activation is called *gene expression*, and the expression level provides a value indicating the number of intermediary molecules (messenger ribonucleic acid and transfer ribonucleic acid) created in this process.

Microarray technology can simultaneously measure the relative gene expression level of thousands of genes in tissue or cell samples. There are two main types of microarray technology: cDNA microarrays and synthetic oligonucleotide microarrays. In both of these methods, a target (extracted from tissue or cell) is hybridized to a probe (genes of known identity or small sequences of DNA). The target is tagged with fluorescent dye before being hybridized to the probe, and a digital image is formed of the chemical reaction. The intensity of the signal then has to be converted to a quantitative value from the image. As one might expect, this involves various image processing techniques, and it could be a major source of error.

A data set containing gene expression levels has information on genes (rows of the matrix) from several experiments (columns of the matrix). Typically, the columns correspond to patients, tumors, time steps, etc. We note that with the analysis of gene expression data, either the rows (genes) or columns (experiments/samples) could correspond to the dimensionality (or sample size), depending on the goal of the analysis. Some of the questions that might be addressed through this technology include:

- What genes are expressed (or not expressed) in a tumor cell versus a normal cell?
- Can we predict the best treatment for a cancer?
- Are there genes that characterize a specific tumor?
- Are we able to cluster cells based on their gene expression level?
- Can we discover subclasses of cancer or tumors?

For more background information on gene expression data, we refer the reader to Schena et al. [1995], Chee et al. [1996], and Lander [1999]. Many gene expression data sets are freely available on the internet, and there are also many articles on the statistical analysis of this type of data. We refer the interested reader to a recent issue of *Statistical Science* (Volume 18, Number 1,

February 2003) for a special section on microarray analysis. One can also look at the *National Academy of Science* website (**http://www.pnas.org**) for articles, many of which have the data available for download. We include three gene expression data sets with this book, and we describe them below.

Yeast Data Set

This data set was originally described in Cho et al. [1998], and it showed the gene expression levels of around 6000 genes over two cell cycles and five phases. The two cell cycles provide 17 time points (columns of the matrix). The subset of the data we provide was obtained by Yeung and Ruzzo [2001] and is available at

http://www.cs.washington.edu/homes/kayee/model.

A full description of the process they used to get the subset can also be found there. First, they extracted all genes that were found to peak in only one of the five phases; those that peaked in multiple phases were not used. Then they removed any rows with negative entries, yielding a total of 384 genes.

The data set is called **yeast.mat**, and it contains two variables: **data** and **classlabs**. The **data** matrix has 384 rows and 17 columns. The variable **classlabs** is a vector containing 384 class labels for the genes indicating whether the gene peaks in phase 1 through phase 5.

Leukemia Data Set

The **leukemia** data set was first discussed in Golub et al., [1999], where the authors measured the gene expressions of human acute leukemia. Their study included prediction of the type of leukemia using supervised learning and the discovery of new classes of leukemia via unsupervised learning. The motivation for this work was to improve cancer treatment by distinguishing between sub-classes of cancer or tumors.

They first classified the leukemias into two groups: (1) those that arise from lymphoid precursors or (2) from myeloid precursors. The first one is called acute lymphoblastic leukemia (ALL), and the second is called acute myeloid leukemia (AML). The distinction between these two classes of leukemia is well known, but a single test to sufficiently establish a diagnosis does not exist [Golub et al., 1999]. As one might imagine, a proper diagnosis is critical to successful treatment and to avoid unnecessary toxicities. The authors turned to microarray technology and statistical pattern recognition to address this problem.

Their initial data set had 38 bone marrow samples taken at the time of diagnosis; 27 came from patients with ALL, and 11 patients had AML. They used oligonucleotide microarrays containing probes for 6,817 human genes to obtain the gene expression information. Their first goal was to construct a classifier using the gene expression values that would predict the type of leukemia. So, one could consider this as building a classifier where the observations have 6,817 dimensions, and the sample size is 38. They had to reduce the dimensionality, so they chose the 50 genes that have the highest correlation with the class of leukemia. They used an independent test set of leukemia samples to evaluate the classifier. This set of data consists of 34 samples, where 24 of them came from bone marrow and 10 came from peripheral blood samples. It also included samples from children and from different laboratories using different protocols.

They also looked at class discovery or unsupervised learning, where they wanted to determine if the patients could be clustered into two groups corresponding to the types of leukemia. They used the method called self-organizing maps (Chapter 3), employing the full set of 6,817 genes. Another aspect of class discovery is to look for subgroups within known classes. For example, the patients with ALL can be further subdivided into patients with B-cell or T-cell lineage.

We decided to include only the 50 genes, rather than the full set. The **leukemia.mat** file has four variables. The variable **leukemia** has 50 genes (rows) and 72 patients (columns). The first 38 columns correspond to the initial training set of patients, and the rest of the columns contain data for the independent testing set. The variables **btcell** and **cancertype** are cell arrays of strings containing the label for B-cell, T-cell, or NA and ALL or AML, respectively. Finally, the variable **geneinfo** is a cell array where the first column provides the gene description, and the second column contains the gene number.

Example 1.1

We show a plot of the 50 genes in Figure 1.1, but only the first 38 samples (i.e., columns) are shown. This is similar to Figure 3B in Golub et al., [1999]. We standardized each gene, so the mean across each row is 0 and the standard deviation is 1. The first 27 columns of the picture correspond to ALL leukemia, and the last 11 columns pertain to the AML leukemia. We can see by the color that the first 25 genes tend to be more highly expressed in ALL, while the last 25 genes are highly expressed in AML. The MATLAB code to construct this plot is given below.

```
% First standardize the data such that each row
% has mean 0 and standard deviation 1.
load leukemia
x = leukemia(:,1:38);
[n,p] = size(x);
y = zeros(n,p);
for i = 1:n
    sig = std(x(i,:));
    mu = mean(x(i,:));
    y(i,:)= (x(i,:)-mu)/sig;
end
% Now do the image of the data.
```

```
pcolor(y)
colormap(gray(256))
colorbar
title('Gene Expression for Leukemia')
xlabel('ALL (1-27) or AML (28-38)')
ylabel('Gene')
```

The results shown in Figure 1.1 indicate that we might be able to distinguish between AML and ALL leukemia using these data.
❏

FIGURE 1.1
This shows the gene expression for the **leukemia** data set. Each row corresponds to a gene, and each column corresponds to a cancer sample. The rows have been standardized such that the mean is 0 and the standard deviation is 1. We can see that the ALL leukemia is highly expressed in the first set of 25 genes, and the AML leukemia is highly expressed in the second set of 25 genes.

Lung Data Set

Traditionally, the classification of lung cancer is based on clinicopathological features. An understanding of the molecular basis and a possible molecular classification of lung carcinomas could yield better therapies targeted to the type of cancer, superior prediction of patient treatment, and the identification of new targets for chemotherapy. We provide two data sets that were originally downloaded from **http://www.genome.mit.edu/MPR/lung** and described in Bhattacharjee et al. [2001]. The authors applied hierarchical and probabilistic clustering to find subclasses of lung adenocarcinoma, and they showed the diagnostic potential of analyzing gene expression data by demonstrating the ability to separate primary lung adenocarcinomas from metastases of extra-pulmonary origin.

A preliminary classification of lung carcinomas comprises two groups: small-cell lung carcinomas (SCLC) or nonsmall-cell lung carcinomas (NSCLC). The NSCLC category can be further subdivided into 3 groups: adenocarcinomas (AD), squamous cell carcinomas (SQ), and large-cell carcinomas (COID). The most common type is adenocarcinomas. The data were obtained from 203 specimens, where 186 were cancerous and 17 were normal lung. The cancer samples contained 139 lung adenocarcinomas, 21 squamous cell lung carcinomas, 20 pulmonary carcinoids, and 6 small-cell lung carcinomas. This is called Dataset A in Bhattacharjee et al. [2001]; the full data set included 12,600 genes. The authors reduced this to 3,312 by selecting the most variable genes, using a standard deviation threshold of 50 expression units. We provide these data in **lungA.mat**. This file includes two variables: **lungA** and **labA**. The variable **lungA** is a 3312×203 matrix, and **labA** is a vector containing the 203 class labels.

The authors also looked at adenocarcinomas separately trying to discover subclasses. To this end, they separated the 139 adenocarcinomas and the 17 normal samples and called it Dataset B. They took fewer gene transcript sequences for this data set by selecting only 675 genes according to other statistical pre-processing steps. These data are provided in **lungB.mat**, which contains two variables: **lungB** (675×156) and **labB** (156 class labels). We summarize these data sets in Table 1.4.

TABLE 1.4

Description of Lung Cancer Data Set

Cancer Type	Label	Number of Data Points
*Dataset A (***lungA.mat***): 3,312 rows, 203 columns*		
Nonsmall cell lung carcinomas		
Adenocarcinomas	AD	139
Pulmonary carcinoids	COID	20
Squamous cell	SQ	21
Normal	NL	17
Small-cell lung carcinomas	SCLC	6
*Dataset B (***lungB.mat***): 675 rows, 156 columns*		
Adenocarcinomas	AD	139
Normal	NL	17

For those who need to analyze gene expression data, we recommend the Bioinformatics Toolbox from The MathWorks. The toolbox provides an integrated environment for solving problems in genomics and proteomics, genetic engineering, and biological research. Some capabilities include the ability to calculate the statistical characteristics of the data, to manipulate sequences, to construct models of biological sequences using Hidden Markov Models, and to visualize microarray data.

1.4.3 Oronsay Data Set

This data set consists of particle size measurements originally presented in Timmins [1981] and analyzed by Olbricht [1982], Fieller, Gilbertson, and Olbricht [1984], and Fieller, Flenley, and Olbricht [1992]. An extensive analysis from a graphical EDA point of view was conducted by Wilhelm, Wegman, and Symanzik [1999]. The measurement and analysis of particle sizes is often used in archaeology, fuel technology (droplets of propellant), medicine (blood cells), and geology (grains of sand). The usual objective is to determine the distribution of particle sizes because this characterizes the environment where the measurements were taken or the process of interest.

The Oronsay particle size data were gathered for a geological application, where the goal was to discover different characteristics between dune sands and beach sands. This characterization would be used to determine whether or not midden sands were dune or beach. The middens were near places where prehistoric man lived, and geologists are interested in whether these middens were beach or dune because that would be an indication of how the coastline has shifted.

There are 226 samples of sand, with 77 belonging to an unknown type of sand (from the middens) and 149 samples of known type (beach or dune). The known samples were taken from *Cnoc Coig* (CC - 119 observations, 90 beach and 29 dune) and *Caisteal nan Gillean* (CG - 30 observations, 20 beach and 10 dune). See Wilhelm, Wegman, and Symanzik [1999] for a map showing these sites on Oronsay island. This reference also shows a more detailed classification of the sands based on transects and levels of sand.

Each observation is obtained in the following manner. Approximately 60g or 70g of sand is put through a stack of 11 sieves of sizes 0.063mm, 0.09mm, 0.125mm, 0.18mm, 0.25mm, 0.355mm, 0.5mm, 0.71mm, 1.0mm, 1.4mm, and 2.0mm. The sand that remains on each of the sieves is weighed, along with the sand that went through completely. This yields 12 weight measurements, and each corresponds to a class of particle size. Note that there are two extreme classes: particle sizes less than 0.063mm (what went through the smallest sieve) and particle sizes larger than 2.0mm (what is in the largest sieve).

Flenley and Olbricht [1993] consider the classes as outlined above, and they apply various multivariate and exploratory data analysis techniques such as principal component analysis and projection pursuit. The **oronsay** data set was downloaded from:

```
http://www.math.usu.edu/
     symanzik/papers/1998_computstat/oronsay.html.
```

More information on the original data set can be found at this website. We chose to label observations first with respect to midden, beach, or dune (in variable **beachdune**):

- Class 0: midden (77 observations)
- Class 1: beach (110 observations)
- Class 2: dune (39 observations)

We then classify observations according to the sampling site (in variable **midden**), as follows

- Class 0: midden (77 observations)
- Class 1: *Cnoc Coig* - CC (119 observations)
- Class 2: *Caisteal nan Gillean* - CG (30 observations)

The data set is in the **oronsay.mat** file. The data are in a 226×12 matrix called **oronsay**, and the data are in raw format; i.e., untransformed and unstandardized. Also included is a cell array of strings called **labcol** that contains the names (i.e., sieve sizes) of the columns.

1.4.4 Software Inspection

The data described in this section were collected in response to efforts for process improvement in software testing. Many systems today rely on complex software that might consist of several modules programmed by different programmers; so ensuring that the software works correctly and as expected is important.

One way to test the software is by inspection, where software engineers inspect the code in a formal way. First they look for inconsistencies, logical errors, etc., and then they all meet with the programmer to discuss what they perceive as defects. The programmer is familiar with the code and can help determine whether or not it is a defect in the software.

The data are saved in a file called **software**. The variables are normalized by the size of the inspection (the number of pages or SLOC – single lines of code). The file **software.mat** contains the preparation time in minutes (**prepage, prepsloc**), the total work hours in minutes for the meeting (**mtgsloc**), and the number of defects found (**defpage, defsloc**). Software engineers and managers would be interested in understanding the relationship between the inspection time and the number of defects found. One of the goals might be to find an optimal time for inspection, where one gets the most payoff (number of defects found) for the amount of time spent reviewing the code. We show an example of these data in Figure 1.2. The defect types include compatibility, design, human-factors, standards, and others.

FIGURE 1.2
This is a scatterplot of the software inspection data. The relationship between the variables is difficult to see.

1.5 Transforming Data

In many real-world applications, the data analyst will have to deal with raw data that are not in the most convenient form. The data might need to be re-expressed to produce effective visualization or an easier, more informative analysis. Some of the types of problems that can arise include data that exhibit nonlinearity or asymmetry, contain outliers, change spread with different levels, etc. We can transform the data by applying a single mathematical function to all of the observations.

In the first subsection below, we discuss the general power transformations that can be used to change the shape of the data distribution. This arises in situations when we are concerned with formal inference methods where the shape of the distribution is important (e.g., statistical hypothesis testing or confidence intervals). In EDA, we might want to change the shape to facilitate visualization, smoothing, and other analyses. Next we cover linear transformations of the data that leave the shape alone. These are typically changes in scale and origin and can be important in dimensionality reduction, clustering, and visualization.

1.5.1 Power Transformations

A *transformation* of a set of data points x_1, x_2, ..., x_n is a function T that substitutes each observation x_i with a new value $T(x_i)$ [Emerson and Stoto, 1983]. Transformations should have the following desirable properties:

1. The order of the data is preserved by the transformation. Because of this, statistics based on order, such as medians are preserved; i.e., medians are transformed to medians.

2. They are continuous functions guaranteeing that points that are close together in raw form are also close together using their transformed values, relative to the scale used.

3. They are smooth functions that have derivatives of all orders, and they are specified by elementary functions.

Some common transformations include taking roots (square root, cube root, etc.), finding reciprocals, calculating logarithms, and raising variables to positive integral powers. These transformations provide adequate flexibility for most situations in data analysis.

Example 1.2

This example uses the software inspection data shown in Figure 1.2. We see that the data are skewed, and the relationship between the variables is difficult to understand. We apply a log transform to both variables using the following MATLAB code, and show the results in Figure 1.3.

```
load software
% First transform the data.
X = log(prepsloc);
Y = log(defsloc);
% Plot the transformed data.
plot(X,Y,'.')
xlabel('Log PrepTime/SLOC')
ylabel('Log Defects/SLOC')
```

We now have a better idea of the relationship between these two variables, which will be examined further in Chapter 7.
❑

Some transformations of the data may lead to insights or discovery of structures that we might not otherwise see. However, as with any analysis, we should be careful about creating something that is not really there, but is just an artifact of the processing. Thus, in any application of EDA, the analyst should go back to the subject area and consult domain experts to verify and help interpret the results.

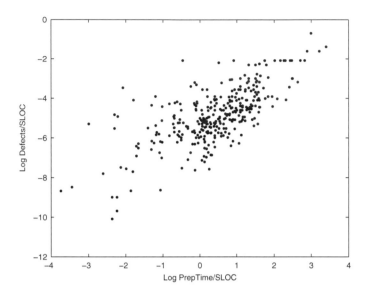

FIGURE 1.3
Each variate was transformed using the logarithm. The relationship between preparation
time per SLOC and number of defects found per SLOC is now easier to see.

1.5.2 Standardization

If the variables are measurements along a different scale or if the standard
deviations for the variables are different from one another, then one variable
might dominate the distance (or some other similar calculation) used in the
analysis. We will make extensive use of interpoint distances throughout the
text in applications such as clustering, multidimensional scaling, and
nonlinear dimensionality reduction. We discuss several 1–D standardization
methods below. However, we note that in some multivariate contexts, the
1–D transformations may be applied to each variable (i.e., on the column of
X) separately.

Transformation Using the Standard Deviation

The first standardization we discuss is called the sample ***z-score***, and it
should be familiar to most readers who have taken an introductory statistics
class. The transformed variates are found using

$$z = \frac{(x - \bar{x})}{s},$$

(1.1)

where x is the original observed data value, $\bar{x}$ is the sample mean, and s is the sample standard deviation. In this standardization, the new variate z will have a mean of zero and a variance of one.

When the z-score transformation is used in a clustering context, it is important that it be applied in a global manner across all observations. If standardization is done within clusters, then false and misleading clustering solutions can result [Milligan and Cooper, 1988].

If we do not center the data at zero by removing the sample mean, then we have the following

$$z = \frac{x}{s}.$$ (1.2)

This transformed variable will have a variance of one and a transformed mean equal to $\bar{x}/s$. The standardizations in Equations 1.1 and 1.2 are linear functions of each other; so Euclidean distances (see Appendix A) calculated on data that have been transformed using the two formulas result in identical dissimilarity values.

For robust versions of Equations 1.1 and 1.2, we can substitute the median and the interquartile range for the sample mean and sample standard deviation respectively. This will be explored in the exercises.

Transformation Using the Range

Instead of dividing by the standard deviation, as above, we can use the range of the variable as the divisor. This yields the following two forms of standardization

$$z = \frac{x}{\max(x) - \min(x)},$$ (1.3)

and

$$z = \frac{x - \min(x)}{\max(x) - \min(x)}.$$ (1.4)

The standardization in Equation 1.4 is bounded by zero and one, with at least one observed value at each of the end points. The transformed variate given by Equation 1.3 is a linear function of the one determined by Equation 1.4; so data standardized using these transformations will result in identical Euclidean distances.

1.5.3 Sphering the Data

This type of standardization called *sphering* pertains to multivariate data, and it serves a similar purpose as the 1–D standardization methods given above. The transformed variables will have a p-dimensional mean of $\mathbf{0}$ and a covariance matrix given by the identity matrix.

We start off with the p-dimensional sample mean given by

$$\bar{\mathbf{x}} = \frac{1}{n}\sum_{i=1}^{n} \mathbf{x}_i.$$

We then find the sample covariance matrix given by the following

$$\mathbf{S} = \frac{1}{n-1}\sum_{i=1}^{n} (\mathbf{x}_i - \bar{\mathbf{x}})(\mathbf{x}_i - \bar{\mathbf{x}})^T,$$

where we see that the covariance matrix can be written as the sum of n matrices. Each of these rank one matrices is the outer product of the centered observations [Duda and Hart, 1973].

We sphere the data using the following transformation

$$\mathbf{Z}_i = \Lambda^{-1/2}\mathbf{Q}^T(\mathbf{x}_i - \bar{\mathbf{x}}) \qquad i = 1, \dots, n,$$

where the columns of $\mathbf{Q}$ are the eigenvectors obtained from $\mathbf{S}$, Λ is a diagonal matrix of corresponding eigenvalues, and $\mathbf{x}_i$ is the i-th observation.

Example 1.3

We now provide the MATLAB code to obtain sphered data. First, we generate 2–D multivariate normal random variables that have the following parameters:

$$\mu = \begin{bmatrix} -2 \\ 2 \end{bmatrix},$$

and

$$\Sigma = \begin{bmatrix} 1 & 0.5 \\ 0.5 & 1 \end{bmatrix},$$

where Σ is the covariance matrix. A scatterplot of these data is shown in Figure 1.4 (top).

```
% First generate some 2-D multivariate normal
% random variables, with mean MU and
% covariance SIGMA. This uses a Statistics
% Toolbox function, but similar functionality
% is available in the EDA Toolbox.
n = 100;
mu = [-2, 2];
sigma = [1,.5;.5,1];
X = mvnrnd(mu,sigma,n);
plot(X(:,1),X(:,2),'.')
```

We now apply the steps to sphere the data, and show the transformed data in Figure 1.4 (bottom).

```
% Now sphere the data.
xbar = mean(X);
% Get the eigenvectors and eigenvalues of the
% covariance matrix.
[V,D] = eig(cov(X));
% Center the data.
Xc = X - ones(n,1)*xbar;
% Sphere the data.
Z = ((D)^(-1/2)*V'*Xc')';
plot(Z(:,1),Z(:,2),'.')
```

By comparing these two plots, we see that the transformed data are sphered and are now centered at the origin.
❑

1.6 Further Reading

There are several books that will provide the reader with more information and other perspectives on EDA. Most of these books do not offer software and algorithms, but they are excellent resources for the student or practitioner of exploratory data analysis.

As we stated in the beginning of this chapter, the seminal book on EDA is Tukey [1977], but the text does not include the more up-to-date view based on current computational resources and methodology. Similarly, the short book on EDA by Hartwig and Dearing [1979] is an excellent introduction to the topic and a quick read, but it is somewhat dated. For the graphical approach, the reader is referred to du Toit, Steyn, and Stumpf [1986], where the authors use SAS to illustrate the ideas. They include other EDA methods

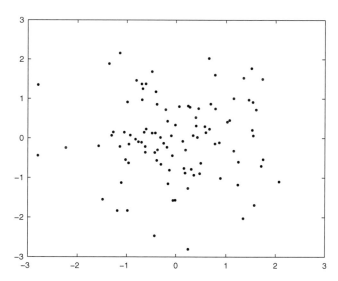

FIGURE 1.4

The top figure shows a scatterplot of the 2–D multivariate normal random variables. Note that these are not centered at the origin, and the cloud is not spherical. The sphered data are shown in the bottom panel. We see that they are now centered at the origin with a spherical spread. This is similar to the z-score standardization in 1–D.

such as multidimensional scaling and cluster analysis. Hoaglin, Mosteller, and Tukey [1983] edited an excellent book on robust and exploratory data analysis. It includes several chapters on transforming data, and we recommend the one by Emerson and Stoto [1983]. The chapter includes a discussion of power transformations, as well as plots to assist the data analyst in choosing an appropriate one.

For a more contemporary resource that explains data mining approaches, of which EDA is a part, Hand, Mannila, and Smyth [2001] is highly recommended. It does not include computer code, but it is very readable. The authors cover the major parts of data mining: EDA, descriptive modeling, classification and regression, discovering patterns and rules, and retrieval by content. Finally, the reader could also investigate the book by Hastie, Tibshirani, and Friedman [2009]. These authors cover a wide variety of topics of interest to exploratory data analysts, such as clustering, nonparametric probability density estimation, multidimensional scaling, and projection pursuit.

As was stated previously, EDA is sometimes defined as an attitude of flexibility and discovery in the data analysis process. There is an excellent article by Good [1983] outlining the philosophy of EDA, where he states that "EDA is more an art, or even a bag of tricks, than a science." While we do not think there is anything "tricky" about the EDA techniques, it is somewhat of an art in that the analyst must try various methods in the discovery process, keeping an open mind and being prepared for surprises! Finally, other summaries of EDA were written by Diaconis [1985] and Weihs [1993]. Weihs describes EDA mostly from a graphical viewpoint and includes descriptions of dimensionality reduction, grand tours, prediction models, and variable selection. Diaconis discusses the difference between exploratory methods and the techniques of classical mathematical statistics. In his discussion of EDA, he considers Monte Carlo techniques such as the bootstrap [Efron and Tibshirani, 1993].

Exercises

1.1   What is exploratory data analysis? What is confirmatory data analysis? How do these analyses fit together?

1.2   Repeat Example 1.1 using the remaining columns (39 – 72) of the **leukemia** data set. Does this follow the same pattern as the others?

1.3   Repeat Example 1.1 using the **lungB** gene expression data set. Is there a pattern?

1.4   Generate some 1–D normally distributed random variables with $\mu = 5$ and $\sigma = 2$ using **normrnd** or **randn** (must transform the results to have the required mean and standard deviation if you use this function). Apply the various standardization procedures described in this chapter and verify the comments regarding the location and spread of the transformed variables.

1.5　　Write MATLAB functions that implement the standardizations mentioned in this chapter.

1.6　　Using the **mvnrnd** function (see Example 1.3), generate some nonstandard bivariate normal random variables. Sphere the data and verify that the resulting sphered data have mean 0 and identity covariance matrix using the MATLAB functions **mean** and **cov**.

1.7　　We will discuss the quartiles and the interquartile range in Chapter 9, but for now look at the MATLAB **help** files on the **iqr** and **median** functions. We can use these robust measures of location and spread to transform our variables. Using Equations 1.1 and 1.2, substitute the median for the sample mean $\bar{x}$ and the interquartile range for the sample standard deviation s. Write a MATLAB function that does this and try it out on the same data you generated in Problem 1.4.

1.8　　Generate $n = 2$ normally distributed random variables. Find the Euclidean distance between the points after they have been transformed first using Equation 1.1 and then Equation 1.2. Are the distances the same? Hint: Use the **pdist** function from the Statistics Toolbox.

1.9　　Repeat Problem 1.8 using the standardizations given by Equations 1.3 and 1.4.

1.10　Generate $n = 100$ uniform 1–D random variables using the **rand** function. Construct a histogram using the **hist** function. Now transform the data by taking the logarithm (use the **log** function). Construct a histogram of these transformed values. Did the shape of the distribution change? Comment on the results.

1.11　Try the following transformations on the software data:

$$T(x) = \log(1 + \sqrt{x})$$
$$T(x) = \log(\sqrt{x})$$

Construct a scatterplot and compare with the results in Example 1.2.

Part II

EDA as Pattern Discovery

Chapter 2

Dimensionality Reduction — Linear Methods

In this chapter we describe several linear methods for dimensionality reduction. We first discuss some classical approaches such as principal component analysis (PCA), singular value decomposition (SVD), non-negative matrix factorization, factor analysis, linear discriminant analysis, and random projections. We conclude this chapter with a discussion of several methods for determining the intrinsic dimensionality of a data set.

2.1 Introduction

Dimensionality reduction is the process of finding a suitable lower-dimensional space in which to represent the original data. Our hope is that the alternative representation of the data will help us:

- Explore high-dimensional data with the goal of discovering structure or patterns that lead to the formation of statistical hypotheses.
- Visualize the data using scatterplots when dimensionality is reduced to 2–D or 3–D.
- Analyze the data using statistical methods, such as clustering, smoothing, probability density estimation, or classification.

One possible method for dimensionality reduction would be to just select subsets of the variables for processing and analyze them in groups. However, in some cases, that would mean throwing out a lot of useful information. An alternative would be to create new variables that are functions (e.g., linear combinations) of the original variables. The methods we describe in this book are of the second type, where we seek a mapping from the higher-dimensional space to a lower-dimensional one, while keeping information on all of the available variables. In general, this mapping can be linear or nonlinear.

Since some of the methods in this chapter transform the data using projections, we take a moment to describe this concept before going on to explain how they work. A projection will be in the form of a matrix that takes the data from the original space to a lower-dimensional one. We illustrate this concept in Example 2.1.

Example 2.1
In this example, we show how projection works when our data set consists of the two bivariate points

$$\mathbf{x}_1 = \begin{bmatrix} 4 \\ 3 \end{bmatrix} \qquad \mathbf{x}_2 = \begin{bmatrix} -4 \\ 5 \end{bmatrix},$$

and we are projecting onto a line that is θ radians from the horizontal or x-axis. For this example, the projection matrix is given by

$$\mathbf{P} = \begin{bmatrix} (\cos\theta)^2 & \cos\theta\sin\theta \\ \cos\theta\sin\theta & (\sin\theta)^2 \end{bmatrix}.$$

The following MATLAB code is used to enter this information.

```
% Enter the data as rows of our matrix X.
X = [4 3; -4 5];
% Provide a value for theta.
theta = pi/3;
% Now obtain the projection matrix.
c2 = cos(theta)^2;
cs = cos(theta)*sin(theta);
s2 = sin(theta)^2;
P = [c2 cs; cs s2];
```

The coordinates of the projected observations are a weighted sum of the original variables, where the columns of $\mathbf{P}$ provide the weights. We project a single observation (represented as a column vector) as follows

$$\mathbf{y}_i = \mathbf{P}^T\mathbf{x}_i \qquad i = 1, \ldots, n,$$

where the superscript T indicates the matrix transpose. Expanding this out shows the new y coordinates as a weighted sum of the x variables:

$$\begin{aligned} y_{i1} &= P_{11}x_{i1} + P_{21}x_{i2} \\ y_{i2} &= P_{12}x_{i1} + P_{22}x_{i2} \end{aligned} \qquad i = 1, \ldots, n.$$

We have to use some linear algebra to project the data matrix $\mathbf{X}$ because the observations are *rows* of this matrix. Taking the transpose of both sides of our projection equation above, we have

$$\mathbf{y}_i^T = (\mathbf{P}^T \mathbf{x}_i)^T$$
$$= \mathbf{x}_i^T \mathbf{P} \ .$$

Thus, we can project the data using this MATLAB code:

```
% Now project the data onto the theta-line.
% Since the data are arranged as rows in the
% matrix X, we have to use the following to
% project the data.
Xp = X*P;
plot(Xp(:,1),Xp(:,2),'o') % Plot the data.
```

This projects the data onto a 1–D subspace that is at an angle θ with the original coordinate axes. As an example of a projection onto the horizontal coordinate axis, we can use these commands:

```
% We can project onto the 1-D space given by the
% horizontal axis using the projection:
Px = [1;0];
Xpx = X*Px;
```

One could also use the projection matrix $\mathbf{P}$ with $\theta = 0$. These data now have only one coordinate value representing the number of units along the x-axis. The projections are shown in Figure 2.1, where the o's represent the projection of the data onto the θ line and the asterisks denote the projections onto the x-axis.
❏

2.2 Principal Component Analysis — PCA

The main purpose of *principal component analysis* (PCA) is to reduce the dimensionality from p to d, where $d < p$, while at the same time accounting for as much of the variation in the original data set as possible. With PCA, we transform the data to a new set of coordinates or variables that are a linear combination of the original variables. In addition, the observations in the new principal component space are uncorrelated. The hope is that we can gain information and better understand the data set by looking at the observations in the new space.

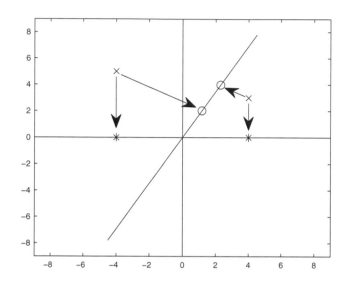

FIGURE 2.1
This figure shows the orthogonal projection of the data points to both the θ line (o's) and the *x*-axis (asterisks).

2.2.1 PCA Using the Sample Covariance Matrix

We start with our centered data matrix $\mathbf{X}_c$ that has dimension $n \times p$. Recall that this matrix contains observations that are centered about the mean; i.e., the sample mean has been subtracted from each row. We then form the sample covariance matrix $\mathbf{S}$ as

$$\mathbf{S} = \frac{1}{n-1}\mathbf{X}_c^T\mathbf{X}_c,$$

where the superscript T denotes the matrix transpose. The jk-th element of $\mathbf{S}$ is given by

$$s_{jk} = \frac{1}{n-1}\sum_{i=1}^{n}\left(x_{ij} - \bar{x}_j\right)\left(x_{ik} - \bar{x}_k\right), \quad j,k = 1,\ldots, p,$$

with

$$\bar{x}_j = \frac{1}{n}\sum_{i=1}^{n}x_{ij}.$$

The next step is to calculate the eigenvectors and eigenvalues of the matrix **S**. The eigenvalues are found by solving the following equation for each l_j, $j = 1, \ldots , p$

$$|\mathbf{S} - l\mathbf{I}| = 0, \tag{2.1}$$

where **I** is a $p \times p$ identity matrix and $| \bullet |$ denotes the determinant. Equation 2.1 produces a polynomial equation of degree p.

The eigenvectors are obtained by solving the following set of equations for $\mathbf{a}_j$

$$(\mathbf{S} - l_j\mathbf{I})\mathbf{a}_j = 0 \qquad j = 1, \ldots , p,$$

subject to the condition that the set of eigenvectors is orthonormal. This means that the magnitude of each eigenvector is one, and they are orthogonal to each other:

$$\mathbf{a}_i\mathbf{a}_i^T = 1$$

$$\mathbf{a}_j\mathbf{a}_i^T = 0,$$

for $i, j = 1, \ldots , p$ and $i \neq j$.

A major result in matrix algebra shows that any square, symmetric, nonsingular matrix can be transformed to a diagonal matrix using

$$\mathbf{L} = \mathbf{A}^T\mathbf{S}\mathbf{A},$$

where the columns of **A** contain the eigenvectors of **S**, and **L** is a diagonal matrix with the eigenvalues along the diagonal. By convention, the eigenvalues are ordered in descending order $l_1 \geq l_2 \geq \ldots \geq l_p$, with the same order imposed on the corresponding eigenvectors.

We use the eigenvectors of **S** to obtain new variables called *principal components* (PCs). The j-th PC is given by

$$z_j = \mathbf{a}_j^T(\mathbf{x} - \bar{\mathbf{x}}) \qquad j = 1, \ldots , p, \tag{2.2}$$

and the elements of **a** provide the weights or coefficients of the old variables in the new PC coordinate space. It can be shown that the PCA procedure defines a principal axis rotation of the original variables about their means [Jackson, 1991; Strang, 1988], and the elements of the eigenvector **a** are the direction cosines that relate the two coordinate systems. Equation 2.2 shows that the PCs are linear combinations of the original variables.

Scaling the eigenvectors to have unit length produces PCs that are uncorrelated and whose variances are equal to the corresponding eigenvalue. Other scalings are possible [Jackson, 1991], such as

$$\mathbf{v}_j = \sqrt{l_j}\,\mathbf{a}_j,$$

and

$$\mathbf{w}_j = \frac{\mathbf{a}_j}{\sqrt{l_j}}.$$

The three eigenvectors $\mathbf{a}_j$, $\mathbf{v}_j$, and $\mathbf{w}_j$ differ only by a scale factor. The $\mathbf{a}_j$ are typically needed for hypothesis testing and other diagnostic methods. Since they are scaled to unity, their components will always be between ± 1. The vectors $\mathbf{v}_j$ and their PCs have the same units as the original variables, making the vectors $\mathbf{v}_j$ useful in other applications. Using $\mathbf{w}_j$ in the transformation yields PCs that are uncorrelated with unit variance.

We transform the observations to the PC coordinate system via the following equation

$$\mathbf{Z} = \mathbf{X}_c\mathbf{A}. \tag{2.3}$$

The matrix $\mathbf{Z}$ contains the ***principal component scores***. Note that these PC scores have zero mean (because we are using the data centered about the mean) and are uncorrelated. We could also transform the original observations in $\mathbf{X}$ by a similar transformation, but the PC scores in this case will have mean $\bar{\mathbf{z}}$. We can invert this transformation to get an expression relating the original variables as a function of the PCs, which is given by

$$\mathbf{x} = \bar{\mathbf{x}} + \mathbf{A}\mathbf{z}.$$

To summarize: the *transformed variables* are the PCs and the individual *transformed data values* are the PC scores.

The dimensionality of the principal component scores in Equation 2.3 is still p; so no dimensionality reduction has taken place. We know from results in linear algebra that the sum of the variances of the original variables is equal to the sum of the eigenvalues. The idea of dimensionality reduction with PCA is that one could include in the analysis only those PCs that have the highest eigenvalues, thus accounting for the highest amount of variation with fewer dimensions or PC variables. We can reduce the dimensionality to d with the following

$$\mathbf{Z}_d = \mathbf{X}_c\mathbf{A}_d, \tag{2.4}$$

where $\mathbf{A}_d$ contains the first d eigenvectors or columns of $\mathbf{A}$. We see that $\mathbf{Z}_d$ is an $n \times d$ matrix (each observation now has only d elements), and $\mathbf{A}_d$ is a $p \times d$ matrix.

2.2.2 PCA Using the Sample Correlation Matrix

We can scale the data first to have standard units, as described in Chapter 1. This means we have the j-th element of $\mathbf{x}^*$ given by

$$x_j^* = \frac{(x_j - \bar{x}_j)}{\sqrt{s_{jj}}}; \qquad j = 1, \dots, p,$$

where s_{jj} is the variance of x_j (i.e., the jj-th element of the sample covariance matrix $\mathbf{S}$). The standardized data $\mathbf{x}^*$ are then treated as observations in the PCA process.

The covariance of this standardized data set is the same as the correlation matrix. The ij-th element of the sample correlation matrix $\mathbf{R}$ is given by

$$r_{ij} = \frac{s_{ij}}{\sqrt{s_{ii}}\sqrt{s_{jj}}},$$

where s_{ij} is the ij-th element of $\mathbf{S}$, and s_{ii} is the i-th diagonal element of $\mathbf{S}$. The rest of the results from PCA hold for the $\mathbf{x}^*$. For example, we can transform the standardized data using Equation 2.3 or reduce the dimensionality using Equation 2.4, where now the matrices $\mathbf{A}$ and $\mathbf{A}_d$ contain the eigenvectors of the correlation matrix.

The correlation matrix should be used for PCA when the variances along the original dimensions are very different; i.e., if some variables have variances that are much greater than the others. In this case, the first few PCs will be dominated by those same variables and will not be very informative. This is often the case when the variables are of different units. Another benefit of using the correlation matrix rather than the covariance matrix arises when one wants to compare the results of PCA among different analyses.

PCA based on covariance matrices does have certain advantages, too. Methods for statistical inference based on the sample PCs from covariance matrices are easier and are available in the literature. It is important to note that the PCs obtained from the correlation and covariance matrices do not provide equivalent information. Additionally, the eigenvectors and eigenvalues from one process do not have simple relationships or correspondence with those from the other one [Jolliffe, 1986]. Since this text is primarily concerned with exploratory data analysis, not inferential methods, we do not discuss this further. In any event, in the spirit of EDA, the analyst should take advantage of both methods to describe and explore different aspects of the data.

2.2.3 How Many Dimensions Should We Keep?

One of the key questions that needs to be answered at this point is how many PCs to keep. We offer the following possible ways to address this question. More details and other options, such as hypothesis tests for equality of eigenvalues, cross-validation, and correlation procedures, can be found in Jackson [1991]. All of the following techniques will be explored in Example 2.2.

Cumulative Percentage of Variance Explained

This is a popular method of determining the number of PCs to use in PCA dimensionality reduction and seems to be implemented in many computer packages. The idea is to select those d PCs that contribute a specified cumulative percentage of total variation in the data, which is calculated using

$$t_d = 100 \frac{\sum_{i=1}^{d} l_i}{\sum_{j=1}^{p} l_j}.$$

If the correlation matrix is used for PCA, then this is simplified to

$$t_d = \frac{100}{p} \sum_{i=1}^{d} l_i.$$

Choosing a value for t_d can be problematic, but typical values range between 70% and 95%. We note that Jackson [1991] does not recommend using this method.

Scree Plot

A graphical way of determining the number of PCs to retain is called the *scree plot*. The original name and idea is from Cattell [1966], and it is a plot of l_k (the eigenvalue) versus k (the index of the eigenvalue). In some cases, we might plot the log of the eigenvalues when the first eigenvalues are very large. This type of plot is called a *log-eigenvalue* or LEV plot. To use the scree plot, one looks for the 'elbow' in the curve or the place where the curve levels off and becomes almost flat. The value of k at this elbow is an estimate for how many PCs to retain. Another way to look at this is by the slopes of the lines connecting the points. When the slopes start to level off and become less steep, that is the number of PCs one should keep.

The Broken Stick

In this method, we choose the number of PCs based on the size of the eigenvalue or the proportion of the variance explained by the individual PC. If we take a line segment and randomly divide it into p segments, then the expected length of the k-th longest segment is

$$g_k = \frac{1}{p} \sum_{i=k}^{p} \frac{1}{i}.$$

If the proportion of the variance explained by the k-th PC is greater than g_k, then that PC is kept. We can say that these PCs account for more variance than would be expected by chance alone.

Size of Variance

One rule that can be used for correlation-based PCA is due to Kaiser [1960], although it is more commonly used in factor analysis. Using this rule, we would retain PCs whose variances are greater than 1 ($l_k \geq 1$). Some suggest that this is too high [Jolliffe, 1972]; so a better rule would be to keep PCs whose variances are greater than 0.7 ($l_k \geq 0.7$). We can use something similar for covariance-based PCA, where we use the average of the eigenvalues rather than 1. In this case, we would keep PCs if

$$l_k \geq 0.7\bar{l} \quad \text{or} \quad l_k \geq \bar{l},$$

where

$$\bar{l} = \frac{1}{p} \sum_{j=1}^{p} l_j.$$

Example 2.2

We show how to perform PCA using the **yeast** cell cycle data set. Recall from Chapter 1 that these contain 384 genes corresponding to five phases, measured at 17 time points. We first load the data and center each row.

```
load yeast
[n,p] = size(data);
% Center the data.
datac = data - repmat(sum(data)/n,n,1);
```

```
% Find the covariance matrix.
covm = cov(datac);
```

We are going to use the covariance matrix in PCA since the variables have common units. The reader is asked to explore the correlation matrix approach in the exercises. The **eig** function in the main MATLAB package is used to calculate the eigenvalues and eigenvectors. MATLAB returns the eigenvalues in a diagonal matrix, and they are in ascending order; so they must be flipped to get the scree plot.

```
[eigvec,eigval] = eig(covm);
eigval = diag(eigval);  % Extract the diagonal elements
% Order in descending order
eigval = flipud(eigval);
eigvec = eigvec(:,p:-1:1);
% Do a scree plot.
figure, plot(1:length(eigval),eigval,'ko-')
title('Scree Plot')
xlabel('Eigenvalue Index - k')
ylabel('Eigenvalue')
```

We see from the scree plot in Figure 2.2 that keeping four PCs seems reasonable. Next we calculate the cumulative percentage of variance explained.

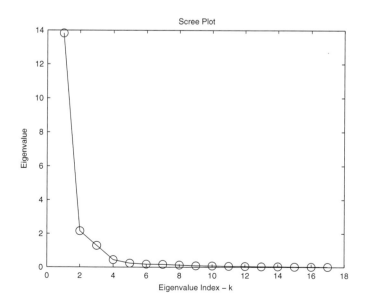

FIGURE 2.2
This is the scree plot for the **yeast** data. The elbow in the curve seems to occur at $k = 4$.

```
% Now for the percentage of variance explained.
pervar = 100*cumsum(eigval)/sum(eigval);
```

The first several values are:

```
73.5923    85.0875    91.9656    94.3217    95.5616
```

Depending on the cutoff value, we would keep four to five PCs (if we are using the higher end of the range of t_d). Now we show how to do the broken stick test.

```
% First get the expected sizes of the eigenvalues.
g = zeros(1,p);
for k = 1:p
    for i = k:p
    g(k) = g(k) + 1/i;
end
end
g = g/p;
```

The next step is to find the proportion of the variance explained.

```
propvar = eigval/sum(eigval);
```

Looking only at the first several values, we get the following for g_k and the proportion of variance explained by each PC:

```
g(1:4) =            0.2023    0.1435    0.1141    0.0945
propvar(1:4) = 0.7359    0.1150    0.0688    0.0236
```

Thus, we see that only the first PC would be retained using this method. Finally, we look at the size of the variance.

```
% Now for the size of the variance.
avgeig = mean(eigval);
% Find the length of ind:
ind = find(eigval > avgeig);
length(ind)
```

According to this test, the first three PCs would be retained. So, we see that different values of *d* are obtained using the various procedures. Because we want to easily visualize the data, we will use the first three PCs to reduce the dimensionality of the data, as follows.

```
% Using d = 3, we will reduce the dimensionality.
P = eigvec(:,1:3);
Xp = datac*P;
figure,plot3(Xp(:,1),Xp(:,2),Xp(:,3),'k*')
xlabel('PC 1'),ylabel('PC 2'),zlabel('PC 3')
```

The results are shown in Figure 2.3.

❑

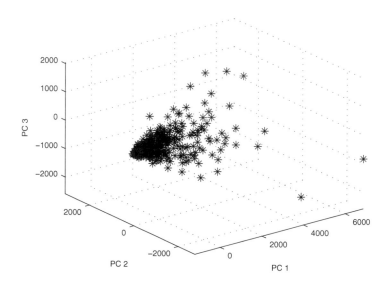

FIGURE 2.3
This shows the results of projecting the **yeast** data onto the first three PCs.

We illustrated the use of the **eig** function that comes in the main MATLAB package. There is another useful function called **eigs** that can be used to find the PCs and eigenvalues of sparse matrices. For those who have the Statistics Toolbox, there is a function called **princomp**. It returns the PCs, the PC scores, and other useful information for making inferences regarding the eigenvalues. For PCA using the covariance matrix, the **pcacov** function is provided.

Before moving on to the next topic, we recall that the PCA methodology described in this book is based on the sample covariance or sample correlation matrix. The procedure is similar if the population version of these matrices is used. Many interesting and useful properties are known about principal components and the associated data transformations, but these are beyond the scope and purpose of this book. We provide references for further reading in the last section.

2.3 Singular Value Decomposition — SVD

Singular value decomposition (SVD) is an important method from linear algebra and is related to PCA. In fact, it provides a way to find the PCs without explicitly calculating the covariance matrix [Gentle, 2002]. It also enjoys widespread use in the analysis of gene expression data [Alter, Brown, and Botstein, 2000; Wall, Dyck, and Brettin, 2001; Wall, Rechtsteiner, and

Rocha, 2003] and in information retrieval applications [Deerwester et al., 1990; Berry, Dumais, and O'Brien, 1995; Berry, Drmac, and Jessup, 1999]; so it is an important technique in its own right.

As before, we start with our data matrix $\mathbf{X}$, where in some cases, we will center the data about their mean to get $\mathbf{X}_c$. We use the noncentered form in the explanation that follows, but the technique is valid for an arbitrary matrix; i.e., the matrix does not have to be square. The SVD of $\mathbf{X}$ is given by

$$\mathbf{X} = \mathbf{UDV}^T, \tag{2.5}$$

where $\mathbf{U}$ is an $n \times n$ matrix, $\mathbf{D}$ is a diagonal matrix with n rows and p columns, and $\mathbf{V}$ has dimensions $p \times p$. The columns of $\mathbf{U}$ and $\mathbf{V}$ are orthonormal. $\mathbf{D}$ is a matrix containing the *singular values* along its diagonal, which are the square roots of the eigenvalues of $\mathbf{X}^T\mathbf{X}$, and zeros everywhere else.

The columns of $\mathbf{U}$ are called the *left singular vectors* and are calculated as the eigenvectors of $\mathbf{XX}^T$ (this is an $n \times n$ matrix). Similarly, the columns of $\mathbf{V}$ are called the *right singular vectors*, and these are the eigenvectors of $\mathbf{X}^T\mathbf{X}$ (this is a $p \times p$ matrix). An additional point of interest is that the singular values are the square roots of the eigenvalues of $\mathbf{X}^T\mathbf{X}$ and $\mathbf{XX}^T$.

Let the rank of the matrix $\mathbf{X}$ be given by r, where

$$r \leq \min(n, p).$$

Then the first r columns of $\mathbf{V}$ form an orthonormal basis for the column space of $\mathbf{X}$, and the first r columns of $\mathbf{U}$ form a basis for the row space of $\mathbf{X}$ [Strang, 1993]. As with PCA, we order the singular values largest to smallest and impose the same order on the columns of $\mathbf{U}$ and $\mathbf{V}$. A lower rank approximation to the original matrix $\mathbf{X}$ is obtained via

$$\mathbf{X}_k = \mathbf{U}_k\mathbf{D}_k\mathbf{V}_k^T, \tag{2.6}$$

where $\mathbf{U}_k$ is an $n \times k$ matrix containing the first k columns of $\mathbf{U}$, $\mathbf{V}_k$ is the $p \times k$ matrix whose columns are the first k columns of $\mathbf{V}$, and $\mathbf{D}_k$ is a $k \times k$ diagonal matrix whose diagonal elements are the k largest singular values of $\mathbf{X}$. It can be shown that the approximation given in Equation 2.6 is the best one in a least squares sense.

To illustrate the SVD approach, we look at an example from information retrieval called *latent semantic indexing* or LSI [Deerwester et al., 1990]. Many applications of information retrieval (IR) rely on lexical matching, where words are matched between a user's query and those words assigned to documents in a corpus or database. However, the words people use to describe documents can be diverse and imprecise; so the results of the queries are often less than perfect. LSI uses SVD to derive vectors that are more robust at representing the meaning of words and documents.

TABLE 2.1

Document Information for Example 2.3

Number	Title
Doc 1	How to *Bake Bread* Without *Recipes*
Doc 2	The Classic Art of Viennese *Pastry*
Doc 3	Numerical *Recipes*: The Art of Scientific Computing
Doc 4	*Breads, Pastries, Pies* and *Cakes*: Quantity *Baking Recipes*
Doc 5	*Pastry*: A Book of Best French *Recipes*
Term 1	bak (e, ing)
Term 2	recipes
Term 3	bread
Term 4	cake
Term 5	pastr (y, ies)
Term 6	pie

Example 2.3

This illustrative example of SVD applied to information retrieval (IR) is taken from Berry, Drmac, and Jessup [1999]. The documents in the corpus comprise a small set of book titles, and a subset of words have been used in the analysis, where some have been replaced by their root words (e.g., *bake* and *baking* are both *bake*). The documents and terms are shown in Table 2.1. We start with a data matrix, where each row corresponds to a term, and each column corresponds to a document in the corpus. The elements of the term-document matrix $\mathbf{X}$ denote the number of times the word appears in the document. In this application, we are not going to center the observations, but we do pre-process the matrix by normalizing each column such that the magnitude of the column is 1. This is done to ensure that relevance between the document and the query is not measured by the absolute count of the terms.[1] The following MATLAB code starts the process:

```
load lsiex
% Loads up variables: X, termdoc, docs and words.
% Convert the matrix to one that has columns
% with a magnitude of 1.
[n,p] = size(termdoc);
for i = 1:p
    termdoc(:,i) = X(:,i)/norm(X(:,i));
end
```

Say we want to find books about *baking bread*. Then the vector that represents this query is given by a column vector with a 1 in the first and third positions:

---

[1] Other term weights can be found in the literature [Berry and Browne, 2005].

```
q1 = [1 0 1 0 0 0]';
```

If we are seeking books that pertain only to *baking*, then the query vector is:

```
q2 = [1 0 0 0 0 0]';
```

We can find the most relevant documents using the original term-document matrix by finding the cosine of the angle between the query vectors and the columns (i.e., the vectors representing documents or books); the higher cosine values indicate greater similarity between the query and the document. This would be a straightforward application of lexical matching. Recall from matrix algebra that the cosine of the angle between two vectors $\mathbf{x}$ and $\mathbf{y}$ is given by

$$\cos\theta_{x,y} = \frac{\mathbf{x}^T\mathbf{y}}{\sqrt{\mathbf{x}^T\mathbf{x}}\sqrt{\mathbf{y}^T\mathbf{y}}}.$$

The MATLAB code to find the cosines for the query vectors in our example is:

```
% Find the cosine of the angle between
% columns of termdoc and a query vector.
% Note that the magnitude of q1 is not 1.
m1 = norm(q1);
cosq1a = q1'*termdoc/m1;
% The magnitude of q2 happens to be 1.
cosq2a = q2'*termdoc;
```

The resulting cosine values are:

```
cosq1a = 0.8165, 0, 0, 0.5774, 0
cosq2a = 0.5774, 0, 0, 0.4082, 0
```

If we use a cutoff value of 0.5, then the relevant books for our first query are the first and the fourth ones, which are those that describe *baking bread*. On the other hand, the second query matches with the first book, but misses the fourth one, which would be relevant. Researchers in the area of IR have applied several techniques to alleviate this problem, one of which is LSI. One of the powerful uses of LSI is in matching a user's query with existing documents in the corpus whose representations have been reduced to lower rank approximations via the SVD. The idea is that some of the dimensions represented by the full term-document matrix are noise and that documents will have closer semantic structure after dimensionality reduction using SVD. So, we now find the singular value decomposition using the function **svd**, which is part of the main MATLAB software.

```
% Find the singular value decomposition.
[u,d,v] = svd(termdoc);
```

We then find the representation of the query vector in the reduced space given by the first k columns of $\mathbf{U}$ in the following manner

$$\mathbf{q}_k = \mathbf{U}_k^T \mathbf{q},$$

which is a vector with k elements. We note that in some applications of LSI, the following is used as the reduced query

$$\mathbf{q}_k = \mathbf{D}_k^{-1} \mathbf{U}_k^T \mathbf{q}.$$

This is simply a scaling by the singular values, since $\mathbf{D}$ is diagonal. The following code projects the query into the reduced space and also finds the cosine of the angle between the query vector and the columns. Berry, Drmac and Jessup show that we do not have to form the full reduced matrix $\mathbf{X}_k$. Instead, we can use the columns of $\mathbf{V}_k$, saving storage space.

```
% Project the query vectors.
q1t = u(:,1:3)'*q1;
q2t = u(:,1:3)'*q2;
% Now find the cosine of the angle between the query
% vector and the columns of the reduced rank matrix,
% scaled by D.
for i = 1:5
    sj = d(1:3,1:3)*v(i,1:3)';
    m3 = norm(sj);
    cosq1b(i) = sj'*q1t/(m3*m1);
    cosq2b(i) = sj'*q2t/(m3);
end
```

From this we have

```
cosq1b = 0.7327, -0.0469, 0.0330, 0.7161, -0.0097
cosq2b = 0.5181, -0.0332, 0.0233, 0.5064, -0.0069
```

Using a cutoff value of 0.5, we now correctly have documents 1 and 4 as being relevant to our queries on *baking bread* and *baking*.
❑

   Note that in the above loop, we are using the magnitudes of the original query vectors as in Berry, Drmac, and Jessup [Equation 6.1, 1999]. This saves on computations and also improves precision (disregarding irrelevant information). We could divide by the magnitudes of the reduced query vectors (`q1t` and `q2t`) in the loop to improve recall (retrieving relevant information) at the expense of precision.
   Before going on to the next topic, we point out that much of the literature describing SVD as it is applied to LSI and gene expression data defines the matrices of Equation 2.5 in a different way. The decomposition is the same,

but the difference is in the size of the matrices **U** and **D**. Some definitions have the dimensions of **U** as $n \times p$ and **D** with dimensions $p \times p$ [Golub and Van Loan, 1996]. We follow the definition in Strang [1988, 1993], which is also used in MATLAB.

2.4 Nonnegative Matrix Factorization

One of the issues with the application of SVD within a text processing framework and other similar applications is that the entries within the original data matrix are all zero or greater. For instance, the elements of the term-document matrix will always be nonnegative, since they represent counts.[2] It would be desirable in these situations to have a method for reducing the dimensionality that is guaranteed to produce nonnegative features. This is not always the case with more classical methods, such as principal component analysis or singular value decomposition. For example, negative entries in the SVD factor matrices **U** and $\mathbf{V}^T$ cause them to not be as easily interpreted as the entries in the original matrix.

An alternative to SVD is called ***nonnegative matrix factorization*** (NMF) [Berry et al., 2007]. NMF casts matrix factorization as a constrained optimization problem that seeks to factor the original matrix into the product of two nonnegative matrices. A nonnegative matrix is one whose entries are all constrained to be nonnegative. Besides being easier to interpret, this type of matrix factorization has been shown to provide better results in information retrieval, clustering, and other applications [Xu, Liu, and Gong, 2003; Berry and Browne, 2005].

We follow Berry et al. [2007] in our development of the mathematical formalism of NMF for dimensionality reduction. Let's say we have our data matrix **X**, which is an $n \times p$ matrix. We seek a rank k approximation to **X** given by the product **WH**, where **W** is a nonnegative $n \times k$ matrix, and **H** is a nonnegative $k \times p$ matrix. We find these factor matrices by minimizing the following mean squared error objective function:

$$f(\mathbf{W}, \mathbf{H}) = \frac{1}{2}\|\mathbf{X} - \mathbf{WH}\|^2. \tag{2.7}$$

The product of the matrices **W** and **H** is called a factorization, but it is important to note that **X** is not necessarily equal to this product. Rather, it is an approximate factorization with rank less than or equal to k.

It turns out that we get some interesting outputs from this factorization. The columns of **W** represent a transformation of the observations in a k-dimensional space, without the need for any further matrix manipulations.

---

[2] This includes counts that might be normalized.

Also, the k rows of **H** contain the coefficients in the linear combination of the original variables in **X**.

Some of the issues that arise in solving the NMF optimization problem are the existence of local minima in the solution space, the choice of an effective initialization for the algorithms, and the lack of a unique solution. These issues have not stopped the development of methods for finding matrices **W** and **H** nor the use of NMF in many data mining and EDA applications.

Berry et al. describe three general classes of algorithms for constructing a nonnegative matrix factorization. These are called the multiplicative update, alternating least squares, and gradient descent algorithms. They note that this taxonomy is somewhat fuzzy, since some approaches can belong to more than one of these categories. In what follows, we will describe the first two algorithms, since they are the ones that are implemented in the MATLAB Statistics Toolbox. However, Cichocki and Zdunek [2006] have a toolbox called NMFLAB that provides functions for all of these approaches.

Most of the standard NMF algorithms start off with matrices **W** and **H** that are initialized with nonnegative entries. This is the case with the original multiplicative update algorithm with the mean squared error objective function from Lee and Seung [2001]. Their procedure is outlined below.

Procedure – Multiplicative Update Algorithm

1. Initialize **W** as an $n \times k$ matrix with nonnegative entries that are randomly generated between zero and one.
2. Initialize **H** as a $k \times p$ random matrix with nonnegative entries between zero and one.
3. Update **H** using

$$\mathbf{H} = \mathbf{H} \mathbin{.*} (\mathbf{W}^T \mathbf{X}) \mathbin{./} (\mathbf{W}^T \mathbf{W} \mathbf{H} + 10^{-9}).$$

4. Update **W** using

$$\mathbf{W} = \mathbf{W} \mathbin{.*} (\mathbf{X} \mathbf{H}^T) \mathbin{./} (\mathbf{W} \mathbf{H} \mathbf{H}^T + 10^{-9}).$$

5. Repeat steps 3 and 4 until convergence or up to some maximum number of iterations.

The procedure outlined above uses the MATLAB operator notation, where '.*' indicates multiplication of matrices element-by-element and './' denotes element-by-element division. The term 10^{-9} is included to prevent a division by zero.

We can see from this algorithm that the results will be dependent on the initial random matrices; so the analyst should obtain the factorization for different starting values and explore the results. The multiplicative update algorithm tends to be more sensitive to initialization than the alternating least squares approach, which is covered next. It has also been shown that the multiplicative update procedure is slow to converge [Berry et al., 2007].

Procedure – Alternating Least Squares

1. Initialize **W** as an $n \times k$ matrix with nonnegative entries that are randomly generated between zero and one.
2. Solve for **H** in the equation

$$\mathbf{W}^T\mathbf{W}\mathbf{H} = \mathbf{W}^T\mathbf{X}.$$

3. Set all negative elements in **H** to 0.
4. Solve for **W** in the equation

$$\mathbf{H}\mathbf{H}^T\mathbf{W}^T = \mathbf{H}\mathbf{X}^T.$$

5. Set all negative elements in **W** to 0.
6. Repeat steps 2 through 5 until convergence or up to some maximum number of iterations.

We can see from the steps above that we have a least squares step, where we solve for one of the factor matrices, followed by another least squares step to solve for the other one. In between, we ensure nonnegativity by setting any negative elements to zero.

The first step in the multiplicative update procedure and the alternating least squares procedure is to initialize the matrices with random values. Other methods for initializing matrix **W** can be used, and a good reference for these is Langville et al., 2006. We explore the use of NMF in the next example.

Example 2.4

We return to the data set used in the previous example to illustrate the use of nonnegative matrix factorization for information retrieval. First, we load the data and normalize the columns.

```
% Loads up variable: X, termdoc, docs, and words
load lsiex
[n,p] = size(termdoc);
% Normalize columns to be unit norm
for i = 1:p
    termdoc(:,i) = X(:,i)/norm(X(:,i));
end
```

Next, we are going to factor the **termdoc** matrix into a nonnegative product of two matrices **W** and **H**, where **W** is 6×3 and **H** is 3×5. The following code utilizes the multiplicative update option of the NMF function included in the Statistics Toolbox.

```
[W,H] = nnmf(termdoc,3,'algorithm','mult');
```

Recall that we had the following queries and are looking for documents that match them.

```
q1 = [1 0 1 0 0 0]';
q2 = [1 0 0 0 0 0]';
```

Now we compute the rank k approximation to our term-document matrix that we obtained using nonnegative matrix factorization.

```
termdocnmfk = W * H;
```

We use the cosine measure and the columns of our approximated term-document matrix to find the closest matches to our queries.

```
for i = 1:5
    m1 = norm(q1);
    m2 = norm(termdocnmfk(:,i));
    cosq1c(i) = (q1' * termdocnmfk(:,i))/(m1*m2);
    m1 = norm(q2);
    m2 = norm(termdocnmfk(:,i));
    cosq2c(i) = (q2' * termdocnmfk(:,i))/(m1*m2);
end
```

For this run, our results are

```
cosq1c = 0.7449 0.0000 0.0100 0.7185 0.0056
cosq2c = 0.5268 0.0000 0.0075 0.5080 0.0043
```

If we use a threshold of 0.5, then we see that the first and fourth documents match our queries.
❑

We used the multiplicative update algorithm with only one replicate in the previous example. The reader is asked to explore this further by running the MATLAB **nnmf** function with several replicates and to also try the alternating least squares algorithm. This has been shown to converge faster and to better solutions. We will return to the topic of nonnegative matrix factorization in Chapter 5, where we show how it can be used for clustering.

2.5 Factor Analysis

There is much confusion in the literature as to the exact definition of the technique called *factor analysis* [Jackson, 1981], but we follow the commonly used definition given in Jolliffe [1986]. In the past, this method has also been confused with PCA, mostly because PCA is sometimes provided in software packages as a special case of factor analysis. Both of these techniques attempt to reduce the dimensionality of a data set, but they are different from one another in many ways. We describe these differences at the end of the discussion of factor analysis.

The idea underlying factor analysis is that the p observed random variables can be written as linear functions of $d < p$ unobserved *latent variables* or *common factors* f_j, as follows:

$$x_1 = \lambda_{11} f_1 + \ldots + \lambda_{1d} f_d + \varepsilon_1$$
$$\ldots \tag{2.8}$$
$$x_p = \lambda_{p1} f_1 + \ldots + \lambda_{pd} f_d + \varepsilon_p.$$

The λ_{ij} ($i = 1, \ldots, p$ and $j = 1, \ldots, d$) in the above model are called the *factor loadings*, and the error terms ε_i are called the *specific factors*. Note that the error terms ε_i are specific to *each* of the original variables, while the f_j are common to *all* of the variables. The sum of the squared factor loadings for the i-th variable

$$\lambda_{i1}^2 + \ldots + \lambda_{id}^2$$

is called the *communality* of x_i.

We see from the model in Equation 2.8 that the original variables are written as a linear function of a smaller number of variables or factors, thus reducing the dimensionality. It is hoped that the factors provide a summary or clustering of the original variables (not the observations), such that insights are provided about the underlying structure of the data. While this model is fairly standard, it can be extended to include more error terms, such as measurement error. It can also be made nonlinear [Jolliffe, 1986].

The matrix form of the factor analysis model is

$$\mathbf{x} = \Lambda \mathbf{f} + \mathbf{e}. \tag{2.9}$$

Some assumptions are made regarding this model, which are

$$E[\mathbf{e}] = 0 \qquad E[\mathbf{f}] = 0 \qquad E[\mathbf{x}] = 0,$$

where $E[\bullet]$ denotes the expected value. If the last of these assumptions is violated, the model can be adjusted to accommodate this, yielding

$$\mathbf{x} = \Lambda \mathbf{f} + \mathbf{e} + \mu, \qquad (2.10)$$

where $E[\mathbf{x}] = \mu$. We also assume that the error terms ε_i are uncorrelated with each other, and that the common factors are uncorrelated with the specific factors f_j. Given these assumptions, the sample covariance (or correlation) matrix is of the form

$$\mathbf{S} = \Lambda^T \Lambda + \Psi,$$

where Ψ is a diagonal matrix representing $E[\mathbf{e}\mathbf{e}^T]$. The variance of ε_i is called the *specificity* of x_i; so the matrix Ψ is also called the *specificity matrix*. Factor analysis obtains a reduction in dimensionality by postulating a model that relates the original variables x_i to the d hypothetical variables or factors.

The matrix form of the factor analysis model is reminiscent of a regression problem, but here both Λ and $\mathbf{f}$ are unknown and must be estimated. This leads to one problem with factor analysis: the estimates are not unique. Estimation of the parameters in the factor analysis model is usually accomplished via the matrices Λ and Ψ. The estimation proceeds in stages, where an initial estimate is found by placing conditions on Λ. Once this initial estimate is obtained, other solutions can be found by rotating Λ. The goal of some rotations is to make the structure of Λ more interpretable, by making the λ_{ij} close to one or zero. Several methods that find rotations such that a desired criterion is optimized are described in the literature. Some of the commonly used methods are varimax, equimax, orthomax, quartimax, promax, and procrustes.

These factor rotation methods can either be *orthogonal* or *oblique*. In the case of orthogonal rotations, the axes are kept at 90 degrees. If this constraint is relaxed, then we have oblique rotations. The orthogonal rotation methods include quartimax, varimax, orthomax, and equimax. The promax and procrustes rotations are oblique.

The goal of the quartimax rotation is to simplify the *rows* of the factor matrix by getting a variable with a high loading on one factor and small loadings on all other factors. The varimax rotation focuses on simplifying the *columns* of the factor matrix. For the varimax approach, perfect simplification is obtained if there are only ones and zeros in a single column. The output from this method tends to have high loadings close to ± 1 and some near zero in each column. The equimax rotation is a compromise between these two methods, where both the rows and the columns of the factor matrix are simplified as much as possible.

Just as we did in PCA, we might want to transform the observations using the estimated factor analysis model either for plotting purposes or for further analysis methods, such as clustering or classification. We could think of these observations as being transformed to the 'factor space.' These are called *factor scores*, similarly to PCA. However, unlike PCA, there is no single method for finding the factor scores, and the analyst must keep in mind that the factor scores are really *estimates* and depend on the method that is used.

An in-depth discussion of the many methods for estimating the factor loadings, the variances Ψ, and the factor scores, as well as the rotations, is beyond the scope of this book. For more information on the details of the methods for estimation and rotation in factor analysis, see Jackson [1991], Lawley and Maxwell [1971], or Cattell [1978]. Before we go on to an example of factor analysis, we note that the MATLAB Statistics Toolbox uses the maximum likelihood method to obtain the factor loadings, and it also implements some of the various rotation methods mentioned earlier.

Example 2.5

In this example, we examine some data provided with the Statistics Toolbox, called **stockreturns**. An alternative analysis of these data is provided in the Statistics Toolbox User's Guide. The data set consists of 100 observations, representing the percent change in stock prices for 10 companies. Thus, the data set has $n = 100$ observations and $p = 10$ variables. It turns out that the first four companies can be classified as technology, the next three as financial, and the last three as retail. We can use factor analysis to see if there is any structure in the data that supports this grouping. We first load up the data set and perform factor analysis using the function **factoran**.

```
load stockreturns
% Loads up a variable called stocks.
% Perform factor analysis:3 factors,default rotation.
[LamVrot,PsiVrot] = factoran(stocks,3);
```

This is the basic syntax for **factoran**, where the user must specify the number of factors (3 in this case), and the default is to use the **varimax** rotation, which optimizes a criterion based on the variance of the loadings. See the MATLAB **help** on **factoran** for more details on the rotations. Next, we specify no rotation, and we plot the matrix **Lam** (the factor loadings) in Figure 2.4.

```
[Lam,Psi] = factoran(stocks,3,'rotate','none');
```

These plots show the pairwise factor loadings, and we can see that the factor loadings are not close to one of the factor axes, making it more difficult to interpret the factors. We can try rotating the matrix next using one of the oblique (nonorthogonal) rotations called **promax**, and we plot these results in Figure 2.5.

```
% Now try the promax rotation.
[LProt,PProt]=factoran(stocks,3,'rotate','promax');
```

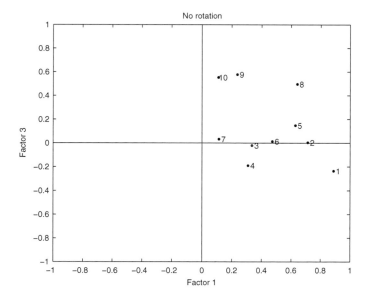

FIGURE 2.4
These show the factor loadings in their unrotated form. We see that the loadings are not grouped around the factor axes, although it is interesting to note that we have the three financial companies (points 5, 6, & 7) grouped together in the upper plot (factors 1 and 2), while the three retail companies (points 8, 9, & 10) are grouped together in the lower plot (factors 1 and 3).

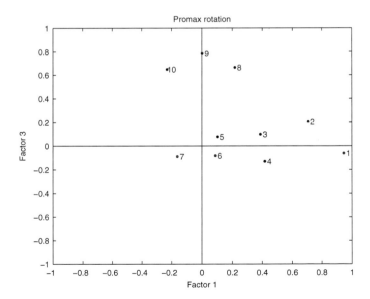

FIGURE 2.5
These plots show the factor loadings after the promax rotation. We see that the stocks can be grouped as technology companies {1, 2, 3, 4}, financial {5, 6, 7}, and retail {8, 9, 10}. The rotation makes the factors somewhat easier to interpret.

Note that we now have a more interpretable structure with the factor loadings, and we are able to group the stocks. We might also be interested in estimating the factor scores. The user is asked to explore this aspect of it in the exercises.
❏

It can be very confusing trying to decide whether to use PCA or factor analysis. Since the objective of this book is to describe exploratory data analysis techniques, we suggest that both methods be used to explore the data, because they take different approaches to the problem and can uncover different facets of the data. We now outline the differences between PCA and factor analysis; a more detailed discussion of which method to use can be found in Velicer and Jackson [1990].

- Both factor analysis and PCA try to represent the structure of the data set based on the covariance or correlation matrix. Factor analysis tries to explain the off-diagonal elements, while PCA explains the variance or diagonal elements of the matrix.

- Factor analysis is typically performed using the correlation matrix, and PCA can be used with either the correlation or the covariance matrix.

- Factor analysis has a model, as given in Equation 2.9 and 2.10, but PCA does not have an explicit model associated with it (unless one is interested in inferential methods associated with the eigenvalues and PCs, in which case, distributional assumptions are made).

- If one changes the number of PCs to keep, then the existing PCs do not change. If we change the number of factors, then the entire solution changes; i.e., existing factors must be re-estimated.

- PCA has a unique solution, while factor analysis does not.

- The PC scores are found in an exact manner, and the factor scores are estimates.

2.6 Fisher's Linear Discriminant

Linear discriminant analysis (LDA) has been used in the statistics community since the time of Fisher [1936]. Thus, similar to PCA, it can be classified as one of the traditional approaches for dimensionality reduction. It is also known as Fisher's linear discriminant or mapping (FLD) [Duda and Hart, 1973] and is one of the tools used for pattern recognition and supervised learning. Thus, linear discriminant analysis is a supervised method, unlike PCA and most other dimensionality reduction techniques.

LDA differs from PCA in other aspects, too. The goal of LDA is to reduce the dimensionality to 1–D, so we will be projecting onto a line (see Figure 2.1). In PCA, we can project to a lower dimension d, such that $1 \leq d < p$. Also, the goal of PCA is to find a d-dimensional representation that explains as much of the variance in the data as possible. In contrast, the objective of LDA is to find a linear projection where the projected observations are well separated.

When we say that LDA is a *supervised method*, we mean that we have class or group labels attached to each of our observations. The classical LDA approach deals with the simplest case, where we have two classes (e.g., diseased and healthy, fraud and not fraud, etc.).

Given a set of two such groups of data in a high-dimensional space, it remains an open problem as to the best way to build a classifier that will distinguish between these observations and that will correctly classify future data. This problem is complicated in the case of high dimensions by Bellman's *curse of dimensionality* [1961]. A discussion of the many ways that have been developed over the years to build classifiers (regardless of how many dimensions there are in the data) is beyond the scope of this text, and the reader is referred to Duda, Hart, and Stork [2001] for a comprehensive treatment of this topic.

One approach to building a classifier with high-dimensional data is to project the observations onto a line in such a way that the observations are well-separated in this lower-dimensional space. The linear separability (and thus the classification) of the observations is greatly affected by the position or orientation of this line. This is illustrated in Figure 2.6, where we have two possible projections of the data. It is clear that the projection to the horizontal axis will produce a representation of the data where the points overlap, which will make building a classifier difficult. The other projection is much better, because it yields a representation with widely separated classes.

In LDA we seek a linear mapping that maximizes the linear class separability in the new representation of the data. We follow Duda and Hart [1973] in our discussion of LDA (or FLD in Duda and Hart). We consider a set of n p-dimensional observations $\mathbf{x}_1, ..., \mathbf{x}_n$, with n_1 samples labeled as belonging to class 1 (λ_1) and n_2 samples as belonging to class 2 (λ_2). We will denote the set of observations in the i-th class as Λ_i.

Given a vector $\mathbf{w}$ with unit norm, we may form a projection of the $\mathbf{x}_i$ onto a line in the direction of $\mathbf{w}$ using

$$y = \mathbf{w}^T \mathbf{x}. \tag{2.11}$$

We want to choose $\mathbf{w}$ in order provide a linear mapping that yields maximal separation of the two classes.

One natural measure of separation between the projected points y_i is the difference between their means. We may calculate the p-dimensional sample mean for each class using

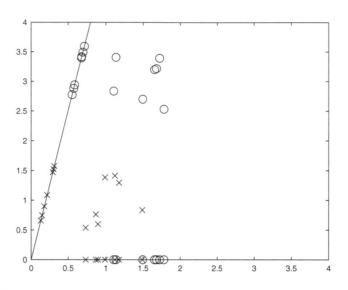

FIGURE 2.6
This shows how different projections of two-class data can produce results that either allow us to linearly separate the classes or not. Here we have two classes represented by **x**'s and **o**'s. We could project to the horizontal axis, but that produces a representation of the data where there is significant overlap for values a little greater than one. On the other hand, we could project the data onto a line such as the one shown here where we see clear separation between the classes.

$$\mathbf{m}_i = \frac{1}{n_i} \sum_{\mathbf{x} \in \Lambda_i} \mathbf{x}. \tag{2.12}$$

The sample mean for the projected points is given by

$$
\begin{aligned}
\tilde{m}_i &= \frac{1}{n_i} \sum_{y \in \Lambda_i} y \\
&= \frac{1}{n_i} \sum_{\mathbf{x} \in \Lambda_i} \mathbf{w}^T \mathbf{x}.
\end{aligned}
\tag{2.13}
$$

Combining Equations 2.12 and 2.13 yields

$$\tilde{m}_i = \mathbf{w}^T \mathbf{m}_i. \tag{2.14}$$

We can use Equation 2.14 to measure the separation of the means for the two classes:

$$\left| \tilde{m}_1 - \tilde{m}_2 \right| = \left| \mathbf{w}^T (\mathbf{m}_1 - \mathbf{m}_2) \right|. \tag{2.15}$$

It is possible to make the difference in Equation 2.15 as large as possible just by scaling $\mathbf{w}$. However, the magnitude of $\mathbf{w}$ is not really important for our desired mapping; it just scales the y. As we saw in Figure 2.6, it is the orientation or direction of the projection line that is significant.

To obtain good class separation, and hence good classification performance, we want the separation of the means to be as large as possible relative to some measure of standard deviation for the observations in each class. We will use the scatter as our measure of the standard deviations.

The scatter for the i-th class of projected data points is given by

$$\tilde{s}_i^2 = \sum_{y \in \Lambda_i} \left(y - \tilde{m}_i \right)^2.$$

We define the total within-class scatter to be $\tilde{s}_1^2 + \tilde{s}_2^2$. The LDA is defined as the vector $\mathbf{w}$ that maximizes the function

$$J(\mathbf{w}) = \frac{\left| \tilde{m}_1 - \tilde{m}_2 \right|^2}{\tilde{s}_1^2 + \tilde{s}_2^2}. \tag{2.16}$$

A little bit of linear algebra manipulation (see Duda and Hart [1973] for details) allows us to write the solution to the maximization of Equation 2.16 as

$$\mathbf{w} = \mathbf{S}_W^{-1}(\mathbf{m}_1 - \mathbf{m}_2), \tag{2.17}$$

where $\mathbf{S}_W$ is the within-class scatter matrix defined by

$$\mathbf{S}_W = \mathbf{S}_1 + \mathbf{S}_2,$$

and

$$\mathbf{S}_i = \sum_{x \in \Lambda_i} \left(\mathbf{x} - \mathbf{m}_i \right) \left(\mathbf{x} - \mathbf{m}_i \right)^T.$$

Note that the matrix $\mathbf{S}_W$ is proportional to the sample covariance matrix for the pooled p-dimensional data. Interestingly, it turns out that the LDA is the linear mapping with the maximum ratio of between-class scatter $\mathbf{S}_B$ to within-class scatter, where

$$S_B = (m_1 - m_2)(m_1 - m_2)^T.$$

Example 2.6

We will implement the LDA approach in this example. First we generate some observations that are multivariate normal using the **mvnrnd** function and plot them as points in the top panel of Figure 2.7.

```
n1 = 100;
n2 = 100;
cov1 = eye(2);
cov2 = [1 .9; .9 1];
% The mvnrnd function is in the Statistics Toolbox.³
dat1 = mvnrnd([-2 2],cov1,n1);
dat2 = mvnrnd([2 -2],cov2,n2);
plot(dat1(:,1),dat1(:,2),'x',...
     dat2(:,1),dat2(:,2),'o')
```

Now, we estimate the optimal linear mapping according to LDA. The first step is to get the within-class scatter matrix.

```
% Calculate the scatter matrices, using the fact that
% they are proportional
% to the sample covariance matrices.
scat1 = (n1-1)*cov(dat1);
scat2 = (n2-1)*cov(dat2);
% Now we compute the within-class scatter matrix.
Sw = scat1 + scat2;
```

Next, we have to get the sample means for each class. Note that these means are for the data represented in the original space.

```
% Next, we need the means for each class in the
% original space.
mu1 = mean(dat1);
mu2 = mean(dat2);
```

Now we can calculate the LDA vector **w** in Equation 2.17.

```
% The next steps calculate the LDA vector w.
% Note that we need to take the transpose of
% the means, since they need to be column vectors.
w = inv(Sw)*(mu1' - mu2');
% Normalize the vector w.
w = w/norm(w);
```

---

[3] The **mvnrnd** and **ksdensity** functions used in this example are from the Statistics Toolbox, but similar functionality is available in the free Computational Statistics Toolbox. See Appendix B for more information.

We use the **w** vector to project the data into a 1–D space (Equation 2.11).

```
% We can now calculate the projected data.
pdat1 = w'*dat1';
pdat2 = w'*dat2';
```

The data now reside in one dimension. So, we will construct a kernel density estimate using the **ksdensity** function to help us visualize them.

```
[ksd1,x1] = ksdensity(pdat1);
[ksd2,x2] = ksdensity(pdat2);
plot(pdat1,zeros(1,100),'x',pdat2,zeros(1,100),'o')
hold on
plot(x1,ksd1,x2,ksd2)
hold off
```

The results are shown in Figure 2.7 (bottom), where we can see that the data are well separated in 1–D when we use the linear mapping obtained from LDA. See Chapter 6 for more information on discriminant analysis.
❑

Duda and Hart [1973] also describe a similar approach for the c-class problem, where $c > 2$. In that case, LDA involves $c - 1$ discriminant functions; so the projection reduces the dimensionality from p dimensions to $c - 1$. The assumption here, of course, is that $p \geq c$.

2.7 Random Projections

In many modern data analysis situations, the dimensionality of the data is so large that the use of standard linear dimensionality reduction methods based on singular value decomposition or principal component analysis is either infeasible or intractable. However, we might want to explore and render our data in a lower dimensional space, such that the mapping is linear. We usually want to have this rendering or projection to be performed in such a manner as to preserve the interpoint distance structures between the data points. This is especially critical if subsequent analysis involves the interpoint distances (e.g., clustering). It turns out, in a rather counterintuitive manner, that one can actually preserve the interpoint distances via *random projections*.

We follow the treatment of random projections given in Vempala [2004]. Let $\mathbf{u} = (u_1, ..., u_p)^T$ be a column vector in $\mathbf{R}^p$. Consider an integer $d < p$ as the size of our smaller dimensional subspace. A d-dimensional subspace can be represented by a $p \times d$ matrix, where the columns are orthonormal. This means they are orthogonal to each other and have a norm equal to one. We will denote this matrix as **R**. From earlier in the chapter (Section 2.1), we see

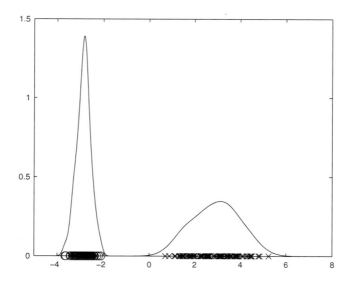

FIGURE 2.7

The top panel shows the scatterplot of the data in the original 2–D space (see Example 2.6). Class one observations are shown as x's, and class two data are shown as o's. The bottom panel shows the data as points on the horizontal axis after they have been mapped to a 1–D space that was found using LDA. We also construct and plot a kernel density estimate of each class and show them on the plot. It is clear that, in this case, the data can be easily separated in 1–D.

that this matrix is similar to the one containing the eigenvectors obtained through PCA.

One approach to obtain such a subspace is to choose the elements of **R** as a random sample from a continuous uniform distribution bounded by zero and one, so $r_{ij} \sim U(0, 1)$. Given this formulation, we may write a vector projected to this subspace as

$$\mathbf{x}_k = \sqrt{\frac{p}{k}}\mathbf{R}^T\mathbf{u}, \tag{2.18}$$

where the scalar coefficient in Equation 2.18 ensures that the squared length of $\mathbf{x}_k$ matches that of $\mathbf{u}$.

It might seem that requiring orthonormality of the columns of **R** would be burdensome, but this can be relaxed. Johnson and Lindenstrauss [1984] show that the entries of **R** can be chosen from a standard normal distribution. They also show that this type of projection ensures (with high probability) that the interpoint distance in the d-dimensional subspace matches approximately those in the original p-dimensional space. If the random entries of **R** are from a standard normal distribution, then we write

$$\mathbf{x}_k = \sqrt{\frac{1}{k}}\mathbf{R}^T\mathbf{u}. \tag{2.19}$$

Text data mining is one of the areas where the large size of the data matrix (i.e., the term–document matrix) makes random projections a viable option to reduce the dimensionality of the data [Bingham and Mannila, 2001; Boutsidis, et al., 2010; Ding, et al., 2013].

A famous data set for text analysis is the *20 Newsgroup* data. This data set was first formulated by Lang [1995]. The *20 Newsgroup* data has around 20,000 documents, chosen from twenty categories. This data set is available at the UCI Machine Learning Repository:

archive.ics.uci.edu/ml/datasets/Twenty+Newsgroups

and

http://qwone.com/~jason/20Newsgroups/

The second web site has a useful link to a compressed file containing a format that is easily read into MATLAB. We will use this data set in the next example to illustrate random projections.

Example 2.7

There are several data sets included with the 20 *Newsgroup* data. We will use the one called **test.data** in this example. Once it is extracted, you can load it, as follows.

```
% This has been downloaded from
%    http://qwone.com/~jason/20Newsgroups/
% Choose the MATLAB/Octave link and
% extract to your current directory.
load('test.data')
```

We can easily convert this to a sparse matrix object, as shown here.

```
% We convert this to a sparse version.
testsp = sparse(test(:,1),...
    test(:,2),test(:,3));

% Determine the size of the data matrix.
[n,p] = size(testsp);
```

The size function returns $n = 7505$ documents and $p = 61188$ words. Thus, we are in a very high-dimensional space. The cosine of the angle between two document vectors is typically used to determine how similar they are. Recall that the cosine between two (column) vectors **x** and **y** is given by

$$\cos(\theta_{x,y}) = \frac{\mathbf{x}^T \mathbf{y}}{\|\mathbf{x}\|\|\mathbf{y}\|},$$

where the numerator is the dot product, and $\|*\|$ is the magnitude of a vector. This is a similarity measure. We can convert to a distance by taking

$$D_{\cos(\theta_{x,y})} = 1 - \cos(\theta_{x,y}).$$

The following code finds the cosine distance between all pairs of documents.

```
% Here we calculate the cosine similarity:
%    cosine(theta) = (x * y)/(||x||*||y||)
% First, the x*y part of the cosine.
tmp1 = testsp * testsp';

% Here is the ||x||*||y|| part.
tmp2 = sqrt(diag(tmp1))*sqrt((diag(tmp1))');

% This is the cosine similarity.
costheta = tmp1./tmp2;

% We convert to a distance matrix.
% This has the distances in the full space.
Df = 1 - costheta;
```

The matrix **Df** is an $n \times n$ matrix of pairwise cosine distances. The diagonal elements of this matrix have to be zero. Some might not be exactly zero due to numerical errors. Thus, we first set those diagonal elements to zero.

```
Df(logical(eye(size(Df)))) = 0;
```

Next, we create the random projection matrix **R** and project to that subspace.

```
% We choose the dimensionality we are projecting to.
d = 300;
% Create the random projection matrix.
R = sparse(randn(p,d));

% We produce a matrix which is 7505 by 300.
testspproj = ((1/sqrt(d))*R'*testsp')';
```

We now calculate the pairwise cosine distances in the subspace.

```
% Calculate the cosine distance for the projected data.
tmp1proj = testspproj * testspproj';
tmp2proj = ...
    sqrt(diag(tmp1proj))*sqrt((diag(tmp1proj))');
costhetaproj = tmp1proj./tmp2proj;
Dproj = 1 - costhetaproj;
```

There are several ways one might compare the distances before and after the dimensionality has been reduced. We use the difference between the distance before and after the reduction.

```
% Compare using the difference in the distances.
D = Df - Dproj;
```

We repeated the same projection steps using a value of $d = 1000$. We show the distribution of the differences in the distances for both values of d in Figure 2.8. As expected, the average distance is close to zero, and we also get less variation when d is larger.
❑

2.8 Intrinsic Dimensionality

Knowing the *intrinsic dimensionality* (sometimes called the *effective dimensionality*) of a data set is useful information in exploratory data analysis. This is defined as the smallest number of dimensions or variables needed to model the data without loss [Kirby, 2001]. Fukunaga [1990] provides a similar definition of intrinsic dimensionality as "the minimum number of parameters needed to account for the observed properties of the data."

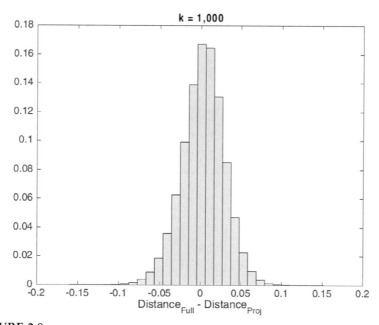

FIGURE 2.8

These plots show the distribution of the distances before and after projecting to a lower-dimensional space. The first one projects to 300 dimensions and the second to 1,000. Both show that the average difference is close to zero. We also see that the distances tend to be closer (less variance) when we use more dimensions.

Several approaches to estimating the intrinsic dimensionality of a data set have been proposed in the literature. Trunk [1968, 1976] describes a statistical approach using hypothesis testing regarding the most likely local dimensionality. Fukunaga and Olsen [1971] present an algorithm where the data are divided into small subregions, and the eigenvalues of the local covariance matrix are computed for each region. The intrinsic dimensionality is then obtained using the size of the eigenvalues.

More methods for estimating the intrinsic dimensionality have been developed in the recent literature, largely due to the increased interest in dimensionality reduction and manifold learning in the machine learning community. These newer approaches are now often categorized as being *local* or *global* [Camastra, 2003]. Local approaches estimate the dimensionality using information in neighborhoods of the observations, while global methods analyze the global properties of the data, making use of the whole data set. We have already discussed one global method for estimating the dimensionality when we looked at the amount of variance explained by the eigenvalues in principal component analysis.

In what follows, we describe several local estimators: nearest neighbor, correlation dimension, and maximum likelihood. These are followed by a global method based on packing numbers.

2.8.1 Nearest Neighbor Approach

Pettis et al. [1979] developed an algorithm for estimating intrinsic dimensionality that is based on nearest neighbor information and a density estimator. We choose to implement the original Pettis algorithm as outlined below; details of the derivation can be found in the original paper. We first set some notation, using 1-D representation for the observations as in the original paper. However, this method works with the *distance* between observations, and it easily extends to the multi-dimensional case. Let $r_{k,x}$ represent the distance from x to the k-th nearest neighbor of x. The average k-th nearest neighbor distance is given by

$$\overline{r}_k = \frac{1}{n} \sum_{i=1}^{n} r_{k,x_i}. \tag{2.20}$$

Pettis et al. [1979] show that the expected value of the average distance in Equation 2.20 is

$$E(\overline{r}_k) = \frac{1}{G_{k,d}} k^{1/d} C_n, \tag{2.21}$$

where

$$G_{k,d} = \frac{k^{1/d}\Gamma(k)}{\Gamma(k+1\div d)},$$

and C_n is independent of k. If we take logarithms of Equation 2.21 and do some rearranging, then we obtain the following

$$\log(G_{k,d}) + \log E(\bar{r}_k) = (1/d)\log(k) + \log(C_n). \tag{2.22}$$

We can get an estimate for $E(\bar{r}_k)$ by taking the observed value of $\bar{r}_k$ based on the data in our sample, yielding the following estimator $\hat{d}$ for the intrinsic dimensionality d

$$\log(G_{k,\hat{d}}) + \log(\bar{r}_k) = (1/\hat{d})\log(k) + \log(C_n). \tag{2.23}$$

This is similar to a regression problem, where the slope is given by $(1/\hat{d})$. The term $\log(C_n)$ affects the intercept, not the slope; so we can disregard this in the estimation process.

The estimation procedure must be iterative since $\hat{d}$ also appears on the response (left) side of Equation 2.23. To get started, we set the term $\log(G_{k,\hat{d}})$ equal to zero and find the slope using least squares. Here the predictor values are given by $\log(k)$, and the responses are $\log(\bar{r}_k)$, $k = 1, ..., K$. Once we have this initial value for $\hat{d}$, we then find $\log(G_{k,\hat{d}})$ using Equation 2.23. Using this value, we find the slope where the responses are now $\log(G_{k,\hat{d}}) + \log(\bar{r}_k)$. The algorithm continues in this manner until the estimates of intrinsic dimensionality converge.

We outline the algorithm shortly, but we need to discuss the effect of outliers first. Pettis et al. [1979] found that their algorithm works better if potential outliers are removed before estimating the intrinsic dimensionality. To define outliers in this context, we first define a measure of the maximum average distance given by

$$m_{\max} = \frac{1}{n}\sum_{i=1}^{n} r_{K,x_i},$$

where r_{K,x_i} represents the i-th K nearest neighbor distance. A measure of the spread is found by

$$s_{\max}^2 = \frac{1}{n-1}\sum_{i=1}^{n}\left(r_{K,x_i} - m_{\max}\right)^2.$$

The data points x_i for which

$$r_{K, x_i} \leq m_{max} + s_{max} \tag{2.24}$$

are used in the nearest neighbor estimate of intrinsic dimensionality. We are now ready to describe the method.

Procedure – Intrinsic Dimensionality

1. Set a value for the maximum number of nearest neighbors K.
2. Determine all of the distances r_{k, x_i}.
3. Remove outliers by keeping the points that satisfy Equation 2.24.
4. Calculate $\log(\bar{r}_k)$.
5. Get the initial estimate $\hat{d}_0$ by fitting a line to

$$\log(\bar{r}_k) = (1/\hat{d})\log(k),$$

and taking the inverse of the slope.
6. Calculate $\log(G_{k, \hat{d}_j})$ using Equation 2.23.
7. Update the estimate of intrinsic dimensionality by fitting a line to

$$\log(G_{k, \hat{d}_j}) + \log(\bar{r}_k) = (1/\hat{d})\log(k).$$

8. Repeat steps 6 and 7 until the estimates converge.

Example 2.8

The following MATLAB code implements the Pettis algorithm for estimating intrinsic dimensionality, which is part of the EDA Toolbox function called **idpettis**. We first generate some data to illustrate the functionality of the algorithm. The helix is described by the following equations, and points are randomly chosen along this path:

$$x = \cos\theta$$
$$y = \sin\theta$$
$$z = 0.1\theta$$

for $0 \leq \theta \leq 4\pi$. For this data set, the dimensionality is 3, but the intrinsic dimensionality is 1. We show a picture of the helix in Figure 2.9. We obtain the data by generating uniform random numbers in the interval $0 \leq \theta \leq 4\pi$.

```
% Generate the random numbers
% unifrnd is from the Statistics Toolbox.
n = 500;
```

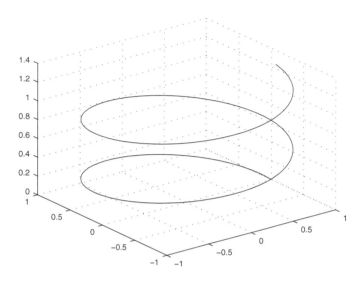

FIGURE 2.9
This shows the helix used in Example 2.8. This is a 1–D structure embedded in 3–D.

```
theta = unifrnd(0,4*pi,1,n);
% Use in the equations for a helix.
x = cos(theta);y = sin(theta);z = 0.1*(theta);
% Put into a data matrix.
X = [x(:),y(:),z(:)];
```

We note that the function **unifrnd** is from the Statistics Toolbox, but users can also employ the **rand** function and scale to the correct interval [Martinez and Martinez, 2015]. In the interest of space and clarity, we show only the code for the algorithm after the outliers have been removed. Thus, you cannot run the code given below as it stands. We refer curious readers to the MATLAB function **idpettis** for the implementation of the rest of the procedure.

```
% Get initial value for d. Values n, k, & logrk
% are defined in the idpettis function. This is
% an extract from that function.
logk = log(k);
[p,s] = polyfit(logk,logrk,1);
dhat = 1/p(1);
dhatold = realmax;
maxiter = 100;
epstol = 0.01;
i = 0;
```

```
while abs(dhatold - dhat) >= epstol & i < maxiter
    % Adjust the y values by adding logGRk
    logGRk = (1/dhat)*log(k)+...
             gammaln(k)-gammaln(k+1/dhat);
    [p,s] = polyfit(logk,logrk + logGRk,1);
    dhatold = dhat;
    dhat = 1/p(1);
    i = i+1;
end
idhat = dhat;
```

As an example of using the **idpettis** function, we offer the following:

```
% Get the distances using the pdist function.
% This returns the interpoint distances.
ydist = pdist(X);
idhat = idpettis(ydist,n);
```

where the input variable **X** is the helix data. The resulting estimate of the intrinsic dimensionality is 1.14. We see from this result that in most cases the estimate will need to be rounded to the nearest integer. Thus, the estimate in this case is the correct one: 1–D.

❑

2.8.2 Correlation Dimension

The *correlation dimension* estimator of intrinsic dimensionality is based on the assumption that the number of observations in a hypersphere with radius r is proportional to r^d [Grassberger and Procaccia, 1983; Camastra and Vinciarelli, 2002]. Given a set of observations $S_n = \{x_1, ..., x_n\}$ in a metric space, we can calculate the relative number of observations that are contained within a hypersphere of radius r by

$$C(r) = \frac{2}{n(n-1)} \sum_{i=1}^{n} \sum_{j=i+1}^{n} c, \qquad (2.25)$$

where

$$c = \begin{cases} 1, & \text{if } \left\| x_i - x_j \right\| \leq r \\ 0, & \text{if } \left\| x_i - x_j \right\| > r. \end{cases}$$

The value of $C(r)$ in Equation 2.25 is proportional to r^d; so we can use this to estimate the intrinsic dimensionality d. Thus, the correlation dimension is given by

$$d = \lim_{r \to 0} \frac{\log C(r)}{\log r}. \tag{2.26}$$

Since we have a finite sample, arriving at the limit of zero in Equation 2.26 is not possible. So, we plot $\log C(r)$ against $\log r$ and then estimate the slope of the linear part. Van der Maaten [2007] shows that this slope, and thus the intrinsic dimensionality, can be estimated by calculating $C(r)$ for two values of r and then finding the ratio:

$$\hat{d}_{corr}(r_1, r_2) = \frac{\log C(r_2) - \log C(r_1)}{\log r_2 - \log r_1}.$$

Fortunately, several local and global estimators of intrinsic dimensionality have been implemented in the Dimensionality Reduction Toolbox by L. J. P. van der Maaten.[4] The toolbox has a function that produces estimates based on the approaches described in this section and the following ones. So, we will wait until a later section to illustrate the use of the correlation dimension approach and other methods.

2.8.3 Maximum Likelihood Approach

As we have seen from the description of the previous two methods, local intrinsic dimensionality estimators are often based on the idea that the number of observations that are covered by a hypersphere of radius r around a given data point grows in proportion to r^d, where d is the intrinsic dimensionality of the manifold around that point. Levina and Bickel [2004] also use this idea to derive the *maximum likelihood estimator*, but instead they model the number of data points inside the hypersphere as a Poisson process [van der Maaten, 2007].

Consider a set of independent and identically distributed observations denoted by $x_1, \ldots, x_n$ residing in a p-dimensional space, R^p. We assume that the observations come from an embedding in a d-dimensional space, where $d \le p$. Thus, the $x_i = g(y_i)$, where the y_i are sampled from an unknown smooth density f with support on R^d. We also make the necessary mathematical assumptions to ensure that observations that are close together in the higher dimensional space are close together in the lower-dimensional embedding space.

---

[4] See Appendix B for the website address.

We choose a point x and assume that $f(x)$ is approximately constant within a small hypersphere $S_x(r)$ of radius r around x. We can treat these observations as a homogeneous Poisson process within $S_x(r)$.

Next, we consider the inhomogeneous process $\{N_x(t), 0 \le t \le r\}$,

$$N_x(t) = \sum_{i=1}^{n} 1\{x_i \in S_x(t)\}. \tag{2.27}$$

Equation 2.27 counts the observations that are within distance t from x. This can be approximated by a Poisson process, and we can write the rate $\lambda(t)$ of the process $N_x(t)$ at dimensionality d as

$$\lambda_x(t) = \frac{f(x)\pi^{d/2} dt^{d-1}}{\Gamma(d/2 + 1)},$$

where $\Gamma(\bullet)$ is the gamma function.

Levina and Bickel provide the relevant log-likelihood function and show that one obtains the following maximum likelihood estimator of the intrinsic dimensionality

$$\hat{d}_r(x) - \left[\frac{1}{N(r,x)} \sum_{j=1}^{N(r,x)} \log \frac{r}{T_j(x)} \right]^{(-1)}, \tag{2.28}$$

where $T_j(x)$ is the Euclidean distance from the point x to the j-th nearest neighbor within the hypersphere centered at x. Alternatively, one could view $T_j(x)$ as the radius of the smallest hypersphere centered at x that contains j observations.

The authors provide a more convenient way to arrive at the estimate by fixing the number of neighbors k instead of the radius r. The maximum likelihood estimate given in Equation 2.26 then becomes

$$\hat{d}_k(x) = \left[\frac{1}{k-1} \sum_{j=1}^{k-1} \log \frac{T_k(x)}{T_j(x)} \right]^{(-1)}. \tag{2.29}$$

It is clear from Equation 2.29 that the estimate depends on the parameter k (or the radius r of the hypersphere), and it also depends on the point x. Sometimes, the intrinsic dimensionality varies as a function of the location (x) in the data set, as well as the scale (k or r). Thus, it is a good idea to have estimates at different locations and scales. Levina and Bickel note that the hypersphere should be small and at the same time contain enough points for the estimator to work well.

They recommend that one obtain estimates of the intrinsic dimensionality at each observation x_i and over different scales k of small to moderate values. The final estimate is obtained by averaging these results, as shown here:

$$\hat{d}_{MLE} = \frac{1}{k_2 - k_1 + 1} \sum_{k=k_1}^{k_2} \overline{d}_k, \qquad (2.30)$$

where

$$\overline{d}_k = \frac{1}{n} \sum_{i=1}^{n} \hat{d}_k(x_i).$$

Levina and Bickel recommend values of $k_1 = 10$ and $k_2 = 20$.

MacKay and Ghahramani [2005] noticed the tendency for the maximum likelihood estimator in Equation 2.30 to exhibit strong bias in the estimate for small k, even when n is large. They advocate averaging the inverses of the estimators to alleviate this problem.

2.8.4 Estimation Using Packing Numbers

Kegl [2003] developed a method for estimating the intrinsic dimensionality that is based on the geometric properties of the data, so it does not use a data generating model. Another nice aspect of this approach is that it does not have parameters to set, as we saw with some of the previous methods. Kegl's estimate is called the *packing number estimator.*

Given a metric space, the *r-covering number* $N(r)$ of a set $S = \{x_1, ..., x_n\}$ in this space is the smallest number of hyperspheres with radius r that is needed to cover all observations x_i in the data set S. The r-covering number $N(r)$ of a d-dimensional data set is proportional to r^{-d}, and we can define the *capacity dimension* as

$$d = -\lim_{r \to 0} \frac{\log N(r)}{\log r}. \qquad (2.31)$$

It is not possible to find the r-covering number $N(r)$ for most data sets. So, Kegl uses the *r-packing number* $M(r)$ instead. This is defined as the maximum number of observations x_i in the data set S that can be covered by a single hypersphere of radius r [van der Maaten, 2007].

Using the following inequality between packing and covering numbers:

$$N(r) \le M(r) \le N(r/2),$$

we can find the intrinsic dimensionality of our data set by using the packing number in Equation 2.31:

$$d = -\lim_{r \to 0} \frac{\log M(r)}{\log r}. \qquad (2.32)$$

Since we have a finite sample, we cannot achieve the limit in Equation 2.32; so the packing estimate for intrinsic dimensionality is found using

$$\hat{d}_{pack}(r_1, r_2) = -\frac{\log M(r_2) - \log M(r_1)}{\log r_2 - \log r_1}.$$

Example 2.9

We are now ready to examine the outputs of the intrinsic dimensionality estimators we have just described. To do this, we will use the function from the Dimensionality Reduction Toolbox[5] called **intrinsic_dim**. The interested reader should download the toolbox and follow the installation instructions in the **readme.txt** file. We will also use another function from this same toolbox called **generate_data** that will allow us to generate data based on a helix, which will be different from our previous example.

```
% First, generate a helix data set and plot it.
% The first argument specifies the type of data set.
% The second argument is the number of points, and
% the third governs the amount of noise.
[X] = generate_data('helix',2000,0.05);
plot3(X(:,1),X(:,2),X(:,3),'.')
grid on
```

The data are plotted in Figure 2.10, where we see that the data fall on a 1–D manifold along a helix. Next, we estimate the intrinsic dimensionality using the correlation dimension, the maximum likelihood estimator, and the packing number.

```
% Now find the estimates of intrinsic dimensionality
% that were described in the last three sections.
d_corr = intrinsic_dim(X,'CorrDim');
d_mle = intrinsic_dim(X,'MLE');
d_pack = intrinsic_dim(X,'PackingNumbers');
```

The results are given here:

```
d_corr = 1.6706
d_mle = 1.6951
d_pack = 1.2320
```

---

[5] See Appendix B for information on downloading the toolbox.

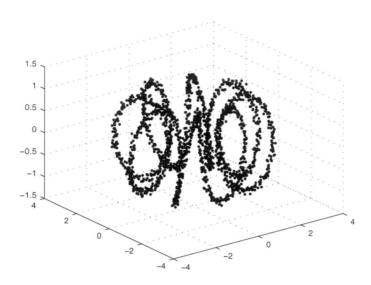

FIGURE 2.10
This is a scatterplot of the helix data generated in Example 2.9.

We see some differences in the results, with the packing number estimator providing the best answer with this artificial data set.
❑

2.8.5 Estimation of Local Dimension

What we presented previously are examples of estimating the global intrinsic dimension. However, there are applications where the data might lie on several manifolds with different dimensions. Some examples include assessing the complexity of a similarity search, detecting outliers, and estimating the local densities in classification [Amsaleg et al. 2015; Carter et al. 2010].

A natural approach to this problem is to find the intrinsic dimensionality in the neighborhood of each observation [Costa et al. 2005]. The process is simple and intuitive. First, we find the k–nearest neighbors for each observation. We then find the intrinsic dimensionality of this neighborhood using one of the methods discussed already. This will give us a distribution of the n local dimensions. We explore this idea in the next example.

Example 2.10

We generated a data set consisting of a cube (3–D), a spherical surface (2–D), and two lines (1–D). The data can be generated using a function we include in the toolbox called **genLDdata**. It will generate a data set similar to the one shown in Figure 2.11. Note that the Dimensionality Reduction Toolbox has to

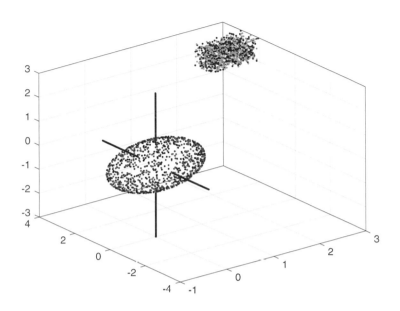

FIGURE 2.11
This data set has three sub-manifolds. We have lines (1–D), a spherical surface (2–D), and a cube (3–D). The global intrinsic dimension is 1.5229.

be downloaded and installed in an appropriate MATLAB directory [van der Maaten, 2007] to run the code in this example

```
% Generate the data set.
A = genLDdata;
% Estimate the global intrinsic dimensionality.
% The default is the MLE.
ID = intrinsic_dim(A);
```

We estimate the global intrinsic dimensionality using the **intrinsic_dim** function, and we get a value of 1.5229. We can estimate the intrinsic dimensionality at each observation with the following steps using a neighborhood size of $k = 100$.

```
% Get the pairwise interpoint distances.
% We use the default Euclidean distance.
Ad = squareform(pdist(A));
% Get the dimensions of the data.
[nr,nc] = size(A);
Ldim = zeros(nr,1);
Ldim2 = Ldim;
[Ads,J] = sort(Ad,2);
% Set the neighborhood size.
```

```
k = 100;
for m = 1 : nr
    Ldim(m,1) = ...
        intrinsic_dim(A(J(m,1:k),:));
end
```

The default method in **intrinsic_dim** is the MLE, which is used here. The maximum local intrinsic dimension is very large (5.3×10^5). The smallest has a value of 0.876. We know the local dimension has to be an integer, so we do a little conversion. This includes converting local dimension values more than three to four, which will indicate an error.

```
% We map all local dimensions greater
% than 3 to 4 and convert to integers.
Ldim(Ldim > 3) = 4;
Ldim = ceil(Ldim);
```

It is interesting to get a frequency distribution of the local dimension values.

```
% Tabulate them.
tabulate(Ldim)
```

Value	Count	Percent
1	3119	51.98%
2	1586	26.43%
3	1089	18.15%
4	206	3.43%

This table indicates at least 206 observations have incorrect estimates of the local dimension. We can visualize our results by coloring observations with their estimated local dimension in a scatterplot. This is shown in Figure 2.12, and the code used to construct the plot is given here.

```
% The following code constructs a
% scatterplot with colors mapped to
% the estimated local dimension.
ind1 = find(Ldim == 1);
ind2 = find(Ldim == 2);
ind3 = find(Ldim == 3);
ind4 = find(Ldim == 4);
scatter3(A(ind1,1),A(ind1,2),A(ind1,3),'r.')
hold on
scatter3(A(ind2,1),A(ind2,2),A(ind2,3),'g.')
scatter3(A(ind3,1),A(ind3,2),A(ind3,3),'b.')
scatter3(A(ind4,1),A(ind4,2),A(ind4,3),'k.')
hold off
```

❑

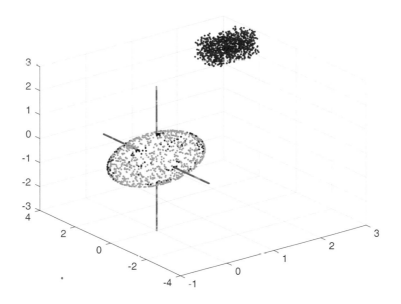

FIGURE 2.12
This is a scatterplot of the data in Figure 2.11 where we colored the points based on their estimated local dimension. We see that some points on the sphere and line are assigned an incorrect intrinsic dimension. (**SEE COLOR INSERT.**)

2.9 Summary and Further Reading

In this chapter, we introduced the concept of dimensionality reduction by presenting several methods that find a linear mapping from a high-dimensional space to a lower one. These methods include principal component analysis, singular value decomposition, and factor analysis. We also addressed the issue of determining the intrinsic dimensionality of a data set.

For a general treatment of linear and matrix algebra, we recommend Strang [1988, 1993] or Golub and van Loan [1996]. Discussions of PCA and factor analysis can be found in most multivariate analysis books; some examples are Manly [1994] or Seber [1984]. Readers who would like a more detailed treatment of these subjects are referred to Jackson [1991] and Jolliffe [1986]. These texts provide extensive discussions (including many extensions and applications) of PCA, SVD, and factor analysis. There are many applications of PCA and SVD to microarray analysis; some examples include gene shaving [Hastie et al., 2000], predicting the clinical status of human breast cancer [West et al., 2001], obtaining a global picture of the dynamics of gene expression [Alter, Brown, and Botstein, 2000], and clustering [Yeung and Ruzzo, 2001].

There have been many applications of nonnegative matrix factorization described in the literature. Some of these include blind source separation [Cichocki et al., 2006], face recognition [Guillamet and Vitria, 2002], music transcription [Smaragdis and Brown, 2003], and the analysis of microarray data [Fogel et al., 2007]. Bucak and Gunsel [2009] introduce an incremental framework for nonnegative matrix factorization that is suitable for online processing of large data sets.

The Fukunaga-Olsen method of estimating intrinsic dimensionality is further explored in Verveer and Duin [1995] and Bruske and Sommer [1998]. The Pettis algorithm is also described in Fukunaga [1990] and is expanded in Verveer and Duin [1995]. Verveer and Duin [1995] revise both the Pettis and the Fukunaga-Olsen algorithms and provide an evaluation of their performance. Costa and Hero [2004] propose a geometric approach based on entropic graph methods. Their method is called the geodesic minimal spanning tree (GMST), and it yields estimates of the manifold dimension and the α-entropy of the sample density on the manifold.

Raginsky and Lazebnik [2006] propose a way to estimate the intrinsic dimensionality using the notion of the quantization dimension and high-rate vector quantization. Costa et al. [2005] devise a global method based on the *k*-nearest neighbor graphs. Li et al. [2007] develop a new method that also uses a nearest neighbor graph, but it is based on the convex hull. Fan et al. [2008] derive an estimator by finding the exponential relationship between the radius of an incising ball and the number of observations within the ball. Little et al. [2009] describe a global estimate of the intrinsic dimensionality that uses a multiscale version of the SVD. Camastra [2003] provides a survey of some dimensionality estimation methods, where his focus is on fractal-based approaches. A good paper that describes local intrinsic dimensionality estimation methods and provides some interesting applications (network anomaly detection, clustering, image segmentation) is Carter et al. [2010].

In this chapter, we discussed ways to use random matrices to reduce the dimensionality of our data. Randomization can also play an important role in classification or supervised learning. To learn more about this, the reader is referred to several papers by Rahimi and Recht [2007, 2008, 2009] detailing their work on the ***random kitchen sinks*** (RKS) algorithm. Their claim is that RKS can perform fast approximate Gaussian process regression. Le et al. [2013] used randomization based methods to develop a procedure called Fastfood, which they claim can compute n non-linear basis functions in $O(n\log p)$ time, where p is the number of dimensions. Finally, we refer the mathematically inclined reader to the recent work of Wang [2015] who provides in-depth discussions on random matrix calculations within MATLAB.

Exercises

2.1   Generate $n - 50$, $p = 3$ normally distributed random variables that have high variance in one dimension. For example, you might use the following MATLAB code to do this:

```
x1 = randn(50,1)*100;
x2 = randn(50,2);
X = [x1,x2];
```

Try PCA using both correlation and covariance matrices. Is the one with the covariance matrix very informative? Which one would be better to use in this case?

2.2   Do a scree plot of the data set used in Example 2.2. Generate multivariate, uncorrelated normal random variables for the same n and p. (Hint: use **randn(n,p)**.) Perform PCA for this data set, and construct its scree plot. It should look approximately like a straight line. Plot both scree plots together; where they intersect is an estimate of the number of PCs to keep. How does this compare with previous results? [Jackson, 1991, p. 46].

2.3   Generate a set of $n = 30$ trivariate normal random variables using **randn(30,3)**.

   a. Subtract the mean from the observations and find the covariance matrix, using **cov**. Get the eigenvectors and eigenvalues based on the covariance matrix of the centered observations. Is the total variance of the original data equal to the sum of the eigenvalues? Verify that the eigenvectors are orthonormal. Find the PC scores and their mean.

   b. Impose a mean of $[2, 2, 2]^T$ on the original variables. Perform PCA using the covariance of these noncentered data. Find the PC scores and their mean.

   c. Center and scale the original variables, so that each one has zero mean and a standard deviation of one. Find the covariance of these transformed data. Find the correlation matrix of the original, non-transformed data. Are these two matrices the same?

   d. Verify that the PC scores produce data that are uncorrelated by finding the correlation matrix of the PC scores.

2.4   Repeat Example 2.2 using the correlation matrix. Compare with the previous results.

2.5   Generate multivariate normal random variables, centered at the origin (i.e., centered about 0). Construct the matrix $X^T X$ **and** find the eigenvectors. Next, construct the matrix XX^T and find the eigenvectors. Now use the SVD on X. Compare the columns of U and V with the eigenvectors. Are they the same? Note that the columns of

U and **V** are unique up to a sign change. What can you say about the eigenvalues and singular values?

2.6   Generate a set of multivariate normal random variables, centered at the origin. Obtain the eigenvalue decomposition of the covariance matrix. Multiply them to get the original matrix back, and also perform the multiplication to get the diagonal matrix **L**.

2.7   Construct a plot based on the slopes of the lines connecting the points in the scree plot. Apply to the **yeast** data. Does this help in the analysis? Hint: use the **diff** function.

2.8   Apply PCA to the following data sets. What value of d would you get using the various methods for choosing the number of dimensions?
        a. Other gene expression data sets.

        b. **oronsay** data.

        c. **sparrow** data.

2.9   Repeat Example 2.3 for the SVD - LSI using $k = 2$. Comment on the results and compare to the document retrieval using $k = 3$.

2.10   Using the term-document matrix in Example 2.3, find the cosine of the angle between the reduced query vectors and the reduced document vectors. However, this time use the magnitude of the reduced query vectors in the loop, not the magnitudes of the original queries. Discuss how this affects your results in terms of precision and recall.

2.11   Generate a bivariate data set using either **rand** or **randn**. Verify that the singular values are the square roots of the eigenvalues of X^TX and XX^T.

2.12   Repeat Example 2.5, using other rotations and methods implemented in MATLAB for estimating the factor loadings. Do they change the results?

2.13   Plot factors 2 and 3 in Example 2.5 for no rotation and promax rotation. Discuss any groupings that might be evident.

2.14   Try Example 2.5 with different values for the number of factors—say two and four. How do the results change?

2.15   The MATLAB function **factoran** includes the option of obtaining estimates of the factor scores. Using the data in Example 2.5, try the following code

```
[lam,psi,T,stats,F]=...
      factoran(stocks,3,'rotate','promax');
```

The factor scores are contained in the variable **F**. These can be viewed using **plotmatrix**.

2.16   Try factor analysis on some of the gene expression data sets to cluster the patients or experiments.

2.17   Generate 2–D standard bivariate random variables using **randn** for increasing values of n. Use the **idpettis** function to estimate the intrinsic dimensionality, repeating for several Monte Carlo trials for each value of n. The true intrinsic dimensionality of these data is 2. How does the algorithm perform on these data sets?

2.18   This data set is taken from Fukunaga [1990], where he describes a Gaussian pulse characterized by three parameters **a**, **m**, and σ. The waveform is given by

$$\mathbf{x}(t) = \mathbf{a}\exp[-(t-\mathbf{m})^2 \div (2\sigma^2)].$$

These parameters are generated randomly in the following ranges:

$$0.7 \leq \mathbf{a} \leq 1.3$$
$$0.3 \leq \mathbf{m} \leq 0.7$$
$$0.2 \leq \sigma \leq 0.4.$$

The signals are generated at 8 time steps in the range $0 \leq t \leq 1.05$; so each of the signals is an 8–D random vector. The intrinsic dimensionality of these data is 3. Generate random vectors for various values of n, and apply **idpettis** to estimate the intrinsic dimensionality and assess the results.

2.19   Estimate the intrinsic dimensionality of the **yeast** data. Does this agree with the results in Example 2.2?

2.20   Estimate the intrinsic dimensionality of the following data sets. Where possible, compare with the results from PCA.

a. All BPM data sets.

b. **oronsay** data.

c. **sparrow** data

d. Other gene expression data.

2.21   The Statistics Toolbox has a function called **rotatefactors** that can be used with either factor analysis or principal component analysis loadings. Do a **help** on this function. Apply it to the factor loadings from the **stockreturns** data. Apply it to the PCA results of the **yeast** data in Example 2.2.

2.22   Consider a set of data that contains two groups, where each group is drawn from different multivariate normal distributions. Use the following code to generate 100 points from each group.

```
sigma = eye(3);
r0 = mvnrnd([7,7,7],sigma,100);
r1 = mvnrnd([5,5,5],sigma,100);
% Put the two groups into a data matrix.
X = [r0;r1];
```

Note that the **mvnrnd** function is from the Statistics Toolbox, but one could also use the **genmix** GUI that is included in the EDA Toolbox.

Use SVD and nonnegative matrix factorization to reduce the data to two dimensions. Visualize the results using the **plot** function, where the first group (i.e., the first 100 rows of the data matrix) is displayed using one color and the second group is displayed with another color. Compare the results from two dimensionality reduction approaches. Is one better for clustering?

2.23   Apply the LDA projection to the Fisher's iris data, taking two classes at a time. Repeat the example in the book and show the data in 1–D, along with the kernel density estimate. Discuss how good the mappings are for each case.

2.24   Repeat Example 2.4 using different replicates and compare the results. Also, try the same example using the alternating least squares approach and compare the results.

2.25   Use some of the alternative ways, such as the correlation dimension, the maximum likelihood approach, and the packing number to estimate the intrinsic dimensionality of the helix data of Example 2.8, and compare the results.

2.26   Apply the nearest neighbor estimator (**idpettis**) to the helix data in Example 2.9 and discuss the results.

2.27   The **intrinsic_dim** function in the Dimensionality Reduction Toolbox will also return the geodesic minimum spanning tree estimator (GMST) of Costa and Hero [2004]. The GMST is based on the idea that the length function of a geodesic minimum spanning tree is dependent on the intrinsic dimensionality. Find the GMST estimator of the helix data sets in Examples 2.8 and 2.9. Discuss your results.

2.28   Estimate the local dimensionalities of the helix data sets in Examples 2.8 and 2.9.

Chapter 3

Dimensionality Reduction — Nonlinear Methods

This chapter covers various methods for nonlinear dimensionality reduction, where the nonlinear aspect refers to the mapping between the high-dimensional space and the low-dimensional space. We start off by discussing a method that has been around for many years called multidimensional scaling. We follow this with several more recently developed nonlinear dimensionality reduction techniques called locally linear embedding, isometric feature mapping, and Hessian eigenmaps. We conclude this chapter by discussing some methods from the machine learning and neural network communities, such as self-organizing maps, generative topographic maps, curvilinear component analysis, autoencoders, and stochastic neighbor embedding.

3.1 Multidimensional Scaling — MDS

In general, *multidimensional scaling* (MDS) is a set of techniques for the analysis of proximity data measured on a set of objects in order to reveal hidden structure. The purpose of MDS is to find a configuration of the data points in a low-dimensional space such that the proximity between objects in the full-dimensional space is represented with some degree of fidelity by the distances between points in the low-dimensional space. This means that observations that are close together in a high-dimensional space should be close in the low-dimensional space. Many aspects of MDS were originally developed by researchers in the social science community, and the method is now widely available in most statistical packages, including the MATLAB Statistics Toolbox.

We first provide some definitions and notation before we go on to describe the various categories of MDS [Cox and Cox, 2001]. As before, we assume that we have a data set with n observations. In general, MDS starts with measures of proximity that quantify how close objects are to one another or how similar they are. They can be of two types: those that indicate similarity or dissimilarity. A measure of dissimilarity between objects r and s is denoted by δ_{rs}, and the similarity is denoted by s_{rs}. For most definitions of these proximity measures, we have

$$\delta_{rs} \geq 0 \qquad \delta_{rr} = 0$$

and

$$0 \leq s_{rs} \leq 1 \qquad s_{rr} = 1.$$

Thus, we see that for a dissimilarity measure δ_{rs}, small values correspond to observations that are close together, while the opposite is true for similarity measures s_{rs}. These two main types of proximity measures are easily converted into the other one when necessary (see Appendix A for more information on this issue). Thus, for the rest of this chapter, we will assume that *proximity* measures *dissimilarity* between objects. We also assume that the dissimilarities are arranged in matrix form, which will be denoted by Δ. In many cases, this will be a symmetric $n \times n$ matrix (sometimes given in either lower or upper triangular form).

We denote the *distance* between observation r and s in the lower-dimensional space by d_{rs}. It should also be noted that in the MDS literature the matrix of coordinate values in the *lower-dimensional space* is denoted by $\mathbf{X}$. We follow that convention here, knowing that it might be rather confusing with our prior use of $\mathbf{X}$ as representing the original set of n p-dimensional observations.

In MDS, one often starts with or might only have the dissimilarities Δ, not the original observations. In fact, in the initial formulations of MDS, the experiments typically involved qualitative judgements about differences between subjects, and p-dimensional observations do not make sense in that context. So, to summarize, with MDS we start with dissimilarities Δ and end up with d-dimensional[1] transformed observations $\mathbf{X}$. Usually, $d = 2$ or $d = 3$ is chosen to allow the analyst to view and explore the results for interesting structure, but any $d < p$ is also appropriate.

There are many different techniques and flavors of MDS, most of which fall into two main categories: metric MDS and nonmetric MDS. The main characteristic that divides them arises from the different assumptions of how the dissimilarities δ_{rs} are transformed into the configuration of distances d_{rs} [Cox and Cox, 2001]. *Metric MDS* assumes that the dissimilarities δ_{rs} based on the p-dimensional data and the distances d_{rs} in a lower-dimensional space are related as follows

---

[1] We realize that the use of the notation d as both the lower dimensionality of the data ($d < p$) and the distance d_{rs} between points in the configuration space might be confusing. However, the meaning should be clear from the context.

$$d_{rs} \approx f(\delta_{rs}) \ , \tag{3.1}$$

where f is a continuous monotonic function. The form of $f(\bullet)$ specifies the MDS model. For example, we might use the formula

$$f(\delta_{rs}) = b\delta_{rs} \ . \tag{3.2}$$

Mappings that follow Equation 3.2 are called *ratio* MDS [Borg and Groenen, 1997]. Another common choice is *interval* MDS, with $f(\bullet)$ given by

$$f(\delta_{rs}) = a + b\delta_{rs} \ ,$$

where a and b are free parameters. Other possibilities include higher degree polynomials, logarithmic, and exponential functions.

Nonmetric MDS relaxes the metric properties of $f(\bullet)$ and stipulates only that the rank order of the dissimilarities be preserved. The transformation or scaling function must obey the monotonicity constraint:

$$\delta_{rs} < \delta_{ab} \Rightarrow f\left(\delta_{rs}\right) \leq f\left(\delta_{ab}\right)$$

for all objects. Because of this constraint, nonmetric MDS is also known as *ordinal MDS*.

3.1.1 Metric MDS

Most of the methods in metric MDS start with the fact that we seek a transformation that satisfies Equation 3.1. We can tackle this problem by defining an objective function and then using a method that will optimize it. One way to define the objective function is to use the squared discrepancies between d_{rs} and $f(\delta_{rs})$ as follows

$$S(d_{rs}) = \sqrt{\frac{\sum_r \sum_s \left[f(\delta_{rs}) - d_{rs}\right]^2}{\text{scale factor}}} \ . \tag{3.3}$$

In general, Equation 3.3 is called the **stress**, and different forms for the scale factor give rise to different forms of stress and types of MDS. The scale factor in the denominator of Equation 3.3 that is used most often is

$$\sum_r \sum_s (d_{rs})^2.$$

In this case, we have an expression called **Stress-1** [Kruskal, 1964a].

Thus, in MDS, we would scale the dissimilarities using $f(\bullet)$ and then find a configuration of points in a d-dimensional space such that when we calculate the distances between them the stress is minimized. This can now be solved using numerical methods and operations research techniques (e.g., gradient or steepest descent). These methods are usually iterative and are not guaranteed to find a global solution to the optimization problem. We will expand on some of the details later, but first we describe a case where a closed form solution is possible.

The phrase *metric multidimensional scaling* is often used in the literature to refer to a specific technique called *classical* MDS. However, metric MDS includes more than this one technique, such as least squares scaling and others [Cox and Cox, 2001]. For metric MDS approaches, we first describe classical MDS followed by a method that optimizes a loss function using a majorization technique.

Classical MDS

If the proximity measure in the original space and the distance are taken to be Euclidean, then a closed form solution exists to find the configuration of points in a d-dimensional space. This is the classical MDS approach. The function $f(\bullet)$ relating dissimilarities and distances is the identity function; so we seek a mapping such that

$$d_{rs} = \delta_{rs}.$$

This technique originated with Young and Householder [1938], Torgerson [1952], and Gower [1966]. Gower was the one that showed the importance of classical scaling, and he gave it the name *principal coordinates analysis*, because it uses ideas similar to those in PCA. Principal coordinate analysis and classical scaling are the same thing, and they have become synonymous with metric scaling.

We now describe the steps of the method only, without going into the derivation. Please see any of the following for the derivation of classical MDS: Cox and Cox [2001], Borg and Groenen [1997], or Seber [1984].

Procedure – Classical MDS

1. Using the matrix of dissimilarities, Δ, find matrix $\mathbf{Q}$, where each element of $\mathbf{Q}$ is given by

$$q_{rs} = -\frac{1}{2}\delta_{rs}^2.$$

2. Find the centering matrix $\mathbf{H}$ using

$$\mathbf{H} = \mathbf{I} - n^{-1}\mathbf{11}^T,$$

where $\mathbf{I}$ is the $n \times n$ identity matrix, and $\mathbf{1}$ is a vector of n ones.

3. Find the matrix $\mathbf{B}$, as follows

$$\mathbf{B} = \mathbf{HQH}.$$

4. Determine the eigenvectors and eigenvalues of $\mathbf{B}$:

$$\mathbf{B} = \mathbf{ALA}^T.$$

5. The coordinates in the lower-dimensional space are given by

$$\mathbf{X} = \mathbf{A}_d \mathbf{L}_d^{1/2},$$

where $\mathbf{A}_d$ contains the eigenvectors corresponding to the d largest eigenvalues, and $\mathbf{L}_d^{1/2}$ contains the square root of the d largest eigenvalues along the diagonal.

We use similar ideas from PCA to determine the dimensionality d to use in Step 5 of the algorithm though $d = 2$ is often used in MDS, since the data can then be represented in a scatterplot.

For some data sets, the matrix $\mathbf{B}$ might not be positive semi-definite, in which case some of the eigenvalues will be negative. One could ignore the negative eigenvalues and proceed to step 5 or add an appropriate constant to the dissimilarities to make $\mathbf{B}$ positive semi-definite. We do not address this second option here, but the reader is referred to Cox and Cox [2001] for more details. If the dissimilarities are in fact Euclidean distances, then this problem does not arise.

Since this uses the decomposition of a square matrix, some of the properties and issues discussed about PCA are applicable here. For example, the lower-dimensional representations are nested. The first two dimensions of the 3–D coordinate representation are the same as the 2–D solution. It is also interesting to note that PCA and classical MDS provide equivalent results when the dissimilarities are Euclidean distances [Cox and Cox, 2001].

Example 3.1

For this example, we use the BPM data described in Chapter 1. Recall that one of the applications for these data is to discover different topics or subtopics. For ease of discussion and presentation, we will look at only two of the topics in this example: the comet falling into Jupiter (topic 6) and DNA in the O. J. Simpson trial (topic 9). First, using the match distances, we extract the required observations from the full interpoint distance matrix.

```
% First load the data - use the 'match' interpoint
% distance matrix.
load matchbpm
% Now get the data for topics 9 and 6.
% Find the indices where they are not equal to 6 or 9.
indlab = find(classlab ~= 6 & classlab ~=9);
% Now get rid of these from the distance matrix.
matchbpm(indlab,:) = [];
matchbpm(:,indlab) = [];
classlab(indlab) = [];
```

The following piece of code shows how to implement the steps for classical MDS in MATLAB.

```
% Now implement the steps for classical MDS
% Find the matrix Q:
Q = -0.5*matchbpm.^2;
% Find the centering matrix H:
n = 73;
H = eye(n) - ones(n)/n;
% Find the matrix B:
B = H*Q*H;
% Get the eigenvalues of B.
[A,L] = eig(B);
% Find the ordering largest to smallest.
[vals, inds] = sort(diag(L));
inds = flipud(inds);
vals = flipud(vals);
% Re-sort based on these.
A = A(:,inds);
L = diag(vals);
```

We are going to plot these results using $d = 2$ for ease of visualization, but we can also construct a scree-type plot to look for the 'elbow' in the curve. As in PCA, this can help us determine a good value to use for d. The code for constructing this plot and finding the coordinates in a 2–D space is given next.

```
% First plot a scree-type plot to look for the elbow.
% The following uses a log scale on the y-axis.
semilogy(vals(1:10),'o')
% Using 2-D for visualization purposes,
% find the coordinates in the lower-dimensional space.
X = A(:,1:2)*diag(sqrt(vals(1:2)));
% Now plot in a 2-D scatterplot
ind6 = find(classlab == 6);
ind9 = find(classlab == 9);
plot(X(ind6,1),X(ind6,2),'x',X(ind9,1),X(ind9,2),'o')
```

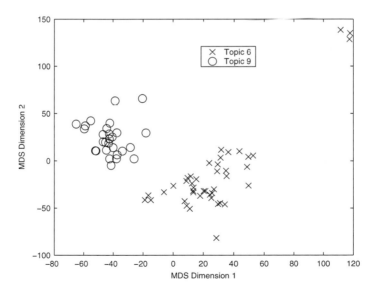

FIGURE 3.1
The top plot shows the logarithm of the eigenvalues. It appears that $d = 3$ might be a good value to use, since there is an elbow there. The bottom plot is a scatterplot of the 2–D coordinates after classical MDS. Note the good separation between the two topics, as well as the appearance of subtopics for topic 6.

```
legend({'Topic 6';'Topic 9'})
```

The scree plot is shown in Figure 3.1 (top), where we see that the elbow looks to be around $d = 3$. The scatterplot of the 2–D coordinates is given in Figure 3.1 (bottom). We see clear separation between topics 6 and 9. However, it is also interesting to note that there seem to be several subtopics in topic 6.
❏

MATLAB provides a function called **cmdscale** for constructing lower-dimensional coordinates using classical MDS. This is available in the Statistics Toolbox.

Metric MDS - SMACOF

The general idea of the *majorization method* for optimizing a function is as follows. Looking only at the one-dimensional case, we want to minimize a complicated function $f(x)$ by using a function $g(x, y)$ that is easily minimized. The function g has to satisfy the following inequality

$$f(x) \le g(x, y),$$

for a given y such that

$$f(y) = g(y, y).$$

Looking at graphs of these function one would see that the function g is always above f, and they coincide at the point $x = y$. The method of minimizing f is iterative. We start with an initial value x_0, and we minimize $g(x, x_0)$ to get x_1. We then minimize $g(x, x_1)$ to get the next value, continuing in this manner until convergence. Upon convergence of the algorithm, we will also have minimized $f(x)$.

The SMACOF (Scaling by Majorizing a Complicated Function) method goes back to de Leeuw [1977], and others have refined it since then, including Groenen [1993]. The method is simple, and it can be used for both metric and nonmetric applications. We now follow the notation of Borg and Groenen [1997] to describe the algorithm for the metric case. They show that the SMACOF algorithm satisfies the requirements for minimizing a function using majorization, as described above. We leave out the derivation, but those who are interested in this and in the nonmetric case can refer to the above references as well as Cox and Cox [2001].

The *raw stress* is written as

$$\sigma(\mathbf{X}) = \sum_{r<s} w_{rs} \left[d_{rs}(\mathbf{X}) - \delta_{rs} \right]^2$$

$$= \sum_{r<s} w_{rs} \delta_{rs}^2 + \sum_{r<s} w_{rs} d_{rs}^2(\mathbf{X}) - 2\sum_{r<s} w_{rs} \delta_{rs} d_{rs}(\mathbf{X}),$$

for all available dissimilarities δ_{rs}. The inequality $r < s$ in the summation means that we only sum over half of the data, since we are assuming that the dissimilarities and distances are symmetric. We might have some missing values; so we define a weight w_{rs} with a value of 1 if the dissimilarity is present and a value of 0 if it is missing. The notation $d_{rs}(\mathbf{X})$ makes it explicit that the distances are a function of $\mathbf{X}$ (the d-dimensional observations), and that we are looking for a configuration $\mathbf{X}$ that minimizes stress, $\sigma(\mathbf{X})$.

Before we describe the algorithm, we need to present some relationships and notation. Let $\mathbf{Z}$ be a possible configuration of points. The matrix $\mathbf{V}$ has elements given by the following

$$v_{ij} = -w_{ij}, \quad i \neq j$$

and

$$v_{ij} = \sum_{j=1,i \neq j}^{n} w_{ij} .$$

This matrix is not of full rank, and we will need the inverse in one of the steps. So we turn to the **Moore-Penrose inverse**, which will be denoted by $\mathbf{V}^+$. [2]
We next define matrix $\mathbf{B}(\mathbf{Z})$ with elements

$$b_{ij} = \begin{cases} -\dfrac{w_{ij}\delta_{ij}}{d_{ij}(\mathbf{Z})}; & d_{ij}(\mathbf{Z}) \neq 0 \\ 0; & d_{ij}(\mathbf{Z}) = 0, \end{cases}$$

for $i \neq j$ and

$$b_{ii} = \sum_{j=1,i \neq j}^{n} b_{ij} .$$

We are now ready to define the **Guttman transform**. The general form of the transform is given by

$$\mathbf{X}^k = \mathbf{V}^+\mathbf{B}(\mathbf{Z})\mathbf{Z} ,$$

where the k represents the iteration number in the algorithm. If all of the weights are one (none of the dissimilarities are missing), then the transform is much simpler:

---

[2] The Moore-Penrose inverse is also called the **pseudoinverse** and can be computed using the singular value decomposition. MATLAB has a function called **pinv** that provides this inverse.

$$\mathbf{X}^k = n^{-1}\mathbf{B}(\mathbf{Z})\mathbf{Z}.$$

Now that we have these definitions, we can describe the SMACOF algorithm below.

SMACOF Algorithm

1. Find an initial configuration of points in R^d. This can either be random or nonrandom (i.e., some regular grid). Call this $\mathbf{X}^0$.
2. Set $\mathbf{Z} = \mathbf{X}^0$ and the counter to $k = 0$.
3. Compute the raw stress $\sigma(\mathbf{X}^0)$.
4. Increase the counter by 1: $k = k + 1$.
5. Obtain the Guttman transform $\mathbf{X}^k$.
6. Compute the stress for this iteration, $\sigma(\mathbf{X}^k)$.
7. Find the difference in the stress values between the two iterations. If this is less than some pre-specified tolerance or if the maximum number of iterations has been reached, then stop.
8. Set $\mathbf{Z} = \mathbf{X}^k$, and go to step 4.

We illustrate this algorithm in the next example.

Example 3.2

We turn to a different data set for this example and look at the **leukemia** data. Recall that we can look at either genes or patients as our observations; in this example we will be looking at the genes. To make things easier, we only implement the case where all the weights are 1, but the more general situation is easily implemented using the above description. First we load the data and get the distances.

```
% Use the Leukemia data, using the genes (columns)
% as the observations.
load leukemia
y = leukemia';
% Get the interpoint distance matrix.
% pdist gets the interpoint distances.
% squareform converts them to a square matrix.
D = squareform(pdist(y,'seuclidean'));
[n,p] = size(D);
% Turn off this warning... :
warning off MATLAB:divideByZero
```

Next we get the required initial configuration and the stress associated with it.

```
% Get the first term of stress.
% This is fixed - does not depend on the configuration.
stress1 = sum(sum(D.^2))/2;
% Now find an initial random configuration.
d = 2;
% This function is part of Statistics Toolbox,
% but one can use the function rand and scale them.
Z = unifrnd(-2,2,n,d);
% Find the stress for this.
DZ = squareform(pdist(Z));
stress2 = sum(sum(DZ.^2))/2;
stress3 = sum(sum(D.*DZ));
oldstress = stress1 + stress2 - stress3;
```

Now we iteratively adjust the configuration until the stress converges.

```
% Iterate until stress converges.
tol = 10^(-6);
dstress = realmax;
numiter = 1;
dstress = oldstress;
while dstress > tol & numiter <= 100000
    numiter = numiter + 1;
    % Now get the update.
    BZ = -D./DZ;
    for i = 1:n
        BZ(i,i) = 0;
        BZ(i,i) = -sum(BZ(:,i));
    end
    X = n^(-1)*BZ*Z;
    Z = X;
    % Now get the distances.
    % Find the stress.
    DZ = squareform(pdist(Z));
    stress2 = sum(sum(DZ.^2))/2;
    stress3 = sum(sum(D.*DZ));
    newstress = stress1 + stress2 - stress3;
    dstress = oldstress - newstress;
    oldstress = newstress;
end
```

A scatterplot of the resulting configuration is shown in Figure 3.2.
❑

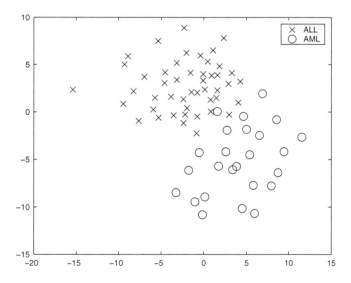

FIGURE 3.2
Here we show the results of using the SMACOF algorithm on the `leukemia` data set. The
top panel shows the data transformed to 2–D and labeled by B-cell and T-cell. The lower
panel illustrates the same data using different symbols based on the ALL or AML labels.
There is some indication of being able to group the genes in a reasonable fashion.

3.1.2 Nonmetric MDS

An algorithm for solving the nonmetric MDS problem was first discussed by Shepard [1962a, 1962b]. However, he did not introduce the idea of using a loss function. That came with Kruskal [1964a, 1964b] who expanded the ideas of Shepard and gave us the concept of minimizing a loss function called *stress*.

Not surprisingly, we first introduce some more notation and terminology for nonmetric MDS. The **disparity** is a measure of how well the distance d_{rs} matches the dissimilarity δ_{rs}. We represent the disparity as $\hat{d}_{rs}$. The r-th point in our configuration $\mathbf{X}$ will have coordinates

$$\mathbf{x}_r = (x_{r1}, \dots, x_{rd})^T.$$

We will use the Minkowski dissimilarity to measure the distance between points in our d-dimensional space. It is defined as

$$d_{rs} = \left\{ \sum_{i=1}^{d} |x_{ri} - x_{si}|^\lambda \right\}^{\frac{1}{\lambda}},$$

where $\lambda > 0$. See Appendix A for more information on this distance and the parameter λ.

We can view the disparities as a function of the distance, such as

$$\hat{d}_{rs} = f(d_{rs}),$$

where

$$\delta_{rs} < \delta_{ab} \Rightarrow \hat{d}_{rs} \leq \hat{d}_{ab}.$$

Thus, the order of the original dissimilarities is preserved by the disparities. Note that this condition allows for possible ties in the disparities.

We define a loss function L, which is really stress, as follows

$$L = S = \sqrt{\frac{\sum_{r<s}\left(d_{rs} - \hat{d}_{rs}\right)^2}{\sum_{r<s} d_{rs}^2}} = \sqrt{\frac{S^*}{T^*}}.$$

It could be the case that we have missing dissimilarities or the values might be meaningless for some reason. If the analyst is faced with this situation, then the summations in the definition of stress are restricted to those pairs (r,s) for which the dissimilarity is available.

As with other forms of MDS, we seek a configuration of points **X**, such that the stress is minimized. Note that the coordinates of the configuration enter into the loss function through the distances d_{rs}. The original dissimilarities enter into the stress by imposing an ordering on the disparities. Thus, the stress is minimized subject to the constraint on the disparities. This constraint is satisfied by using *isotonic regression* (also known as *monotone regression*)[3] to obtain the disparities.

We now pause to describe the isotonic regression procedure. This was first described in Kruskal [1964b], where he developed an up-and-down-blocks algorithm. A nice explanation of this is given in Borg and Groenen [1997], as well as in the original paper by Kruskal. We choose to describe and implement the method for isotonic regression outlined in Cox and Cox [2001]. Isotonic regression of the d_{rs} on the δ_{rs} partitions the dissimilarities into blocks over which the $\hat{d}_{rs}$ are constant. The estimated disparities $\hat{d}_{rs}$ are given by the mean of the d_{rs} values within the block.

Example 3.3

The easiest way to explain (and hopefully to understand) isotonic regression is through an example. We use a simple one given in Cox and Cox [2001]. There are four objects with the following dissimilarities

$$\delta_{12} = 2.1, \delta_{13} = 3.0, \delta_{14} = 2.4, \delta_{23} = 1.7, \delta_{24} = 3.9, \delta_{34} = 3.2.$$

A configuration of points yields the following distances

$$d_{12} = 1.6, d_{13} = 4.5, d_{14} = 5.7, d_{23} = 3.3, d_{24} = 4.3, d_{34} = 1.3.$$

We now order the dissimilarities from smallest to largest and use a single subscript to denote the rank. We also impose the same order (and subscript) on the corresponding distances. This yields

$$\delta_1 = 1.7, \delta_2 = 2.1, \delta_3 = 2.4, \delta_4 = 3.0, \delta_5 = 3.2, \delta_6 = 3.9$$
$$d_1 = 3.3, d_2 = 1.6, d_3 = 5.7, d_4 = 4.5, d_5 = 1.3, d_6 = 4.3.$$

The constraint on the disparities requires these distances to be ordered such that $d_i < d_{i+1}$. If this is what we have from the start, then we need not adjust things further. Since that is not true here, we must use isotonic regression to get $\hat{d}_{rs}$. To do this, we first get the cumulative sums of the distances d_i defined as

---

$$D_i = \sum_{j=1}^{i} d_j, \qquad i = 1, \ldots, N ,$$

where N is the total number of dissimilarities available. In essence, the algorithm finds the greatest convex minorant of the graph of D_i, going through the origin. See Figure 3.3 for an illustration of this. We can think of this procedure as taking a string, attaching it to the origin on one end and the last point on the other end. The points on the greatest convex minorant are those where the string touches the points D_i. These points partition the distances d_i into blocks, over which we will have disparities of constant value. These disparities are the average of the distances that fall into the block. Cox and Cox [2001] give a proof that this method yields the required isotonic regression. We now provide some MATLAB code that illustrates these ideas.

```
% Enter the original data.
dissim = [2.1 3 2.4 1.7 3.9 3.2];
dists = [1.6 4.5 5.7 3.3 4.3 1.3];
N = length(dissim);
% Now re-order the dissimilarities.
[dissim,ind] = sort(dissim);
% Now impose the same order on the distances.
dists = dists(ind);
% Now find the cumulative sums of the distances.
D = cumsum(dists);
% Add the origin as the first point.
D = [0 D];
```

It turns out that we can find the greatest convex minorant by finding the slope of each D_i with respect to the origin. We first find the smallest slope, which defines the first partition (i.e., it is on the greatest convex minorant). We then find the next smallest slope, after removing the first partition from further consideration. We continue in this manner until we reach the end of the points. The following code implements this process.

```
% Now find the slope of these.
slope = D(2:end)./(1:N);
% Find the points on the convex minorant by looking
% for smallest slopes.
i = 1;
k = 1;
while i <= N
    val = min(slope(i:N));
    minpt(k) = find(slope == val);
    i = minpt(k) + 1;
    k = k + 1;
end
```

It turns out that this procedure yields extra points, ones that are *not* on the convex minorant. MATLAB has a function called **convhull** that finds all of the points that are on the convex hull[4] of a set of points. This also yields extra points because we only want those points that are on the "bottom" of the convex hull. To get the desired points, we take the intersection of the two sets.[5]

```
K = convhull(D, 0:N);
minpt = intersect(minpt+1,K) - 1;
```

Now that we have the points that divide the distances into blocks, we find the disparities, which are given by the average distance in that block.

```
% Now that we have all of the minorant points
% that divide into blocks, the disparities are
% the averages of the distances over those blocks.
j = 1;
for i = 1:length(minpt)
    dispars(j:minpt(i)) = mean(dists(j:minpt(i)));
    j = minpt(i) + 1;
end
```

The disparities are given by

$$\hat{d}_1 = 2.45, \hat{d}_2 = 2.45, \hat{d}_3 = 3.83, \hat{d}_4 = 3.83, \hat{d}_5 = 3.83, \hat{d}_6 = 4.3 .$$

The graphs that illustrate these concepts are shown in Figure 3.3, where we see that the disparities fall into three groups, as given above.
❑

Now that we know how to do isotonic regression, we can describe Kruskal's algorithm for nonmetric MDS. This is outlined below.

Procedure – Kruskal's Algorithm

1. Choose an initial configuration X_0 for a given dimensionality d. Set the iteration counter k equal to 0. Set the step size α to some desired starting value.
2. Normalize the configuration to have mean of zero (i.e., the centroid is at the origin) and a mean square distance from the origin equal to 1.
3. Compute the interpoint distances for this configuration.

---

[4] The convex hull of a data set is the smallest convex region that contains the data set.
[5] Please see the M-file for an alternative approach, courtesy of Tom Lane of The MathWorks.

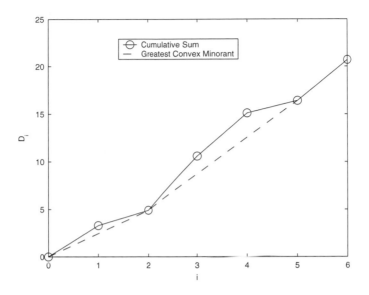

FIGURE 3.3
This shows the idea behind isotonic regression using the greatest convex minorant. The greatest convex minorant is given by the dashed line. The places where it touches the graph of the cumulative sums of the distances partition the distances into blocks. Each of the disparities in the blocks are given by the average of the distances falling in that partition.

4. Check for equal dissimilarities. If some are equal, then order those such that the corresponding distances form an increasing sequence within that block.

5. Find the disparities $\hat{d}_{rs}$ using the current configuration of points and isotonic regression.

6. Append all of the coordinate points into one vector such that

$$\mathbf{x} = (x_{11}, \dots, x_{1d}, \dots, x_{n1}, \dots, x_{nd})^T.$$

7. Compute the gradient for the k-th iteration given by

$$\frac{\partial S}{\partial x_{ui}} = g_{ui} = S \sum_{r<s} \left(\delta^{ur} - \delta^{us} \right) \left(\frac{d_{rs} - \hat{d}_{rs}}{S^*} - \frac{d_{rs}}{T^*} \right) \left(\frac{|x_{ri} - x_{si}|}{d_{rs}} \right)^{\lambda - 1} \mathrm{sgn}\left(x_{ri} - x_{si} \right),$$

where δ^{ur} represents the Kronecker delta function, *not* dissimilarities. This function is defined as follows

$$\delta^{ur} = 1 \qquad \text{if } u = r$$

$$\delta^{ur} = 0 \qquad \text{if } u \neq r$$

The value of the sgn($\bullet$) function is +1 if the argument is positive and -1 if it is negative.

8. Check for convergence. Determine the magnitude of the gradient vector for the current configuration using

$$\text{mag}(\mathbf{g}_k) = \sqrt{\frac{1}{n} \sum_{u,i} g_{ui}^2} \ .$$

If this magnitude is less than some small value ε or if some maximum allowed number of iterations has been reached, then the process stops.

9. Find the step size α

$$\alpha_k = \alpha_{k-1} \times \text{angle factor} \times \text{relaxation factor} \times \text{good luck factor} \ ,$$

where k represents the iteration number. The angle factor $= 4^{(\cos\theta)^3}$, with

$$\cos\theta = \frac{\sum_{r<s} g_{k_{rs}} g_{k_{rs}-1}}{\sqrt{\sum_{r<s} g_{k_{rs}}^2} \sqrt{\sum_{r<s} g_{k_{rs}-1}^2}} \ .$$

The relaxation factor is given by

$$\frac{1.3}{1+\beta^5}, \qquad \beta = \min\left\{1, \frac{S_k}{S_{k-5}}\right\} \ ,$$

and the good luck factor is

$$\min\left\{1, \frac{S_k}{S_{k-1}}\right\} \ .$$

10. The new configuration is given by

$$\mathbf{x}_{k+1} = \mathbf{x}_k - \alpha_k \frac{\mathbf{g}_k}{\text{mag}(\mathbf{g}_k)} \ .$$

11. Increment the counter: $k = k + 1$. Go to step 2.

A couple of programming issues should be noted. When we are in the beginning of the iterative algorithm, we will not have values of the stress for $k - 5$ or $k - 1$ (needed in Step 9). In these cases, we will simply use the value at the first iteration until we have enough. We can use a similar idea for the gradient. Kruskal [1964b] states that this step size provides large steps during the initial iterations and small steps at the end. There is no claim of optimality with this procedure, and the reader should be aware that the final configuration is not guaranteed to be a global solution. It is likely that the configuration will be a local minimum for stress, which is typically the case for greedy iterative algorithms. It is recommended that several different starting configurations be used, and the one with the minimum stress be accepted as the final answer.

This leads to another programming point. How do we find an initial configuration to start the process? We could start with a grid of points that is evenly laid out over a d-dimensional space. Another possibility is to start from the configuration given by classical MDS. Finally, one could also generate random numbers according to a Poisson process in R^d.

Another issue that must be addressed with nonmetric MDS is how to handle ties in the dissimilarities. There are two approaches to this. The primary approach states that if $\delta_{rs} = \delta_{tu}$, then $\hat{d}_{rs}$ is not required to be equal to $\hat{d}_{tu}$. The secondary and more restrictive approach requires the disparities to be equal when dissimilarities are tied. We used the primary approach in the above procedure and in our MATLAB implementation.

Example 3.4

The **nmmds** function (that we include in the EDA Toolbox) that implements Kruskal's nonmetric MDS is quite long, and it involves several helper functions. So, we just show how it would be used rather than repeating all of the code here. We return to our BPM data, but this time we will look at two different topics: topic 8 (the death of the North Korean leader) and topic 11 (Hall's helicopter crash in North Korea). Since these are both about North Korea, we would expect to see some similarity between them. Previous experiments showed that the documents from these two topics were always grouped together, but in several different subgroups [Martinez, 2002]. We apply the nonmetric MDS method to these data using the Ochiai measure of semantic dissimilarity.

```
load ochiaibpm
% First get out the data for topics 8 and 11.
% Find the indices where they are not equal to 8 or 11.
indlab = find(classlab ~= 8 & classlab ~= 11);
% Now get rid of these from the distance matrix.
ochiaibpm(indlab,:) = [];
ochiaibpm(:,indlab)=[];
```

```
classlab(indlab) = [];
% We only need the upper part.
n = length(classlab);
dissim = [];
for i = 1:n
    dissim = [dissim, ochiaibpm(i,(i+1):n)];
end
% Find configuration for R^2.
d = 2;
r = 1;
% The nmmds function is in the EDA Toolbox.
[Xd,stress,dhats] = nmmds(dissim,d,r);
ind8 = find(classlab == 8);
ind11 = find(classlab == 11);
% Plot with symbols mapped to class (topic).
plot(Xd(ind8,1),Xd(ind8,2),'.',...
    Xd(ind11,1),Xd(ind11,2),'o')
legend({'Class 8';'Class 11'})
```

The resulting plot is shown in Figure 3.4. We see that there is no clear separation between the two topics, but we do have some interesting structure in this scatterplot indicating the possible presence of subtopics.
❏

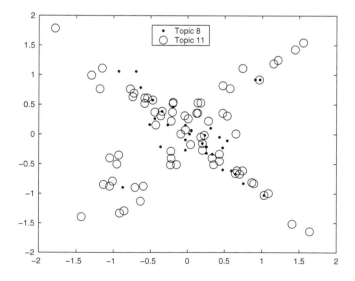

FIGURE 3.4

This shows the results of applying Kruskal's nonmetric MDS to topics 8 and 11, both of which concern North Korea. We see some interesting structure in this configuration and the possible presence of subtopics.

What should the dimensionality be when we are applying MDS to a set of data? This is a rather difficult question to answer for most MDS methods. In the spirit of EDA, one could apply the procedure for several values of d and record the value for stress. We could then construct a plot similar to the scree plot in PCA, where we have the dimension d on the horizontal axis and the stress on the vertical axis. Stress always decreases with the addition of more dimensions, so we would be looking for the point at which the payoff from adding more dimensions decreases (i.e., the value of d where we see an elbow in the curve).

The Statistics Toolbox includes a function for multi-dimensional scaling called **mdscale**. It does both metric and nonmetric MDS, and it includes several choices for the stress. With this function, one can use weights, specify the type of starting configuration, and request replicates for different random initial configurations.

3.2 Manifold Learning

Some recently developed approaches tackle the problem of nonlinear dimensionality reduction by assuming that the data lie on a submanifold of Euclidean space M. The common goal of these methods is to produce coordinates in a lower dimensional space, such that the neighborhood structure of the submanifold is preserved. In other words, points that are neighbors along the submanifold are also neighbors in the reduced parameterization of the data. (See Figure 3.5 and Example 3.5 for an illustration of these concepts.)

These methods are discussed in this section. We will first present locally linear embedding, which is an unsupervised learning algorithm that exploits the local symmetries of linear reconstructions. Next, we cover isometric feature mapping, which is an extension to classical MDS. Finally, we present a novel method called Hessian eigenmaps, which addresses one of the limitations of isometric feature mapping.

All of these methods are implemented in MATLAB, and the code is freely available for download. (We provide the URLs in Appendix B.) Because of this, we will not be including all of the code in the examples, but will only show how to use the existing implementations of the techniques.

3.2.1 Locally Linear Embedding

Locally linear embedding (LLE) was developed by Roweis and Saul [2000]. The method is an eigenvector-based method, and its optimizations do not involve local minima or iterative algorithms. The technique is based on simple geometric concepts. First, we assume that the data are sampled in

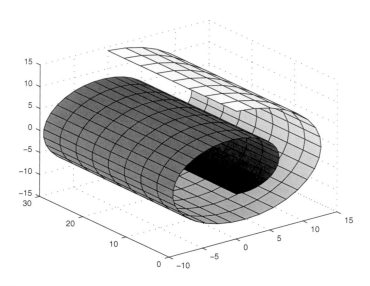

FIGURE 3.5
This shows the submanifold for the Swiss roll data set. We see that this is really a 2–D manifold (or surface) embedded in a 3–D space.

sufficient quantity from a smooth submanifold. We also assume that each data point and its neighbors lie on or close to a locally linear patch of the manifold M.

The LLE algorithm starts by characterizing the local geometry of the patches by finding linear coefficients that reconstruct each data point by using only its k nearest neighbors, with respect to Euclidean distance. There will be errors in the reconstruction, and these are measured by

$$\varepsilon(W) = \sum_i \left| \mathbf{x}_i - \sum_j W_{ij} \mathbf{x}_j \right|^2 , \qquad (3.4)$$

where the subscript j ranges over those points that are in the neighborhood of $\mathbf{x}_i$. The weights are found by optimizing Equation 3.4 subject to the following constraint:

$$\sum_j W_{ij} = 1 .$$

The optimal weights are found using least squares, the details of which are omitted here.

Once the weights W_{ij} are found, we fix these and find a representation $\mathbf{y}_i$ of the original points in a low-dimensional space. This is also done by optimizing a cost function, which in this case is given by

$$\Phi(\mathbf{y}) = \sum_i \left| \mathbf{y}_i - \sum_j W_{ij} \mathbf{y}_j \right|^2 . \qquad (3.5)$$

This defines a quadratic form in $\mathbf{y}_i$, and it can be minimized by solving a sparse eigenvector problem. The d eigenvectors corresponding to the *smallest* nonzero eigenvalues provide a set of orthogonal coordinates centered at the origin. The method is summarized below.

Locally Linear Embedding

1. Determine a value for k and d.
2. For each data point $\mathbf{x}_i$, find the k nearest neighbors.
3. Compute the weights W_{ij} that optimally reconstruct each data point $\mathbf{x}_i$ from its neighbors (Equation 3.4).
4. Find the d-dimensional points $\mathbf{y}_i$ that are optimally reconstructed using the same weights found in step 3 (Equation 3.5).

Note that the algorithm requires a value for k, which governs the size of the neighborhood, and a value for d. Of course, different results can be obtained when we vary these values. We illustrate the use of LLE in Example 3.5.

3.2.2 Isometric Feature Mapping — ISOMAP

Isometric feature mapping (ISOMAP) was developed by Tenenbaum, de Silva, and Langford [2000] as a way of enhancing classical MDS. The basic idea is to use distances along a geodesic path (presumably measured along the manifold M) as measures of dissimilarity. As with LLE, ISOMAP assumes that the data lie on an unknown submanifold M that is embedded in a p-dimensional space. It seeks a mapping $f: X \rightarrow Y$ that preserves the intrinsic metric structure of the observations. That is, the mapping preserves the distances between observations, where the distance is measured along the geodesic path of M. It also assumes that the manifold M is globally isometric to a convex subset of a low-dimensional Euclidean space.

In Figure 3.6, we show an example that illustrates this idea. The Euclidean distance between two points on the manifold is shown by the straight line connecting them. If our goal is to recover the manifold, then a truer indication of the distance between these two points is given by their distance on the manifold, i.e., the geodesic distance along the manifold between the points.

The ISOMAP algorithm has three main steps. The first step is to find the neighboring points based on the interpoint distances d_{ij}. This can be done by either specifying a value of k to define the number of nearest neighbors or a radius ε. The distances are typically taken to be Euclidean, but they can be any valid metric. The neighborhood relations are then represented as a

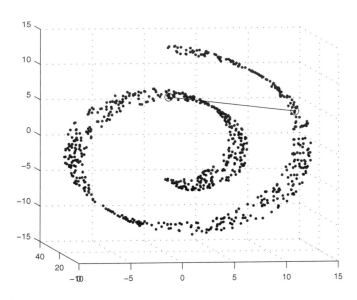

FIGURE 3.6

This is a data set that was randomly generated according to the Swiss roll parametrization [Tenenbaum, de Silva, and Langford, 2000]. The Euclidean distance between two points indicated by the circles is given by the straight line shown here. If we are seeking the neighborhood structure as given by the submanifold M, then it would be better to use the geodesic distance (the distance along the manifold or the roll) between the points. One can think of this as the distance a bug would travel if it took the shortest path between these two points, while walking on the manifold.

weighted graph, where the edges of the graph are given weights equal to the distance d_{ij}. The second step of ISOMAP provides estimates of the geodesic distances between all pairs of points i and j by computing their shortest path distance using the graph obtained in step one. In the third step, classical MDS is applied to the geodesic distances and an embedding is found in d-dimensional space as described in the first section of this chapter.

Procedure – ISOMAP

1. Construct the neighborhood graph over all observations by connecting the ij-th point if point i is one of the k nearest neighbors of j (or if the distance between them is less than ε). Set the lengths of the edges equal to d_{ij}.
2. Calculate the shortest paths between points in the graph.
3. Obtain the d-dimensional embedding by applying classical MDS to the geodesic paths found in step 2.

The input to this algorithm is a value for k or ε and the interpoint distance matrix. It should be noted that different results are obtained when the value for k (or ε) is varied. In fact, it could be the case that ISOMAP will return fewer than n lower-dimensional points. If this happens, then the value of k (or ε) should be increased.

3.2.3 Hessian Eigenmaps

We stated in the description of ISOMAP that the manifold M is assumed to be convex. Donoho and Grimes [2003] developed a method called **Hessian eigenmaps** that will recover a low-dimensional parametrization for data lying on a manifold that is locally isometric to an open, connected subset of Euclidean space. This significantly expands the class of data sets where manifold learning can be accomplished using isometry principles. This method can be viewed as a modification of the LLE method. Thus, it is also called **Hessian locally linear embedding** (HLLE).

We start with a parameter space $\Theta \subset R^d$ and a smooth mapping $\psi : \Theta \to R^p$, where R^p is the embedding space and $d < p$. Further, we assume that Θ is an open, connected subset of R^d, and ψ is a locally isometric embedding of Θ into R^p. The manifold can be written as a function of the parameter space, as follows

$$M = \psi\,(\Theta)\,.$$

One can think of the manifold as the enumeration $m = \psi(\theta)$ of all possible measurements as one varies the parameters θ for a given process.

Let us assume that we have some observations m_i that represent measurements over many different choices of control parameters θ_i ($i = 1,...,n$). These measurements are the same as our observations x_i.[6] The HLLE description given in Donoho and Grimes [2003] considers the case where all data points lie exactly on the manifold M. The goal is to recover the underlying parameterization ψ and the parameter settings θ_i, up to a rigid motion.

We now describe the main steps of the HLLE algorithm. We leave out the derivation and proofs because we just want to provide the general ideas underlying the algorithm. The reader is referred to the original paper and the MATLAB code for more information and implementation details.

The two main assumptions of the HLLE algorithm are:

1. In a small enough neighborhood of each point m, geodesic distances to nearby points m' (both on the manifold M) are identical to Euclidean distances between associated parameter points θ and θ'.

---

[6] We use different notation here to be consistent with the original paper.

This is called *local isometry*. (ISOMAP deals with the case where M assumes a globally isometric parameterization.)

2. The parameter space Θ is an open, connected subset of R^d, which is a weaker condition than the convexity assumption of ISOMAP.

In general, the idea behind HLLE is to define a neighborhood around some m in M and obtain local tangent coordinates. These local coordinates are used to define the Hessian of a smooth function $f: M \to R$. The function f is differentiated in the tangent coordinate system to produce the tangent Hessian. A quadratic form $\mathcal{H}(f)$ is obtained using the tangent Hessian, and the isometric coordinates θ can be recovered by identifying a suitable basis for the null space of $\mathcal{H}(f)$.

The inputs required for the algorithm are a set of n p-dimensional data points, a value for d, and the number of nearest neighbors k to determine the neighborhood. The only constraint on these values is that $\min(k,p) > d$. The algorithm estimates the tangent coordinates by using the SVD on the neighborhood of each point, and it uses the empirical version of the operator $\mathcal{H}(f)$. The output of HLLE is a set of n d-dimensional embedding coordinates.

Example 3.5

We generated some data from an S-curve manifold to be used with LLE and ISOMAP and saved it in a file called **scurve.mat**.[7] We show both the true manifold and the data randomly generated from this manifold in Figure 3.7.

```
load scurve
% The scurve.mat file contains our data matrix X.
% First set up some parameters for LLE.
K = 12;
d = 2;
% Run LLE - note that LLE takes the input data
% as rows corresponding to dimensions and
% columns corresponding to observations. This is
% the transpose of our usual data matrix.
Y = lle(X,K,d);
% Plot results in scatterplot.
scatter(Y(1,:),Y(2,:),12,X(3,:),'+');
```

We show the results of LLE in Figure 3.8. Note by the colors that neighboring points on the manifold are mapped into neighboring points in the 2–D embedding. Now we use ISOMAP on the same data.

```
% Now run the ISOMAP - we need the distances for input.
% We need the data matrix as n x p.
X = X';
dists = squareform(pdist(X));
```

---

[7] Look at the file called **example35.m** for more details on how to generate the data.

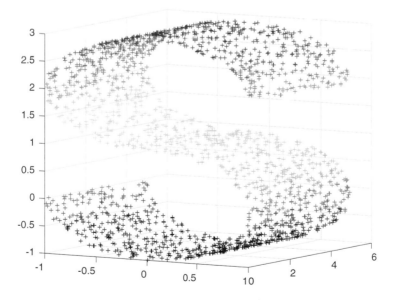

FIGURE 3.7
The top panel shows the true S-curve manifold, which is a 2–D manifold embedded in 3–D. The bottom panel is the data set randomly generated from the manifold. The gray scale values are an indication of the neighborhood. (**SEE COLOR INSERT.**)

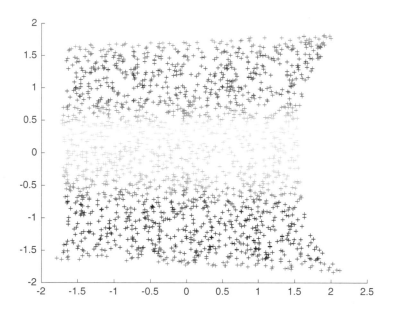

FIGURE 3.8
This is the embedding recovered from LLE. Note that the neighborhood structure is preserved. (**SEE COLOR INSERT.**)

```
options.dims = 1:10;     % These are for ISOMAP.
options.display = 0;
[Yiso, Riso, Eiso] = isomap(dists, 'k', 7, options);
```

Constructing a scatterplot of this embedding to compare with LLE is left as an exercise to the reader. As stated in the text, LLE and ISOMAP have some problems with data sets that are not convex. We show an example here using both ISOMAP and HLLE to discover such an embedding. These data were generated according to the code provided by Donoho and Grimes [2003]. Essentially, we have the Swiss roll manifold with observations removed that fall within a rectangular area along the surface.

```
% Now run the example from Grimes and Donoho.
load swissroll
options.dims = 1:10;
options.display = 0;
dists = squareform(pdist(X'));
[Yiso, Riso, Eiso] = isomap(dists, 'k', 7, options);
% Now for the Hessian LLE.
Y2 = hlle(X,K,d);
scatter(Y2(1,:),Y2(2,:),12,tt,'+');
```

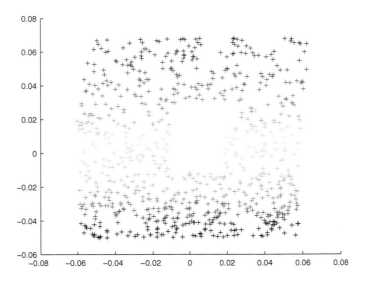

FIGURE 3.9
The top panel contains the 2–D coordinates from ISOMAP. We do see the hole in the embedding, but it looks like an oval rather than a rectangle. The bottom panel shows the 2–D coordinates from HLLE. HLLE was able to recover the correct embedding.

We see in Figure 3.9 that ISOMAP was unable to recover the correct embedding. We did find the hole, but it is distorted. HLLE was able to find the correct embedding with no distortion.
❑

We note a few algorithmic complexity issues. The two locally linear embedding methods, LLE and HLLE, can be used on problems with many data points (large n), because the initial computations are performed on small neighborhoods and they can exploit sparse matrix techniques. On the other hand, ISOMAP requires the calculation of a full matrix of geodesic distances for its initial step. LLE and HLLE are affected by the dimensionality p, because they must estimate a local tangent space at each data point. Also, in the HLLE algorithm, we must estimate second derivatives, which can be difficult with high-dimensional data.

3.3 Artificial Neural Network Approaches

We now discuss two other methods that we categorize as ones based on artificial neural network (ANN) ideas: the self-organizing map, the generative topographic mapping, and curvilinear component analysis. These ANN approaches also look for intrinsically low-dimensional structures that are embedded nonlinearly in a high-dimensional space. As in MDS and the manifold learning approaches, these seek a single global low-dimensional nonlinear model of the observations. In general, these methodologies try to fit a grid or predefined topology (usually 2–D) to the data, using greedy algorithms that first fit the large-scale linear structure of the data and then make small-scale nonlinear refinements.

There are MATLAB toolboxes available for both self-organizing maps and generative topographic maps. They are free, and they come with extensive documentation and examples. Because the code for these methods can be very extensive and would not add to the understanding of the reader, we will not be showing the code in the book. Instead, we will show how to use some of the functions in the examples.

3.3.1 Self-Organizing Maps

The *self-organizing map* (SOM) is a tool for the exploration and visualization of high-dimensional data [Kohonen, 1998]. It derives an orderly mapping of data onto a regular, low-dimensional grid. The dimensionality of the grid is usually $d = 2$ for ease of visualization. It converts complex, nonlinear relationships in the high-dimensional space into simpler geometric relationships such that the important topological and metric relationships are conveyed. The data are organized on the grid in such a way that observations that are close together in the high-dimensional space are also closer to each other on the grid. Thus, this is very similar to the ideas of MDS, except that the positions in the low-dimensional space are restricted to the grid. The grid locations are denoted by $\mathbf{r}_i$.

There are two methods used in SOM: incremental learning and batch mapping. We will describe both of them, and then illustrate some of the functions in the SOM Toolbox[8] in Example 3.6. We will also present a brief discussion of the various methods for visualizing and exploring the results of SOM.

The *incremental* or *sequential learning method* for SOM is an iterative process. We start with the set of observations $\mathbf{x}_i$ and a set of p-dimensional model vectors $\mathbf{m}_i$. These *model vectors* are also called *neurons, prototypes*, or *codebooks*. Each model vector is associated with a vertex ($\mathbf{r}_i$) on a 2–D lattice that can be either hexagonal or rectangular and starts out with some initial value ($\mathbf{m}_i(t = 0)$). This could be done by setting them to random values, by setting them equal to a random subset of the original data points or by using PCA to provide an initial ordering.

At each training step, a vector $\mathbf{x}_i$ is selected and the distance between it and all the model vectors is calculated, where the distance is typically Euclidean. The SOM Toolbox handles missing values by excluding them from the calculation, and variable weights can also be used. The best-matching unit (BMU) or model vector is found and is denoted by $\mathbf{m}_c$. Once the closest model vector $\mathbf{m}_c$ is found, the model vectors are updated so that $\mathbf{m}_c$ is moved closer to the data vector $\mathbf{x}_i$. The neighbors of the BMU $\mathbf{m}_c$ are also updated, in a weighted fashion. The update rule for the model vector $\mathbf{m}_i$ is

$$\mathbf{m}_i(t + 1) \;=\; \mathbf{m}_i(t) + \alpha(t)h_{ci}(t)[\mathbf{x}(t) - \mathbf{m}_i(t)],$$

where t denotes time or iteration number. The learning rate is given by $\alpha(t)$ and $0 < \alpha(t) < 1$, which decreases monotonically as the iteration proceeds.

The neighborhood is governed by the function $h_{ci}(t)$, and a Gaussian centered at the best-matching unit is often used. The SOM Toolbox has other neighborhood functions available, as well as different learning rates $\alpha(t)$. If the Gaussian is used, then we have

$$h_{ci}(t) \;=\; \exp\left[-\frac{\|\mathbf{r}_i - \mathbf{r}_c\|^2}{2\sigma^2(t)}\right],$$

where the symbol $\|\cdot\|$ denotes the distance, and the $\mathbf{r}_i$ are the coordinates on the grid. The size or width of the neighborhood is governed by $\sigma(t)$, and this value decreases monotonically.

---

[8] See Appendix B for the website where one can download the SOM Toolbox.

The training is usually done in two phases. The first phase corresponds to a large learning rate and neighborhood radius σ (*t*), which provides a large-scale approximation to the data. The map is fine tuned in the second phase, where the learning rate and neighborhood radius are small. Once the process finishes, we have the set of prototype vectors over the 2–D coordinates on the map grid.

The *batch training method* or Batch Map is also iterative, but it uses the whole data set before adjustments are made rather than a single vector. At each step of the algorithm, the data set is partitioned such that each observation is associated with its nearest model vector. The updated model vectors are found as a weighted average of the data, where the weight of each observation is the value of the neighborhood function $h_{ic}(t)$ at its BMU *c*.

Several methods exist for visualizing the resulting map and prototype vectors. These methods can have any of the following goals. The first is to get an idea of the overall shape of the data and whether clusters are present. The second goal is to analyze the prototype vectors for characteristics of the clusters and correlations among components or variables. The third task is to see how new observations fit with the map or to discover anomalies. We will focus on one visualization method called the U-matrix that is often used to locate clusters in the data [Ultsch and Siemon, 1990].

The U-matrix is based on distances. First, the distance of each model vector to each of its immediate neighbors is calculated. This distance is then visualized using a color scale. Clusters are seen as those map units that have smaller distances, surrounded by borders indicating larger distances. Another approach is to use the size of the symbol to represent the average distance to its neighbors; so cluster borders would be shown as larger symbols. We illustrate the SOM and the U-matrix visualization method in Example 3.6.

Example 3.6

We turn to the **oronsay** data set to illustrate some of the basic commands in the SOM Toolbox. First we have to load the data and put it into a MATLAB data structure that is recognized by the functions. This toolbox comes with several normalization methods, and it is recommended that they be used before building a map.

```
% Use the oronsay data set.
load oronsay
% Convert labels to cell array of strings.
for i = 1:length(beachdune)
    mid{i} = int2str(midden(i));
    % Use next one in exercises.
    bd{i} = int2str(beachdune(i));
end
% Normalize each variable to have unit variance.
D = som_normalize(oronsay,'var');
```

```
% Convert to a data structure.
sD = som_data_struct(D);
% Add the labels - must be transposed.
sD = som_set(sD,'labels',mid');
```

We can visualize the results in many ways, most of which will be left as an exercise. We show the U-matrix in Figure 3.10 and include labels for the codebook vectors.

```
% Make the SOM
sM = som_make(sD);
sM = som_autolabel(sM,sD,'vote');
% Plot U matrix.
som_show(sM,'umat','all');
% Add labels to an existing plot.
som_show_add('label',sM,'subplot',1);
```

Note that the larger values indicate cluster borders, and low values indicate clusters. By looking at the colors, we see a couple of clusters - one in the lower left corner and one in the top. The labels indicate some separation into groups.
❏

3.3.2 Generative Topographic Maps

The SOM has several limitations [Bishop, Svensén, and Williams, 1996]. First, the SOM is based on heuristics, and the method is not derived from the optimization of an objective function. Preservation of the neighborhood structure is not guaranteed by the SOM method, and there could be problems with convergence of the prototype vectors. The SOM does not define a density model, and the use of the codebook vectors as a model of the distribution of the original data is limited. Finally, the choice of how the neighborhood function should shrink during training is somewhat of an art, so it is difficult to compare different runs of the SOM procedure on the same data set. The *generative topographic mapping* (GTM) was inspired by the SOM and attempts to overcome its limitations.

The GTM is described in terms of a latent variable model (or space) with dimensionality d [Bishop, Svensén, and Williams, 1996, 1998a]. The goal is to find a representation for the distribution $p(\mathbf{x})$ of p-dimensional data, in terms of a smaller number of d latent variables $\mathbf{m} = (m_1, \ldots, m_d)$. As usual, we want $d < p$, and often take $d = 2$ for ease of visualization. We can achieve this goal by finding a mapping $\mathbf{y}(\mathbf{m}; \mathbf{W})$, where $\mathbf{W}$ is a matrix containing weights, that takes a point $\mathbf{m}$ in the latent space and maps it to a point $\mathbf{x}$ in the data space. This might seem somewhat backward from what we considered previously, but we will see later how we can use the inverse of this mapping to view summaries of the observations in the reduced latent space.

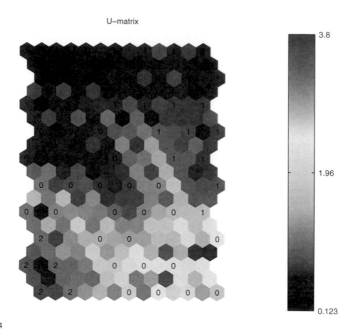

U–matrix

SOM 26–Jun–2004

FIGURE 3.10
This is the SOM for the **oronsay** data set. We can see some cluster structure here by the colors. One is in the upper part of the map, and the other is in the lower left corner. The labels on the map elements also indicate some clustering. Recall that the classes are midden (0), *Cnoc Coig* (1), and *Caisteal nan Gillean* (2). (**SEE COLOR INSERT.**)

We start with a probability distribution $p(\mathbf{m})$ defined on the latent variable space (with d dimensions), which in turn induces a distribution $p(\mathbf{y}\,|\,\mathbf{W})$ in the data space (with p dimensions). For a given $\mathbf{m}$ and $\mathbf{W}$, we choose a Gaussian centered at $\mathbf{y}(\mathbf{m};\mathbf{W})$ as follows:

$$p(\mathbf{x}\,|\,\mathbf{m},\mathbf{W},\beta) = \left[\frac{\beta}{2\pi}\right]^{p/2}\exp\left[\frac{-\beta}{2}\|\mathbf{y}(\mathbf{m};\mathbf{W}) - \mathbf{x}\|^2\right],$$

where the variance is β^{-1} and $\|\bullet\|$ denotes the inner product. Other models can be used, as discussed in Bishop, Svensén, and Williams [1998b]. The matrix $\mathbf{W}$ contains the parameters or weights that govern the mapping. To derive the desired mapping, we must estimate β and the matrix $\mathbf{W}$.

The next step is to assume some form for $p(\mathbf{m})$ defined in the latent space. To make this similar to the SOM, the distribution is taken as a sum of delta functions centered on the nodes of a regular grid in latent space:

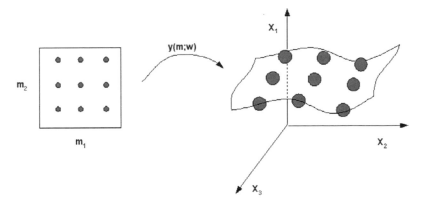

FIGURE 3.11
This illustrates the mapping used in GTM. Our latent space is on the left, and the data space is on the right. A Gaussian is centered at each of the data points, represented by the spheres.

$$p(\mathbf{m}) = \frac{1}{K}\sum_{k=1}^{K}\delta(\mathbf{m}-\mathbf{m}_k),$$

where K is the total number of grid points or delta functions. Each point $\mathbf{m}_k$ in the latent space is mapped to a corresponding point $\mathbf{y}(\mathbf{m}_k;\mathbf{W})$ in the data space, where it becomes the center of a Gaussian density function. This is illustrated in Figure 3.11.

We can use maximum likelihood and the Expectation-Maximization (EM) algorithm (see Chapter 6) to get estimates for β and $\mathbf{W}$. Bishop, Svensén, and Williams [1998a] show that the log-likelihood function for the distribution described here is given by

$$L(\mathbf{W},\beta) = \sum_{i=1}^{n} \ln\left\{\frac{1}{K}\sum_{k=1}^{K}p(\mathbf{x}_i\,|\,\mathbf{m}_k,\mathbf{W},\beta)\right\}. \qquad (3.6)$$

They next choose a model for $\mathbf{y}(\mathbf{m};\mathbf{W})$, which is given by

$$\mathbf{y}(\mathbf{x};\mathbf{W}) = \mathbf{W}\phi(\mathbf{x}),$$

where the elements of $\phi(\mathbf{m})$ have M fixed basis functions $\phi_j(\mathbf{m})$, and $\mathbf{W}$ is of size $p\times M$. For basis functions, they choose Gaussians whose centers are distributed on a uniform grid in latent space. Note that the centers for these basis functions are *not* the same as the grid points $\mathbf{m}_j$. Each of these Gaussian basis functions ϕ has a common width parameter σ. The smoothness of the manifold is determined by the value of σ, the number of basis functions M, and their spacing.

Looking at Equation 3.6, we can view this as a missing-data problem, where we do not know which component k generated each data point x_i. The EM algorithm for estimating β and W consists of two steps that are done iteratively until the value of the log-likelihood function converges. In the E-step, we use the current values of the parameters to evaluate the posterior probabilities of each component k for every data point. This is calculated according to the following

$$\tau_{ki}\left(W_{\text{old}}, \beta_{\text{old}}\right) = \frac{p\left(\mathbf{x}_i \mid \mathbf{m}_k, W_{\text{old}}, \beta_{\text{old}}\right)}{\sum_{c=1}^{K} p\left(\mathbf{x}_i \mid \mathbf{m}_c, W_{\text{old}}, \beta_{\text{old}}\right)}, \qquad (3.7)$$

where the subscript 'old' indicates the current values.

In the M-step, we use the posterior probabilities to find weighted updates for the parameters. First, we calculate a new version of the weight matrix from the following equation

$$\Phi^T G_{\text{old}} \Phi W_{\text{new}}^T = \Phi^T T_{\text{old}} X, \qquad (3.8)$$

where Φ is a $K \times M$ matrix with elements $\Phi_{kj} = \phi_j(\mathbf{m}_k)$, X is the data matrix, T is a $K \times n$ matrix with elements τ_{ki}, and G is a $K \times K$ diagonal matrix where the elements are given by

$$G_{kk} = \sum_{i=1}^{n} \tau_{ki}(W, \beta).$$

Equation 3.8 is solved for W_{new} using standard linear algebra techniques. A time-saving issue related to this update equation is that Φ is constant; so it only needs to be evaluated once.

We now need to determine an update for β that maximizes the log-likelihood. This is given by

$$\frac{1}{\beta_{\text{new}}} = \frac{1}{np} \sum_{i=1}^{n} \sum_{k=1}^{K} \tau_{kn}\left(W_{\text{old}}, \beta_{\text{old}}\right) \left\| W_{\text{old}} \phi\left(\mathbf{m}_k - \mathbf{x}_i\right) \right\|^2. \qquad (3.9)$$

To summarize, the algorithm requires starting points for the matrix W and the inverse variance β. We must also specify a set of points $\mathbf{m}_i$, as well as a set of basis functions $\phi_j(\mathbf{m})$. The parameters W and β define a mixture of Gaussians with centers $W\phi(\mathbf{m}_k)$ and equal covariance matrices given by $\beta^{-1}\mathbf{I}$. Given initial values, the EM algorithm is used to estimate these parameters. The E-step finds the posterior probabilities using Equation 3.7, and the M-step updates the estimates for W and β using Equations 3.8 and 3.9. These steps are repeated until convergence.

So, the GTM gives us a mapping from this latent space to the original p-dimensional data space. For EDA purposes, we are really interested in going the other way—mapping our p-dimensional data into some lower-dimensional space. As we see from the development of the GTM, each datum $\mathbf{x}_i$ provides a posterior distribution in our latent space. Thus, this posterior *distribution* in latent space provides information about a *single* observation. Visualizing all observations in this manner would be too difficult; so each distribution should be summarized in some way. Two summaries that come to mind are the mean and the mode, which are then visualized as individual points in our latent space. The mean for observation $\mathbf{x}_i$ is calculated from

$$\overline{\mathbf{m}}_i = \sum_{k=1}^{K} \tau_{ki} \mathbf{m}_k \, .$$

The mode for the i-th observation (or posterior distribution) is given by the maximum value τ_{ki} over all values of k. The values $\overline{\mathbf{m}}_i$ or modes are shown as symbols in a scatterplot or some other visualization scheme.

Example 3.7

We again turn to the **oronsay** data to show the basic functionality of the GTM Toolbox.[9] The parameters were set according to the example in their documentation.

```
load oronsay
% Initialize parameters for GTM.
noLatPts = 400;
noBasisFn = 81;
sigma = 1.5;
% Initialize required variables for GTM.
[X,MU,FI,W,beta] = gtm_stp2(oronsay,noLatPts,...
    noBasisFn,sigma);
lambda = 0.001;
cycles = 40;
[trndW,trndBeta,llhLog] = gtm_trn(oronsay,FI,W,...
    lambda,cycles,beta,'quiet');
```

The function **gtm_stp2** initializes the required variables, and **gtm_trn** does the training. Each observation gets mapped into a probability distribution in the 2–D map; so we need to find either the mean or mode of each one to show as a point. We can do this as follows:

```
% Get the means in latent space.
mus = gtm_pmn(oronsay,X,FI,trndW,trndBeta);
% Get the modes in latent space.
```

---

[9] This is included with the EDA Toolbox.

```
modes = gtm_pmd(oronsay,X,FI,trndW);
```

We now plot the values in the lower-dimensional space using symbols corresponding to their class.

```
ind0 = find(midden == 0);
ind1 = find(midden == 1);
ind2 = find(midden == 2);
plot(mus(ind0,1),mus(ind0,2),'k.',mus(ind1,1),...
     mus(ind1,2),'kx',mus(ind2,1),mus(ind2,2),'ko')
```

The resulting plot is shown in Figure 3.12, where we can see some separation into the three groups.
❏

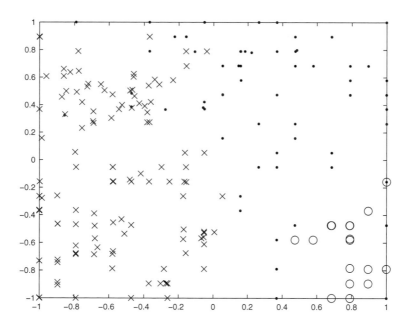

FIGURE 3.12
This shows the map obtained from GTM, where the distribution for each point is summarized by the mean. Each class is displayed using a different symbol: Class 0 is '.', Class 1 is 'x', and Class 2 is 'o.' We can see some separation into groups in this plot.

3.3.3 Curvilinear Component Analysis

Curvilinear component analysis (CCA) was developed by Demartines and Herault [1997] as an improvement on self-organizing maps (SOMs) and Sammon's nonlinear mapping [1969]. The CCA approach is a neural network strategy that produces a nonlinear mapping from the original space (where the data reside) to a lower-dimensional continuous space that can provide the correct shape of the submanifold when other methods such as SOM, MDS, and ISOMAP fail. We provide a comparison of these approaches at the end of this section.

Curvilinear component analysis is a self-organized neural network that encompasses two main steps. First, there is a vector quantization step [Ahalt et al., 1990] where input vectors become prototypes of the distribution. Some examples of vector quantization include a type of k-means clustering[10] and SOM. The next step is to find a nonlinear projection of the quantizing (or codebook) vectors by minimizing a cost function based on interpoint distances in the input and output spaces. As we will see in the following discussion, CCA combines aspects of MDS and SOM to *unfold* the data and reveal interesting structure.

As before, we are given n p-dimensional observations $\mathbf{x}_i$, and we seek a mapping to a reduced d-dimensional space ($d < p$) with the observations denoted as $\mathbf{y}_i$ in that space. The final output of CCA depends on a nonlinear mapping of the inputs or observations after they have been converted to prototypes of the distribution using vector quantization. The position of the output points is governed by the Euclidean distances between the observations in the original space and also in the output space. As in MDS, the goal is to find a mapping that preserves the interpoint distances from the original high-dimensional space.

Here, the distance between the i-th and j-th observations will be denoted as

$$X_{ij} = d(\mathbf{x}_i, \mathbf{x}_j) ;$$

the corresponding distance in the d-dimensional output space is

$$Y_{ij} = d(\mathbf{y}_i, \mathbf{y}_j) ,$$

where $d(\bullet)$ is usually the Euclidean distance.

While we would like to have $X_{ij} = Y_{ij}$, this is not always possible at all scales. So, a weighting function is introduced, and the following quadratic cost function is used:

$$E = \frac{1}{2} \sum_i \sum_{j \neq i} \left(X_{ij} - Y_{ij} \right)^2 F\left(Y_{ij}, \lambda_y \right) , \qquad (3.10)$$

where λ_y is a neighborhood parameter that governs the scale. The goal is to find values for the $\mathbf{y}_i$ such that Equation 3.10 is minimized.

The weight function $F(Y_{ij}, \lambda_y)$ in Equation 3.10 is chosen as a bounded and monotonically decreasing function. The reason for this is to preserve the local

---

[10] This is discussed in Chapter 4.

topology in the new space. Demartines and Herault [1997] used the following in their simulations:

$$F\left(Y_{ij}, \lambda_y\right) = \begin{cases} 1 & \text{if } Y_{ij} \le \lambda_y \\ 0 & \text{if } Y_{ij} > \lambda_y. \end{cases}$$

However, other functions can be used, such as a decreasing exponential or a sigmoid.

The parameter λ_y is similar to the neighborhood radius in SOM, and it generally changes as the minimization of Equation 3.10 proceeds. This is typically driven by an automatic schedule, but the original paper describes how this might also be chosen interactively.

The minimization of the cost function is achieved using a novel variant of gradient descent methods. We do not provide details here, but there are a couple aspects of the optimization approach that should be mentioned. First, their method yields a significant savings in computation time, making it suitable for large data sets. One of the ways this is achieved is with the vector quantization step where the original data are represented by a smaller set of prototype vectors. The other benefit of the optimization method used for CCA is that the cost function can temporarily increase, which allows the algorithm to escape from local minima of E. We will explore the use of CCA in the next example.

Example 3.8
We use a function for CCA that is provided with the SOM Toolbox and is included with the EDA Toolbox for convenience. We are going to generate some points that fall on the unit sphere ($p = 3$), and we are not adding noise so the submanifold is easier to see.

```
% Generate the data using the sphere function
% in the main MATLAB package.
[x,y,z] = sphere;
% Visualize the points in a scatterplot.
colormap(cmap)
scatter3(x(:),y(:),z(:),3,z(:),'filled')
axis equal
view([-34, 16])
```

The scatterplot of the original data is shown in the top panel of Figure 3.13. We now apply CCA to find a nonlinear mapping to 2–D. Note that we have to specify the number of iterations for the optimization algorithm. This is the **epochs** argument in the **cca** function. Demartins and Herault state that a data set of $n = 1000$ and a linear manifold usually requires around 50 iterations. When the submanifold is not linear, then many more steps (in the thousands) might be needed. Our data set is small (n = 21), and we will use 30 iterations.

```
% Apply CCA where we reduce to 2-D.
Pcca = cca([x(:),y(:),z(:)],2,30);
```

A scatterplot of the data in 2–D is shown in the bottom of Figure 3.13. We see that CCA was able to unfold the sphere and find the correct mapping.
❑

Another related method to CCA is called *curvilinear distance analysis* (CDA) [Lee et al., 2000 and Lee, Lendasse, and Verleysen, 2002]. CDA is similar to ISOMAP in that it uses the same geodesic distance to find neighbors, but instead the developers of CDA call it a *curvilinear* distance. The rest of the CDA strategy (e.g., the cost function E and its minimization) is the same as in CCA. Thus, one can view CDA as being the same as CCA, but with the input interpoint distances being determined using a geodesic distance.

We mentioned at the beginning of this section that CCA was developed as an improvement to Sammon's nonlinear mapping. The cost function E in Sammon's algorithm is given by

$$E = \frac{1}{c} \sum_i \sum_{j<i} \left(X_{ij} - Y_{ij} \right)^2 \frac{1}{X_{ij}},$$

where c is a normalization constant

$$c = \sum_i \sum_{j<i} X_{ij}.$$

Demartines and Herault [1997] note that this mapping is biased in favor of preserving small distances of the input space; so a correct unfolding can be difficult to obtain in applications.

Classical MDS and PCA are both linear approaches to dimensionality reduction; so they will produce a linear approximation to the nonlinear submanifold represented by the data. Nonmetric MDS can handle nonlinear data structures, but recall that the rank orders of the interpoint distances are preserved instead of the actual distances. This could result in quantization errors and a poor mapping.

We also stated that CCA is an improvement to self-organizing maps, which makes sense, since SOM or some other vector quantization is used as a first step in CCA. The SOM achieves vector quantization with a predefined neighborhood between neurons, so it can miss the actual shape. In CCA, the neurons find an appropriate position in the output space that preserves the local topology in addition to the shape of the submanifold.

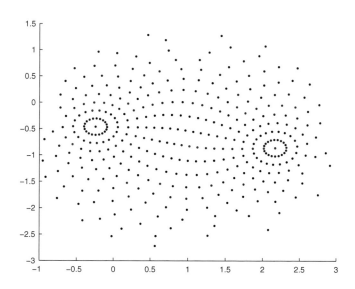

FIGURE 3.13

The plot given in the top panel displays a scatterplot of some observations that fall on the unit sphere. No noise was added to the data so the shape of the sphere would be easier to see. The scatterplot in the lower panel shows the points on the sphere after they have been mapped to 2–D using curvilinear component analysis. Note that the CCA approach was able to unfold the sphere in such a way that we can readily see the correct topology.

3.3.4 Autoencoders

There has been a resurgence of interest in artificial neural networks (ANN) under the guise of deep learning. **Deep learning** involves supervised or unsupervised learning tasks using multiple layered models from machine learning. Besides neural networks, this approach uses concepts from artificial intelligence, stochastic optimization, graphical modeling, statistical pattern recognition, and more [Lewis, 2016].

This renewed interest has been driven by many things, including algorithmic breakthroughs, improved network architectures, and increased computing capacity and speed. This renaissance has allowed these networks to best humans in chess [Lai, 2015], the game go [Zastrow, 2016], and to learn to play Atari games based on visual input [Minh et al. 2013]. The interested reader is referred to the recent paper by LeCun, Bengio, and Hinton [2015] for a very high level introduction to the deep learning field.

In keeping with the focus of this chapter on nonlinear dimensionality reduction, we will limit our discussions to the autoencoder deep learning architecture [Hinton and Salakhutdinov, 2006]. In this article, Hinton states

> "It has been obvious since the 1980s that backpropagation through deep autoencoders would be very effective for nonlinear dimensionality reduction, provided that computers were fast enough, data sets were big enough, and the initial weights were close enough to a good solution. All three conditions are now satisfied."

We follow Lewis [2016] in our description of autoencoders.

Autoencoders seek to learn the identity mapping on a data set. In other words, they are learning their input values. Autoencoders are usually feedforward neural networks with at least one hidden layer. The inputs and the outputs are the same, so it is just reconstructing the inputs.

Suppose we have a data set $\mathbf{X}$ residing in R^{10} (i.e., a ten-dimensional space). We wish to learn the following two mappings

$$\alpha: R^{10} \to R^2$$
$$\beta: R^2 \to R^{10}$$

The function α is an **encoder**, and the mapping β is a **decoder** function. The encoder is a mapping that takes the data from a higher dimensional space to only two dimensions in this example. Thus, it is a mapping that reduces the dimensionality of our data.

The encoder and decoders are chosen to minimize the reconstruction error

$$\|\mathbf{X} - (\beta \circ \alpha)\mathbf{X}\|^2 \tag{3.11}$$

over all possible encoders α and decoders β. Note that reconstruction errors other than the squared error in Equation 3.11 can be used.

If we could learn a network architecture that mapped the data to a lower dimensional space and back to the higher dimensionality space with minimal error, then the mapping (or encoding) has done a good job at capturing our data. We illustrate a single autoencoder network in Figure 3.14.

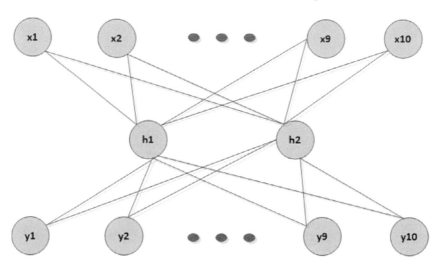

FIGURE 3.14
This is an example of a single hidden layer autoencoder network. The input x_i are equal to the output y_i, and the network estimates a mapping or encoder function that best matches these two layers. The hidden layer nodes are usually nonlinear sigmoid functions.

Because the input values are the same as the outputs, an autoencoder is essentially learning or modeling the identity function $h(x) = y \approx \hat{x}$. It first maps the inputs to the hidden representation (or layer) using the encoder function given by

$$h = f(\mathbf{W}_1\mathbf{x} + \mathbf{b}_1),\qquad\qquad(3.12)$$

where f is a nonlinear function (e.g., sigmoid) [Weisstein, 2016], $\mathbf{W}_1$ is the matrix of weights connecting the input and hidden layers, and $\mathbf{b}_1$ is a vector of biases or offsets for the hidden nodes.

The reconstruction part of the process is achieved using a decoder function

$$y = f(\mathbf{W}_2\mathbf{x} + \mathbf{b}_2),$$

where the $\mathbf{W}_2$ and $\mathbf{b}_2$ are analogous to the entities in Equation 3.12. We have the option to constrain the decoder to use *tied weights* $\mathbf{W}_2 = \mathbf{W}_1^T$.

In the case of an autoencoder network with a single hidden layer, one can initialize the weights randomly and use error backpropagation [Rumelhart et al., 1986] to adjust these weights to minimize the squared reconstruction error (Equation 3.11). The real power of the autoencoder comes when one includes multiple layers as shown in Figure 3.15.

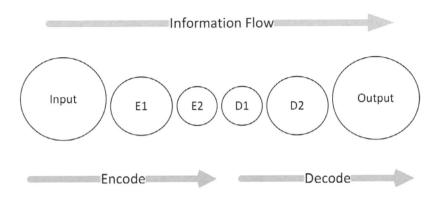

FIGURE 3.15
Here we show a schematic of a multilayer autoencoder.

Referring to Figure 3.15, we ultimately end up with an encoding of the data **X** in a space with greatly reduced dimensionality in the E2 layer. These deep encoding architectures can contain many hidden encoding and decoding layers that can fail under simple forward and backward based propagation training. Fortunately, Hinton and Salakhutdinov [2006] formulated a method to train these networks by using Boltzmann machines [Smolensky, 1986] to pre-train the sublayers in the network architecture.

Example 3.9
We return to the S–curve manifold from Example 3.5. This data set consists of 2,000 data points that lie along a 2–D manifold in three dimensions. The data can be uploaded from the **scurve** file.

```
% Load the scurve data
% Contains 3-D data in X and
% height along z-axis in 'height'.
load scurve
```

This loads the data matrix **X** and the **height** variable. We use the **height** to color code the symbols, which gives us a sense of the neighborhood structure. To better illustrate our description of autoencoders, we will add some noisy dimensions to the data.

```
% Add 7 columns of noise.
Y = [X',randn(2000,7)];
```

The Dimensionality Reduction Toolbox [van der Maaten, 2007] has a function for constructing an autoencoder. The following function call uses two nodes in the hidden layer. Thus, we are reducing the dimensionality to 2–D.

```
% Reduce the dimensionality to 2-D.
hn = 2;
[network, encode2d, decode10d] = ...
    train_autoencoder(Y,hn,0,5000);
```

The output variable **encode2d** contains the 2,000 points in 2–D. We show them in Figure 3.16. We can compare the results to Figures 3.7 and 3.8, where we see that the dimensionality reduction from the autoencoder approach is superior in some aspects.
❏

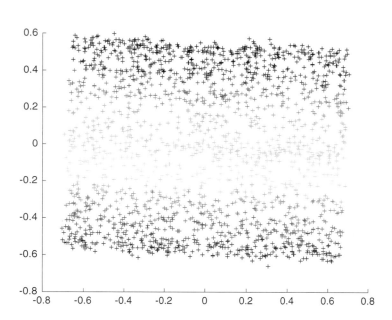

FIGURE 3.16
Here is a scatterplot of the 2–D autoencoder embedding for the S–curve data. The color of the symbols is matched to the height in 3–D, and we see that most of the neighborhood structure has been preserved. Furthermore, this appears to have a better embedding than the one from LLE (see Figure 3.8) in that it reflects the true rectangular boundary of the data in 2–D. (**SEE COLOR INSERT**.)

3.4 Stochastic Neighbor Embedding

The *stochastic neighbor embedding* (SNE) approach for reducing the dimensionality of a data set was developed by Hinton and Roweis [2002]. The goal is to allow us to visualize high-dimensional data in either two or three dimensions. In some ways, SNE has the same goal as MDS, in that it obtains a mapping to a lower-dimensional space such that the interpoint distance relationships in the higher-dimensional space are preserved.

The input to SNE is the interpoint distance matrix, usually obtained using Euclidean distances. The distances are converted to similarities using conditional probabilities. The similarity of an observation $\mathbf{x}_i$ to $\mathbf{x}_j$ is the conditional probability $p_{i|j}$ that $\mathbf{x}_i$ would have $\mathbf{x}_j$ as a neighbor. This choice is proportional to their probability under a Gaussian density centered at $\mathbf{x}_i$ [van der Maaten and Hinton, 2008], as given here

$$p_{j|i} = \frac{\exp(-\|\mathbf{x}_i - \mathbf{x}_j\|^2/(2\sigma_i^2))}{\sum_{k \neq i} \exp(-\|\mathbf{x}_i - \mathbf{x}_k\|^2/(2\sigma_i^2))}, \tag{3.13}$$

where σ_i^2 is the variance of the Gaussian centered at $\mathbf{x}_i$.

Recall, that the goal is to find an embedding of the high-dimensional data in a lower-dimensional space. The corresponding observations in this space are denoted by $\mathbf{y}_i$ and $\mathbf{y}_j$. We use a similar form of conditional probability to represent the similarities between the observations in this space. It is given by

$$q_{j|i} = \frac{\exp(-\|\mathbf{y}_i - \mathbf{y}_j\|^2)}{\sum_{k \neq i} \exp(-\|\mathbf{y}_i - \mathbf{y}_k\|^2)}, \tag{3.14}$$

where the variance has been set to $1/\sqrt{2}$. Note that $p_{i|i} = q_{i|i} = 0$.

We can measure how well these neighborhood probabilities match in the two spaces using the summation of the Kullback-Liebler divergences given by

$$C = \sum_i \sum_j p_{j|i} \log \frac{p_{j|i}}{q_{j|i}}. \tag{3.15}$$

Equation 3.15 is our cost function, and it is minimized using gradient descent.

The Kullback-Liebler divergence is not symmetric, so the errors in pairwise distances in the lower-dimensional space are not weighted equally [van der

Maaten and Hinton, 2008]. The result is that the cost function (Equation 3.15) emphasizes the retention of the local structure in the smaller space.

We have to specify the variance σ_i^2 of the Gaussian used in Equation 3.13. In most applications, there is likely no optimal value for all $i = 1, \ldots, n$. The SNE approach conducts a search for a good value of the variance at a data point $\mathbf{x}_i$ with a fixed **perplexity**, which is specified by the user. One can think of the perplexity as defining the number of neighbors. van der Maaten and Hinton [2008] define it as a "smooth measure of the effective number of neighbors." Values might range from 5 to 50.

The SNE method has some drawbacks. First, it is often difficult and costly to solve the objective function in Equation 3.15. Secondly, it tends to map data points to the same place or near each other in the lower dimensional space. This is sometimes called the "crowding problem." Other nonlinear dimensionality reduction methods suffer from the same issue.

The *t-SNE* or *t*–Distributed Stochastic Neighbor Embedding method was developed to address these problems [van der Maaten and Hinton, 2008]. It uses a symmetrized version of the cost function, and it employs a Student's *t* distribution to find the similarity between points in the lower-dimensional space. The *t* distribution has heavier tails, which reduces crowding.

To symmetrize the cost function, we minimize a single divergence between a joint probability distribution in the high-dimensional space and a joint probability distribution in the lower-dimensional space. The cost function is

$$C = \sum_i \sum_j p_{ij} \log \frac{p_{ij}}{q_{ij}}. \tag{3.16}$$

As before, we have $p_{ii} = q_{jj} = 0$. This is symmetric since $p_{ij} = p_{ji}$ and $q_{ij} = q_{ji}$, for all i and j.

The joint probabilities p_{ij} are obtained from the symmetrized conditional probabilities

$$p_{ij} = \frac{p_{j|i} + p_{i|j}}{2n}. \tag{3.17}$$

We first determine the conditional probabilities using Equation 3.13, and then symmetrize using Equation 3.17. Besides having an easier gradient to use in the optimization, this way of calculating the joint probabilities p_{ij} ensures that each $\mathbf{x}_i$ makes a significant contribution to the objective function [van der Maaten and Hinton, 2008].

The second adjustment that *t*-SNE makes is to use a Student's *t*-distribution with one degree of freedom to determine the joint probabilities q_{ij}. These are

$$q_{ij} = \frac{(1 + \|\mathbf{y}_i - \mathbf{y}_j\|^2)^{-1}}{\displaystyle\sum_{k \neq l}(1 + \|\mathbf{y}_k - \mathbf{y}_l\|^2)^{-1}} . \tag{3.18}$$

This is faster to compute because it does not have the exponential function. It also means that the lower-dimensional map is optimized similarly at most scales.

The *t*-SNE method is implemented in the Dimensionality Reduction Toolbox [van der Maaten, 2007] and in other computing languages at

http://lvdmaaten.github.io/tsne/

This website also provides a fast implementation of the algorithm and visual examples of its application to famous data sets.

Example 3.10

The Dimensionality Reduction Toolbox has a function that will generate 3–D data, and we use that to generate some clusters. Besides the data, the function returns a set of cluster labels that can be used to aid visualization. We show these data in the top panel of Figure 3.17.

```
% First, we generate some 3-D data.
[X,labels] = generate_data('3d_clusters');
scatter3(X(:,1),X(:,2),X(:,3),5,labels)
box on
```

We now apply *t*-SNE, specifying an embedding dimension of 2–D, an initial dimension of 3–D, and perplexity of 30 (the default). The **tsne** function starts off with a preprocessing step that uses PCA to reduce to a smaller dimension before starting *t*-SNE. We specified 3–D for this value because our initial dimensionality is small.

```
d = 2;
dpca = 3;
perp = 30;
ydata = tsne(X, labels, d, dpca, perp);
scatter(ydata(:,1),ydata(:,2),5,labels)
```

The scatterplot of the data is shown in the lower part of Figure 3.17. We see that the interpoint distances seem to be preserved, as is the distance between the clusters. Thus, there is little crowding.

❏

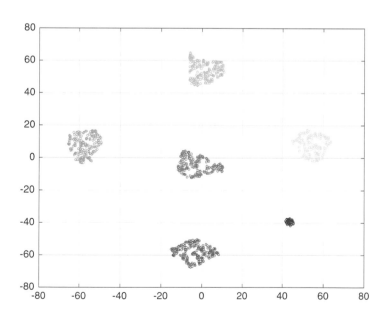

FIGURE 3.17

The top panel shows a scatterplot of the 3–D cluster data generated in Example 3.10. We see that the clusters are rather separated. The lower plot shows the 2–D mapping using *t*-SNE. We see that the clusters are not crowded, and the distance between clusters seems to be reasonably preserved, as is the individual interpoint distances.

3.5 Summary and Further Reading

In this chapter, we discussed several methods for finding a nonlinear mapping from a high-dimensional space to one with lower dimensionality. The first set of methods was grouped under the name multidimensional scaling, and we presented both metric and nonmetric MDS. We note that in some cases, depending on how it is set up, MDS finds a linear mapping. We also presented several methods for learning manifolds, where the emphasis is on nonlinear manifolds. These techniques are locally linear embedding, ISOMAP, and Hessian locally linear embedding. ISOMAP is really an enhanced version of classical MDS, where geodesic distances are used as input to classical MDS. HLLE is similar in spirit to LLE, and its main advantage is that it can handle data sets that are not convex. Finally, we presented two artificial neural network approaches called self-organizing maps and generative topographic maps.

We have already mentioned some of the major MDS references, but we also include them here with more information on their content. Our primary reference for terminology and methods was Cox and Cox [2001]. This is a highly readable book, suitable for students and practitioners with a background in basic statistics. The methods are clearly stated for both classical MDS and nonmetric MDS. Also, the authors provide a CD-ROM with programs (running under DOS) and data sets so the reader can use these tools.

Another book by Borg and Groenen [1997] brings many of the algorithms and techniques of MDS together in a way that can be understood by those with a two-semester course in statistics for social sciences or business. The authors provide the derivation of the methods, along with ways to interpret the results. They do not provide computer software or show how this can be done using existing software packages, but their algorithms are very clear and understandable.

A brief introduction to MDS can be obtained from Kruskal and Wish [1978]. This book is primarily focused on applications in the social sciences, but it would be useful for those who need to come to a quick understanding of the basics of MDS. Computational and algorithmic considerations are not stressed in this book. There are many overview papers on MDS in the various journals and encyclopedias; these include Mead [1992], Siedlecki, Siedlecka, and Sklansky [1988], Steyvers [2002], and Young [1985].

Some of the initial work in MDS was done by Shepard in 1962. The first paper in the series [Shepard, 1962a] describes a computer program to reconstruct a configuration of points in Euclidean space, when the only information that is known about the distance between the points is some unknown monotonic function of the distance. The second paper in the series [Shepard, 1962b] presents two applications of this method to artificial data.

Kruskal continued the development of MDS by introducing an objective function and developing a method for optimizing it [Kruskal, 1964a, 1964b]. We highly recommend reading the original Kruskal papers; they are easily understood and should provide the reader with a nice understanding of the origins of MDS.

A nice overview of manifold learning is given in Saul and Roweis [2002], with an emphasis on LLE. Further issues with ISOMAP are explored in Balasubramanian and Schwartz [2002]. More detailed information regarding HLLE can be found in a technical report written by Donoho and Grimes [2002]. An edited book by Gorban et al. [2008] contains many excellent articles on manifold learning for dimensionality reduction, as well as visualization. There is a text called *Nonlinear Dimensionality Reduction* by Lee and Verleysen [2007] that describes the details of manifold learning and is for the mathematically inclined reader.

There is one book dedicated to SOM written by Kohonen [2001]. Many papers on SOM have appeared in the literature. A 1998 technical report lists 3,043 works based on the SOM [Kangas and Kaski, 1998]. A nice, short overview of SOM can be found in Kohonen [1998]. Some recent applications of SOM have been in the area of document clustering [Kaski et al., 1998; Kohonen et al., 2000] and the analysis of gene microarrays [Tamayo et al., 1999]. Theoretical aspects of the SOM are discussed in Cottrell, Fort, and Pages [1998]. The use of the SOM for clustering is described in Kiang [2001] and Vesanto and Alhoniemi [2000]. Visualization and EDA methods for SOM are discussed in Mao and Jain [1995], Ultsch and Siemon [1990], Deboeck and Kohonen [1998], and Vesanto [1997; 1999]. For those who want to learn more about GTM, we recommend Bishop, Svensén, and Williams [1996, 1997a, 1997b, 1998a, and 1998b], Bishop, Hinton, and Strachan [1997], and Bishop and Tipping [1998].

Additional resources on deep learning and *t*-SNE continue to grow. Those who want to learn more about deep learning are referred to the extensive article by Bengio [2009]. A book on deep learning in draft form is available online at [Goodfellow et al., 2016]

http://www.deeplearningbook.org/

The book by Lewis [2016] is a hands–on guide to deep learning using the R computing environment.

Exercises

3.1   Try the classical MDS approach using the **skull** data set. Plot the results in a scatterplot using the text labels as plotting symbols (see the **text** function). Do you see any separation between the categories of gender [Cox and Cox, 2001]? Try PCA on the **skull** data. How does this compare with the classical MDS results?

3.2 Apply the SMACOF and nonmetric MDS methods to the **skull** data set. Compare your results with the configuration obtained through classical MDS.

3.3 Use **plot3** (similar to **plot**, but in three dimensions) to construct a 3–D scatterplot of the data in Example 3.1. Describe your results.

3.4 Apply the SMACOF method to the **oronsay** data set and comment on the results.

3.5 Repeat Examples 3.2, 3.4, and Problem 3.4 for several values of d. See the **help** on **gplotmatrix** and use it to display the results for $d > 2$. Do a scree-like plot of stress versus d to see which d is best.

3.6 The Shepard diagram for nonmetric MDS is a plot where the ordered dissimilarities are on the horizontal axis. The distances (shown as points) and the disparities (shown as a line) are on the vertical. With large data sets, this is not too useful. However, with small data sets, it can be used to see the shape of the regression curve. Implement this in MATLAB and test it on one of the smaller data sets.

3.7 Try using **som_show(sM)** in Example 3.6. This shows a U-matrix for each variable. Look at each one individually and add the labels to the elements: **som_show(sM,'comp',1)**, etc. See the SOM Toolbox documentation for more information on how these functions work.

3.8 Repeat Example 3.6 and Problem 3.7 using the other labels for the **oronsay** data set. Discuss your results.

3.9 Repeat the plot in Example 3.7 (GTM) using the modes instead of the means. Do you see any difference between them?

3.10 Do a help on the Statistics Toolbox function **mdscale**. Apply the methods (metric and nonmetric MDS) to the **skulls** and **oronsay** data sets.

3.11 Apply the ISOMAP method to the **scurve** data from Example 3.5. Construct a scatterplot of the data and compare to the results from LLE.

3.12 Apply the LLE method to the **swissroll** data of Example 3.5. Construct a scatterplot and compare with HLLE and ISOMAP.

3.13 What is the intrinsic dimensionality of the **swissroll** and **scurve** data sets?

3.14 Where possible, apply MDS, ISOMAP, LLE, HLLE, SOM, CCA, and GTM to the following data sets. Discuss and compare the results.

a. BPM data sets

b. gene expression data sets

c. **iris**

d. **pollen**

e. **posse**

f. **oronsay**

g. **skulls**

3.15 Repeat Example 3.2 with different starting values to search for different structures. Analyze your results.

3.16 Use PCA and ISOMAP to the sphere data in Example 3.8. Compare your results with the output from CCA.

3.17 Lewis [2016] has an example of an autoencoder where the input is an image. He then compares the input and output images to assess the performance of the autoencoder. The following runs a similar test using the autoencoder function from the Dimensionality Reduction Toolbox. Try it out on the small MATLAB logo image using different autoencoder configurations.

```
% Read file and convert it to grayscale.
imdata = imread('matlogo.png','png');
imgray = rgb2gray(imdata);
% Display image.
image(imgray)
colormap(gray(256));

% We convert it to double and scale between
% 0 and 1 for additional processing
matlogo = double(imgray/255);

% We process the transpose so we have
% n = 462 and p = 416.
% We use 100 nodes in a single hidden layer
% and 500 iterations.
[network100, encode100, decode100] = ...
   train_autoencoder(matlogo', 100,0,1000);

% Map the data to gray scale.
mat_gs = 255*decode100';
% Plot the image and compare.
figure
image(mat_gs)
colormap(gray(256))
```

3.18 Use ISOMAP to reduce the data generated in Example 3.10 to 2–D. A value of $k = 200$ is needed. Compare your results with the *t*-SNE.

3.19 Find the distances between all pairs of groups in the 3–D data generated in Example 3.10. Compare these with the intergroup distances in the 2–D embedding. How well are they preserved?

Chapter 4

Data Tours

In previous chapters we searched for lower-dimensional representations of our data that might show interesting structure. However, there are an infinite number of possibilities; so we might try touring through space, looking at many of these representations. This chapter includes a description of several tour methods that can be roughly categorized into the following groups:

1. **Grand Tours**: If the goal is to look at the data from all possible viewpoints to get an idea of the overall distribution of the p-dimensional data, then we might want to look at a random sequence of lower dimensional projections. Typically, there is little user interaction with these methods other than to set the step size for the sequence and maybe to stop the tour when an interesting structure is found. This was the idea behind the original torus grand tour of Asimov [1985].

2. **Interpolated Tours**: In this type of tour, the user chooses a starting plane and an ending plane. The data are projected onto the first plane. The tour then proceeds from one plane to the other by interpolation. At each step of the sequence, the user is presented with a different view of the data, usually in the form of a scatterplot [Hurley and Buja, 1990].

3. **Guided Tours**: These tours can be either partly or completely guided by the data. An example of this type of tour is the EDA projection pursuit method. While this is not usually an interactive or visual tour in the sense of the others, it does look at many projections of the data searching for interesting structure such as holes, clusters, etc.

Each of these methods is described in more detail in the following sections.

4.1 Grand Tour

In the grand tour methods, we want to view the data from 'all' possible perspectives. This is done by projecting the data onto a 2–D subspace and then viewing it as a scatterplot. We do this repeatedly and rapidly, so the user ends up seeing an animated sequence (or movie) of scatterplots. We could also project to spaces with dimensionality greater than 2, but the visualization would have to be done differently (see Chapter 10 for more on this subject). Projections to 1–D (a line) are also possible, where we could show the individual points on the line or as a distribution (e.g., histogram or other estimate of data density). We describe two grand tour methods in this section: the torus winding method and the pseudo grand tour.

In general, grand tours should have the following desirable characteristics:

1. The sequence of planes (or projections) should be dense in the space of all planes, so the tour eventually comes close to any given 2–D projection.

2. The sequence should become dense rapidly; so we need an efficient algorithm to compute the sequence, project the data, and present it to the user.

3. We want our sequence of planes to be uniformly distributed, because we do not want to spend a lot of time in one area.

4. The sequence of planes should be 'continuous' to aid user understanding and to be visually appealing. However, a trade-off between continuity and speed of the tour must be made.

5. The user should be able to reconstruct the sequence of planes after the tour is over. If the user stops the tour at a point where interesting structure is found, then that projection should be recovered easily.

To achieve these, the grand tour algorithm requires a continuous, space-filling path through the set of 2–D subspaces in p-dimensional space.

To summarize, the grand tour provides an overview or tour of a high-dimensional space by visualizing a sequence of 2–D scatterplots, in such a way that the tour is representative of all projections of the data. The tour continues until the analyst sees some interesting structure, at which time it is halted. The output of the grand tour method is a movie or animation with information encoded in the smooth motion of the 2–D scatterplots. A benefit from looking at a moving sequence of scatterplots is that two additional dimensions of information are available in the speed vectors of the data points [Buja and Asimov, 1986]. For example, the further away a point is from the computer screen, the faster the point rotates.

4.1.1 Torus Winding Method

The *torus winding method* was originally proposed by Asimov [1985] and Buja and Asimov [1986] as a way of implementing the grand tour. We let $\{\lambda_1,...,\lambda_N\}$ be a set of real numbers that are linearly independent over the integers. We also define a function a(t) as

$$a(t) = (\lambda_1 t,..., \lambda_N t), \qquad (4.1)$$

where the coordinates $\lambda_i t$ are interpreted modulo 2π. It is known that the mapping in Equation 4.1 defines a space-filling path [Asimov, 1985; Wegman and Solka, 2002] that winds around a torus.

We let the vector $\mathbf{e}_i$ represent the canonical basis vector. It contains zeros everywhere except in the i-th position where it has a one. Next we let $\mathbf{R}_{ij}(\theta)$ denote a $p \times p$ matrix that rotates the $\mathbf{e}_i \mathbf{e}_j$ plane through an angle of size θ. This is given by the identity matrix, but with the following changes:

$$r_{ii} = r_{jj} = \cos(\theta); \; r_{ij} = -\sin(\theta); \; r_{ji} = \sin(\theta).$$

We then define a function f as follows

$$f(\theta_{1,2}, ..., \theta_{p-1,p}) = \mathbf{Q} = \mathbf{R}_{12}(\theta_{12}) \times \mathbf{R}_{13}(\theta_{13}) \times ... \times \mathbf{R}_{p-1,p}(\theta_{p-1,p}). \quad (4.2)$$

Note that we have N arguments (or angles) in the function f, and it is subject to the restrictions that $0 \le \theta_{ij} \le 2\pi$ and $1 \le i < j \le p$. We use a reduced form of this function with fewer terms [Asimov, 1985] in our procedure outlined below.

Procedure – Torus Method

1. The number of factors in Equation 4.2 is given by $N = 2p - 3$.[1] We use only $R_{ij}(\theta_{ij})$ with $i = 1$ or $i = 2$. If $i = 1$, then $2 \le j \le p$. If $i = 2$, then $3 \le j \le p$.
2. Choose real numbers $\{\lambda_1, ... , \lambda_N\}$ and a stepsize t such that the numbers $\{2\pi, \lambda_1 t, ... , \lambda_N t\}$ are linearly independent. The stepsize t should be chosen to yield a continuous sequence of planes.
3. The values $\lambda_1 Kt, ... , \lambda_N Kt$, $K = 1, 2, ...$ are used as the arguments to the function $f(\bullet)$. K is the iteration number.
4. Form the product $\mathbf{Q}_K$ of all the rotations (Equation 4.2).
5. Rotate the first two basis vectors using

---

[1] We are using the reduced form described in the appendix of Asimov [1985]. The complete form of this rotation consists of $(p^2 - p)/2$ possible plane rotations, corresponding to the distinct 2–planes formed by the canonical basis vectors.

$$\mathbf{A}_K = \mathbf{Q}_K\mathbf{E}_{12},$$

where the columns of $\mathbf{E}_{12}$ contain the first two basis vectors, $\mathbf{e}_1$ and $\mathbf{e}_2$.

6. Project the data onto the rotated coordinate system for the K-th step:

$$\mathbf{X}_K = \mathbf{X}\mathbf{A}_K.$$

7. Display the points as a scatterplot.
8. Repeat from step 3 for the next value of K.

We need to choose λ_i and λ_j such that the ratio λ_i/λ_j is irrational for every i and j. Additionally, we must choose these such that no λ_i/λ_j is a rational multiple of any other ratio. It is also recommended that the time step t be a small positive irrational number. Two possible ways to obtain irrational values are the following [Asimov, 1985]:

1. Let $\lambda_i = \sqrt{P_i}$, where P_i is the i-th prime number.
2. Let $\lambda_i = e^i \bmod 1$.

We show how to implement the torus grand tour in Example 4.1.

Example 4.1
We use the **yeast** data in this example of a torus grand tour. First we load the data and set some constants.

```
load yeast
[n,p] = size(data);
% Set up vector of frequencies.
N = 2*p - 3;
% Use second option from above.
lam = mod(exp(1:N),1);
% This is a small irrational number:
delt = exp(-5);
% Get the indices to build the rotations.
% As in step 1 of the torus method.
J = 2:p;
I = ones(1,length(J));
I = [I, 2*ones(1,length(J)-1)];
J = [J, 3:p];
E = eye(p,2);    % Basis vectors
% Just do the tour for some number of iterations.
maxit = 2150;
```

Next we implement the tour itself.

```
% Get an initial plot.
z = zeros(n,2);
ph = plot(z(:,1),z(:,2),'o','erasemode','normal');
axis equal, axis off
% Use some Handle Graphics to remove flicker.
set(gcf,'backingstore','off','renderer',...
    'painters','DoubleBuffer','on')
% Start the tour.
for k = 1:maxit
    % Find the rotation matrix.
    Q = eye(p);
    for j = 1:N
        dum = eye(p);
        dum([I(j),J(j)],[I(j),J(j)]) = ...
         cos(lam(j)*k*delt);
        dum(I(j),J(j)) = -sin(lam(j)*k*delt);
        dum(J(j),I(j)) = sin(lam(j)*k*delt);
        Q = Q*dum;
    end
    % Rotate basis vectors.
    A = Q*E;
    % Project onto the new basis vectors.
    z = data*A;
    % Plot the transformed data.
    set(ph,'xdata',z(:,1),'ydata',z(:,2))
    % Forces Matlab to plot the data.
    pause(0.02)
end
```

We provide a function called **torustour** that implements this code. The configuration obtained at the end of this tour is shown in Figure 4.1.
❑

4.1.2 Pseudo Grand Tour

Asimov [1985] and Buja and Asimov [1986] described other ways of implementing a grand tour called the *at-random method* and the *random-walk method*. These methods, along with the torus grand tour, have some limitations. With the torus method we may end up spending too much time in certain regions, and it can be computationally intensive. Other techniques are better computationally, but cannot be reversed easily (to recover the projection) unless the set of random numbers used to generate the tour is retained.

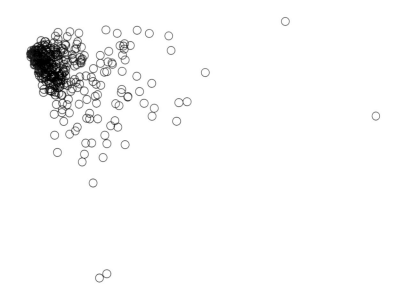

FIGURE 4.1
This shows the end of the torus grand tour using the **yeast** data.

We now discuss the *pseudo grand tour* first described in Wegman and Shen [1993] and later implemented in MATLAB by Martinez and Martinez [2015]. One of the important aspects of the torus grand tour is that it provides a continuous space-filling path through the manifold of planes. The following method does not employ a space-filling curve; thus it is called a pseudo grand tour. Another limitation is that the pseudo grand tour does not generalize to higher dimensional tours like the torus method. In spite of this, the pseudo grand tour has many benefits, such as speed, ease of calculation, uniformity of the tour, and a recoverable projection.

For the tour we need unit vectors that comprise the desired projection. Our first unit vector is denoted as $\alpha(t)$, such that

$$\alpha^T(t)\alpha(t) = \sum_{i=1}^{p} \alpha_i^2(t) = 1,$$

for every t, where t represents the stepsize as before. We need a second unit vector $\beta(t)$ that is orthonormal to $\alpha(t)$, so

$$\beta^T(t)\beta(t) = \sum_{i=1}^{p} \beta_i^2(t) = 1, \qquad \alpha^T(t)\beta(t) = 0.$$

For the pseudo grand tour, $\alpha(t)$ and $\beta(t)$ must be continuous functions of t and should produce 'all' possible orientations of a unit vector.

Before we continue in our development, we consider an observation $\mathbf{x}$. If p is odd, then we augment each data point with a zero, to get an even number of elements. In this case,

$$\mathbf{x} = [x_1, \ldots, x_p, 0]^T, \qquad \text{for } p \text{ odd.}$$

This will not affect the projection. So, without loss of generality, we present the method with the understanding that p is even. We take the vector $\alpha(t)$ to be

$$\alpha(Kt) = \sqrt{\frac{2}{p}} \times [\sin\omega_1 Kt, \cos\omega_1 Kt, \ldots, \sin\omega_{p/2} Kt, \cos\omega_{p/2} Kt]^T, \qquad (4.3)$$

for $K = 1, 2, \ldots$ and the vector $\beta(t)$ as

$$\beta K(t) = \sqrt{\frac{2}{p}} \times [\cos\omega_1 Kt, -\sin\omega_1 Kt, \ldots, \cos\omega_{p/2} Kt, -\sin\omega_{p/2} Kt]^T. \qquad (4.4)$$

We choose ω_i and ω_j in a similar manner to the λ_i and λ_j in the torus grand tour. The steps for implementing the 2–D pseudo grand tour are given here, and the details on how to implement this in MATLAB are given in Example 4.2.

Procedure- Pseudo Grand Tour

1. Set each ω_i to an irrational number. Determine a small positive irrational number for the stepsize t.
2. Find vectors $\alpha(Kt)$ and $\beta(Kt)$ using Equations 4.3 and 4.4.
3. Project the data onto the plane spanned by these vectors.
4. Display the projected points in a 2–D scatterplot.
5. Repeat from step 2 for the next value of K.

Example 4.2

We will use the **oronsay** data set for this example that illustrates the pseudo grand tour. We provide a function in the EDA Toolbox called **pseudotour** that implements the pseudo grand tour, however details are given below. Since the **oronsay** data set has an even number of variables, we do not have to augment the observations with a zero.

```
load oronsay
```

```
x = oronsay;
maxit = 10000;
[n,p] = size(x);
% Set up vector of frequencies as in grand tour.
th = mod(exp(1:p),1);
% This is a small irrational number:
delt = exp(-5);
cof = sqrt(2/p);
% Set up storage space for projection vectors.
a = zeros(p,1); b = zeros(p,1);
z = zeros(n,2);
% Get an initial plot.
ph = plot(z(:,1),z(:,2),'o','erasemode','normal');
axis equal, axis off
set(gcf,'backingstore','off','renderer',...
  'painters','DoubleBuffer','on')
for t = 0:delt:(delt*maxit)
 % Find the transformation vectors.
 for j = 1:p/2
   a(2*(j-1)+1) = cof*sin(th(j)*t);
   a(2*j) = cof*cos(th(j)*t);
   b(2*(j-1)+1) = cof*cos(th(j)*t);
   b(2*j) = cof*(-sin(th(j)*t));
 end
 % Project onto the vectors.
 z(:,1) = x*a;
 z(:,2) = x*b;
 set(ph,'xdata',z(:,1),'ydata',z(:,2))
 drawnow
end
```

A scatterplot showing an interesting configuration of points is shown in Figure 4.2. The reader is encouraged to view this tour as it shows some interesting structure along the way.
❑

4.2 Interpolation Tours

We present a version of the *interpolation tour* described in Young and Rheingans [1991] and Young, Faldowski, and McFarlane [1993]. The mathematics underlying this type of tour were presented in Hurley and Buja [1990] and Asimov and Buja [1994]. The idea behind interpolation tours is that it starts with two subspaces: an initial one and a target subspace. The

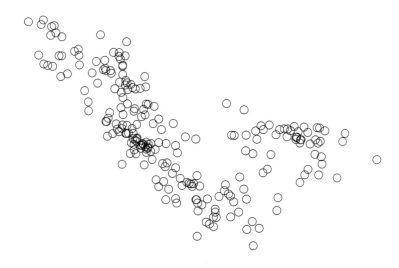

Iteration: 1822

FIGURE 4.2
This shows the scatterplot of points for an interesting projection of the **oronsay** data found during the pseudo grand tour in Example 4.2.

tour proceeds by traveling from one to the other via geodesic interpolation paths between the two spaces. Of course, we also display the projected data in a scatterplot at each step in the path for a movie view, and one can continue to tour the data by going from one target space to another.

We assume that the data matrix $\mathbf{X}$ is column centered; i.e., the centroid of the data space is at the origin. As with the other tours, we must have a visual space to present the data to the user. The visual space will be denoted by $\mathbf{V}_t$, which is an $n \times 2$ matrix of coordinates in 2–D.

One of the difficulties of the interpolation tour is getting the target spaces. Some suggested spaces include those spanned by subsets of the eigenvectors in PCA, which is what we choose to implement here. So, assuming that we have the principal component scores (see Chapter 2), the initial visible space and target space will be $n \times 2$ matrices, whose columns contain different principal components.

The interpolation path is obtained through the following rotation:

$$\mathbf{V}_t = \mathbf{T}_k[\cos \mathbf{U}_t] + \mathbf{T}_{k+1}[\sin \mathbf{U}_t], \tag{4.5}$$

where $\mathbf{V}_t$ is the visible space at the t-th step in the path, $\mathbf{T}_k$ indicates the k-th target in the sequence, and $\mathbf{U}_t$ is a diagonal 2×2 matrix with values θ_k between 0 and $\pi/2$. At each value of t, we increment the value of θ_k for some small stepsize. Note that the subscript k indicates the k-th plane in the target sequence, since we can go from one target plane to another.

Example 4.3

We use the **oronsay** data set to illustrate our function that implements the interpolation tour. The function takes the data matrix as its first argument. The next two inputs to the function contain column indices to the matrix of principal components. The first vector designates the starting plane and the second one corresponds to the target plane.

```
load oronsay
% Set up the vector of indices to the columns spanning
% the starting and target planes.
T1 = [3 4];
T2 = [5 6];
intour(oronsay, T1, T2);
```

We show the scatterplots corresponding to the starting plane and the target plane in Figure 4.3. This function actually completes a full rotation back to the starting plane after pausing at the target. Those readers who are interested in the details of the tour can refer to the M-file **intour** for more information.
❏

4.3 Projection Pursuit

In contrast to the grand tour, the ***projection pursuit*** method performs a directed search based on some index that indicates a type of structure one is looking for. In this sense, the tour is guided by the data, because it keeps touring until a possible structure is found. Like the grand tour method, projection pursuit seeks to find projections of the data that are interesting; i.e., show departures from normality, such as clusters, linear structures, holes, outliers, etc. The objective is to find a projection plane that provides a 2–D view of our data such that the structure (or departure from normality) is maximized over all possible 2–D projections.

Friedman and Tukey [1974] describe projection pursuit as a way of searching for and exploring nonlinear structure in multi-dimensional data by examining many 2–D projections. The idea is that 2–D orthogonal projections of the data should reveal structure in the original data. The projection pursuit technique can also be used to obtain 1–D projections, but we look only at the 2–D case. Extensions to this method are also described in the literature by Friedman [1987], Posse [1995a, 1995b], Huber [1985], and Jones and Sibson [1987]. In our presentation of projection pursuit exploratory data analysis, we follow the method of Posse [1995a, 1995b].

FIGURE 4.3
This shows the start plane (top) and the target plane (bottom) for an interpolation tour using the **oronsay** data set.

Projection pursuit exploratory data analysis (PPEDA) is accomplished by visiting many projections in search of something interesting, where *interesting* is measured by an index. In most cases, the projection pursuit index measures the departure from normality. We use two indexes in our implementation. One is the ***chi-square index*** developed in Posse [1995a, 1995b], and the other is the ***moment index*** of Jones and Sibson [1987].

PPEDA consists of two parts:

1. A projection pursuit index that measures the degree of departure from normality, and

2. A method for finding the projection that yields the highest value for the index.

Posse [1995a, 1995b] uses a random search to locate a plane with an optimal value of the projection index and combines it with the structure removal of Friedman [1987] to get a sequence of interesting 2–D projections. Each projection found in this manner shows a structure that is less important (in terms of the projection index) than the previous one. Before we describe this method for PPEDA, we give a summary of the notation that we use to present the method.

Notation

$\mathbf{Z}$ is the matrix of sphered data.

α, β are orthonormal p-dimensional vectors that span the projection plane.

(α, β) is the projection plane spanned by α and β.

z_i^α, z_i^β are the sphered observations projected onto the vectors α and β.

(α^*, β^*) denotes the plane where the index is at a current maximum.

$PI_{\chi^2}(\alpha, \beta)$ denotes the chi-square projection index evaluated using the data projected onto the plane spanned by α and β.

$PI_M(\alpha, \beta)$ denotes the moment projection index.

c is a scalar that determines the size of the neighborhood around (α^*, β^*) that is visited in the search for planes that provide better values for the projection pursuit index.

v is a vector uniformly distributed on the unit p-dimensional sphere.

half specifies the number of steps without an increase in the projection index, at which time the value of the neighborhood is halved.

m represents the number of searches or random starts to find the best plane.

Finding the Structure

How we calculate the projection pursuit indexes $PI_{\chi^2}(\alpha, \beta)$ and $PI_M(\alpha, \beta)$ for each candidate plane is discussed at the end of this chapter. So, we first turn our attention to the second part of PPEDA, where we must optimize the projection index over all possible projections onto 2–D planes. Posse [1995a] shows that his random search optimization method performs better than the steepest-ascent techniques [Friedman and Tukey, 1974] typically used in optimization problems of this type.

The Posse algorithm starts by randomly selecting a starting plane, which becomes the current best plane (α^*, β^*). The method seeks to improve the current best solution by considering two candidate solutions within its neighborhood. These candidate planes are given by

$$
\begin{aligned}
a_1 &= \frac{\alpha^* + cv_1}{\|\alpha^* + cv_1\|} & b_1 &= \frac{\beta^* - (a_1^T \beta^*)a_1}{\|\beta^* - (a_1^T \beta^*)a_1\|} \\
a_2 &= \frac{\alpha^* - cv_2}{\|\alpha^* - cv_2\|} & b_2 &= \frac{\beta^* - (a_2^T \beta^*)a_2}{\|\beta^* - (a_2^T \beta^*)a_2\|}.
\end{aligned}
\tag{4.6}
$$

We start a global search by looking in large neighborhoods of the current best solution plane (α^*, β^*). We gradually focus in on a plane that yields a maximum index value by decreasing the neighborhood after a specified number of steps with no improvement in the value of the projection pursuit index. The optimization process is terminated when the neighborhood becomes small.

Because this method is a random search, the result could be a locally optimal solution. So, one typically goes through this procedure several times for different starting planes, choosing the final configuration as the one corresponding to the largest value of the projection pursuit index.

A summary of the steps for the exploratory projection pursuit procedure is given here. The complete search for the best plane involves repeating steps 2 through 9 of the procedure m times, using different random starting planes. The best plane (α^*, β^*) chosen is the plane where the projected data exhibit the greatest departure from normality as measured by the projection pursuit index.

Procedure – PPEDA

1. Sphere the data to obtain **Z**. See Chapter 1 for details on sphering data.
2. Generate a random starting plane, (α_0, β_0). This is the current best plane, (α^*, β^*).

3. Evaluate the projection index $PI_{\chi^2}(\alpha_0, \beta_0)$ or $PI_M(\alpha_0, \beta_0)$ for the starting plane.

4. Generate two candidate planes (a_1, b_1) and (a_2, b_2) according to Equation 4.6.

5. Calculate the projection index for these candidate planes.

6. Choose the candidate plane with a higher value of the projection pursuit index as the current best plane (α^*, β^*).

7. Repeat steps 4 through 6 while there are improvements in the projection pursuit index.

8. If the index does not improve for *half* times, then decrease the value of c by half.

9. Repeat steps 4 through 8 until c is some small number.

Structure Removal

We have no reason to assume that there is only one interesting projection, and there might be other views that reveal insights about our data. To locate other views, Friedman [1987] devised a method called **structure removal**. The overall procedure is to perform projection pursuit as outlined above, remove the structure found at that projection, and repeat the projection pursuit process to find a projection that yields another maximum value of the projection pursuit index. Proceeding in this manner will provide a sequence of projections providing informative views of the data.

Structure removal in two dimensions is an iterative process. The procedure repeatedly transforms the projected data to standard normal until they stop becoming more normal as measured by the projection pursuit index. We start with a $p \times p$ matrix $\mathbf{U}^*$, where the first two rows of the matrix are the vectors of the projection obtained from PPEDA. The rest of the rows of $\mathbf{U}^*$ have ones on the diagonal and zero elsewhere. For example, if $p = 4$, then

$$\mathbf{U}^* = \begin{bmatrix} \alpha_1^* & \alpha_2^* & \alpha_3^* & \alpha_4^* \\ \beta_1^* & \beta_2^* & \beta_3^* & \beta_4^* \\ 0 & 0 & 1 & 0 \\ 0 & 0 & 0 & 1 \end{bmatrix}.$$

We use the Gram-Schmidt process [Strang, 1988] to make the rows of $\mathbf{U}^*$ orthonormal. We denote the orthonormal version as $\mathbf{U}$. The next step in the structure removal process is to transform the $\mathbf{Z}$ matrix using the following

$$\mathbf{T} = \mathbf{U}\mathbf{Z}^T. \qquad (4.7)$$

In Equation 4.7, $\mathbf{T}$ is $p \times n$; so each column of the matrix corresponds to a p-dimensional observation. With this transformation, the first two dimensions (the first two rows of $\mathbf{T}$) of every transformed observation are the projection onto the plane given by (α^*, β^*).

We now remove the structure that is represented by the first two dimensions. We let Θ be a transformation that transforms the first two rows of $\mathbf{T}$ to a standard normal and the rest remain unchanged. This is where we actually remove the structure, making the data normal in that projection (the first two rows). Letting $\mathbf{T}_1$ and $\mathbf{T}_2$ represent the first two rows of $\mathbf{T}$, we define the transformation as follows

$$
\begin{aligned}
\Theta(\mathbf{T}_1) &= \Phi^{-1}[F(\mathbf{T}_1)] \\
\Theta(\mathbf{T}_2) &= \Phi^{-1}[F(\mathbf{T}_2)] \\
\Theta(\mathbf{T}_i) &= \mathbf{T}_i \; ; \qquad i = 3, \ldots, p ,
\end{aligned}
\tag{4.8}
$$

where Φ^{-1} is the inverse of the standard normal cumulative distribution function, and F is a function defined below (see Equations 4.9 and 4.10). We see from Equation 4.8 that we will be changing only the first two rows of $\mathbf{T}$.

We now describe the transformation of Equation 4.8 in more detail, working only with $\mathbf{T}_1$ and $\mathbf{T}_2$. First, we note that $\mathbf{T}_1$ can be written as

$$
\mathbf{T}_1 = (z_1^{\alpha^*}, \ldots, z_j^{\alpha^*}, \ldots, z_n^{\alpha^*}) ,
$$

and $\mathbf{T}_2$ as

$$
\mathbf{T}_2 = (z_1^{\beta^*}, \ldots, z_j^{\beta^*}, \ldots, z_n^{\beta^*}) .
$$

Recall that $z_j^{\alpha^*}$ and $z_j^{\beta^*}$ would be coordinates of the j-th observation projected onto the plane spanned by (α^*, β^*).

Next, we define a rotation about the origin through the angle γ as follows

$$
\begin{aligned}
\tilde{z}_j^{1(t)} &= z_j^{1(t)} \cos\gamma + z_j^{2(t)} \sin\gamma \\
\tilde{z}_j^{2(t)} &= z_j^{2(t)} \cos\gamma - z_j^{1(t)} \sin\gamma,
\end{aligned}
\tag{4.9}
$$

where $\gamma = 0, \pi \div 4, \pi \div 8, 3\pi \div 8$, and $z_j^{1(t)}$ represents the j-th element of $\mathbf{T}_1$ at the t-th iteration of the process. We next apply the following transformation to the rotated points,

$$
z_j^{1(t+1)} = \Phi^{-1}\left\{ \frac{r(\tilde{z}_j^{1(t)}) - 0.5}{n} \right\} \qquad z_j^{2(t+1)} = \Phi^{-1}\left\{ \frac{r(\tilde{z}_j^{2(t)}) - 0.5}{n} \right\}, \tag{4.10}
$$

where $r(\tilde{z}_j^{1(t)})$ represents the rank (position in the ordered list) of $\tilde{z}_j^{1(t)}$.

This transformation replaces each rotated observation by its normal score in the projection. With this procedure, we are deflating the projection index by making the data more normal. It is evident in the procedure given below, that this is an iterative process. Friedman [1987] states that the projection index should decrease rapidly in the first few iterations. After approximate normality is obtained, the index might oscillate with small changes. Usually, the process takes between 5 to 15 complete iterations to remove the structure.

Once the structure is removed using this process, we must transform the data back using

$$\mathbf{Z'} = \mathbf{U}^T \Theta (\mathbf{UZ}^T). \tag{4.11}$$

From matrix theory [Strang, 1988], we know that all directions orthogonal to the structure (i.e., all rows of $\mathbf{T}$ other than the first two) have not been changed, whereas the structure has been Gaussianized and then transformed back.

Procedure – Structure Removal

1. Create the orthonormal matrix $\mathbf{U}$, where the first two rows of $\mathbf{U}$ contain the vectors α^*, β^*.

2. Transform the data $\mathbf{Z}$ using Equation 4.7 to get $\mathbf{T}$.

3. Using only the first two rows of $\mathbf{T}$, rotate the observations using Equation 4.9.

4. Normalize each rotated point according to Equation 4.10.

5. For angles of rotation $\gamma = 0, \pi \div 4, \pi \div 8, 3\pi \div 8$, repeat steps 3 through 4.

6. Evaluate the projection index using $z_j^{1(t+1)}$ and $z_j^{2(t+1)}$, after going through an entire cycle of rotation (Equation 4.9) and normalization (Equation 4.10).

7. Repeat steps 3 through 6 until the projection pursuit index stops changing.

8. Transform the data back using Equation 4.11.

Example 4.4

We use the **oronsay** data to illustrate the projection pursuit procedure, which is implemented in the **ppeda** function provided with this text. First we do some preliminaries, such as loading the data and setting the parameter values.

```
load oronsay
X = oronsay;
```

```
[n,p] = size(X);
% For m = 5 random starts, find the N = 2
% best projection planes.
N = 2;
m = 5;
% Set values for other constants.
c = tan(80*pi/180);
half = 30;
% These will store the results for the
% 2 structures.
astar = zeros(p,N);
bstar = zeros(p,N);
ppmax = zeros(1,N);
```

Next we sphere the data to obtain matrix **Z**.

```
% Sphere the data.
muhat = mean(X);
[V,D] = eig(cov(X));
Xc = X - ones(n,1)*muhat;
Z = ((D)^(-1/2)*V'*Xc')';
```

Now we find each of the desired number of structures using the **ppeda** function with the index argument set to the Posse chi-square index.

```
% Now do the PPEDA: Find a structure, remove it,
% and look for another one.
Zt = Z;
for i = 1:N
    % Find one structure
    [astar(:,i),bstar(:,i),ppmax(i)] = ...
        ppeda(Zt,c,half,m,'chi');
    % Now remove the structure.
    % Function comes with text.
    Zt = csppstrtrem(Zt,astar(:,i),bstar(:,i));
end
```

The following MATLAB code shows how to project the data to each of these projection planes and then plot them. The plots are shown in Figure 4.4. The first projection has an index of 9.97, and the second has an index of 5.54.

```
% Now project and see the structure.
proj1 = [astar(:,1), bstar(:,1)];
proj2 = [astar(:,2), bstar(:,2)];
Zp1 = Z*proj1;
Zp2 = Z*proj2;
figure
plot(Zp1(:,1),Zp1(:,2),'k.'),title('Structure 1')
xlabel('\alpha*'),ylabel('\beta*')
```

```
figure
plot(Zp2(:,1),Zp2(:,2),'k.'),title('Structure 2')
xlabel('\alpha*'),ylabel('\beta*')
```

We repeat this **for** loop, but this time use the moment index to the **ppeda** function by replacing **chi** with **mom** in the argument to **ppeda**. The first projection from this procedure has a moment index of 425.71, and the second one yields an index of 424.51. Scatterplots of the projected data onto these two planes are given in Figure 4.5. We see from these plots that the moment index tends to locate projections with outliers.
❏

4.4 Projection Pursuit Indexes

We briefly describe the two projection pursuit indexes (PPIs) that are implemented in the accompanying MATLAB code. Other projection indexes for PPEDA are given in the literature (see some of the articles mentioned in the last section). A summary of these indexes, along with a simulation analysis of their performance, can be found in Posse [1995b].

4.4.1 Posse Chi-Square Index

Posse [1995a, 1995b] developed an index for projection pursuit that is based on the chi-square. We present only the empirical version here, but we first provide some notation.

Notation

ϕ_2 is the standard bivariate normal density.

c_k is the probability evaluated over the k-th region using the standard bivariate normal,

$$c_k = \int\int_{B_k} \phi_2 dz_1 dz_2 .$$

B_k is a box in the projection plane.

I_{B_k} is the indicator function for region B_k.

$\lambda_j = \pi j/36, j = 0, ..., 8$ is the angle by which the data are rotated in the plane before being assigned to regions B_k.

$\alpha(\lambda_j)$ and $\beta(\lambda_j)$ are given by

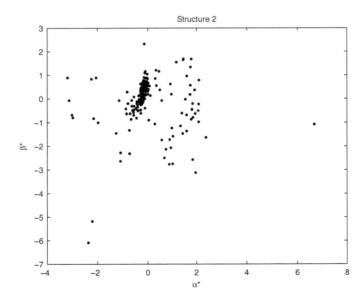

FIGURE 4.4
Here we show the results from applying PPEDA to the **oronsay** data set. The top configuration has a chi-square index of 9.97, and the second one has a chi-square index of 5.54.

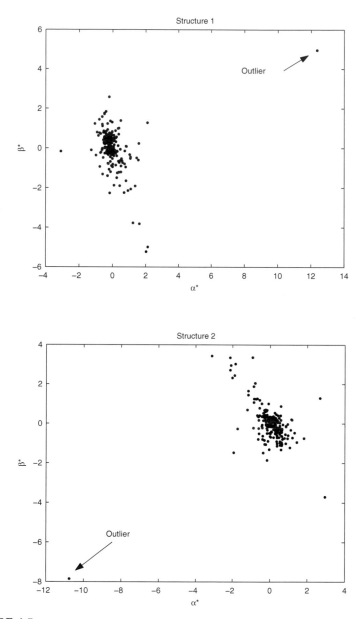

FIGURE 4.5
Here we see scatterplots from two planes found using the moment projection pursuit index. This index tends to locate projections with outliers. In the first structure, there is an outlying point in the upper right corner. In the second one, there is an outlying point in the lower left corner.

$$\alpha(\lambda_j) = \alpha\cos\lambda_j - \beta\sin\lambda_j$$
$$\beta(\lambda_j) = \alpha\sin\lambda_j + \beta\cos\lambda_j .$$

The plane is first divided into 48 regions or boxes B_k that are distributed in rings. See Figure 4.6 for an illustration of how the plane is partitioned. All regions have the same angular width of 45 degrees and the inner regions have the same radial width of $(2\log 6)^{1/2} : 5$. This choice for the radial width provides regions with approximately the same probability for the standard bivariate normal distribution. The regions in the outer ring have probability $1/48$. The regions are constructed in this way to account for the radial symmetry of the bivariate normal distribution. The projection index is given by

$$PI_{\chi^2}(\alpha, \beta) = \frac{1}{9}\sum_{j=0}^{8}\sum_{k=1}^{48}\frac{1}{c_k}\left[\frac{1}{n}\sum_{i=1}^{n}I_{B_k}(z_i^{\alpha(\lambda_j)}, z_i^{\beta(\lambda_j)}) - c_k\right]^2 .$$

The chi-square projection index is not affected by the presence of outliers. It is sensitive to distributions that have a hole in the core, and it will also yield projections that contain clusters. The chi-square projection pursuit index is fast and easy to compute, making it appropriate for large sample sizes. Posse [1995a] provides a formula to approximate the percentiles of the chi-square index so the analyst can assess the significance of the observed value of the projection index.

4.4.2 Moment Index

This index was developed in Jones and Sibson [1987] and is based on bivariate third and fourth moments. This is very fast to compute; so it is useful for large data sets. However, a problem with this index is that it tends to locate structure in the tails of the distribution. It is given by

$$PI_M(\alpha, \beta) = \frac{1}{12}\left\{\kappa_{30}^2 + 3\kappa_{21}^2 + 3\kappa_{12}^2 + \kappa_{03}^2 + \frac{1}{4}(\kappa_{40}^2 + 4\kappa_{31}^2 + 6\kappa_{22}^2 + 4\kappa_{13}^2 + \kappa_{04}^2)\right\},$$

where

$$\kappa_{21} = \frac{n}{(n-1)(n-2)}\sum_{i=1}^{n}(z_i^\alpha)^2 z_i^\beta \qquad \kappa_{12} = \frac{n}{(n-1)(n-2)}\sum_{i=1}^{n}(z_i^\beta)^2 z_i^\alpha$$

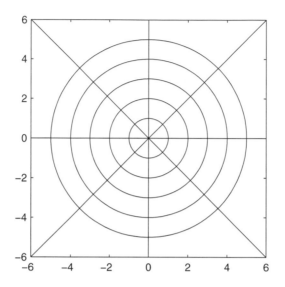

FIGURE 4.6
This shows the layout of the regions B_k for the chi-square projection index [Posse, 1995a].

$$\kappa_{22} = \frac{n(n+1)}{(n-1)(n-2)(n-3)}\left\{\sum_{i=1}^{n}(z_i^\alpha)^2(z_i^\beta)^2 - \frac{(n-1)^3}{n(n+1)}\right\}$$

$$\kappa_{30} = \frac{n}{(n-1)(n-2)}\sum_{i=1}^{n}(z_i^\alpha)^3 \qquad \kappa_{03} = \frac{n}{(n-1)(n-2)}\sum_{i=1}^{n}(z_i^\beta)^3$$

$$\kappa_{31} = \frac{n(n+1)}{(n-1)(n-2)(n-3)}\sum_{i=1}^{n}(z_i^\alpha)^3 z_i^\beta$$

$$\kappa_{13} = \frac{n(n+1)}{(n-1)(n-2)(n-3)}\sum_{i=1}^{n}(z_i^\beta)^3 z_i^\alpha$$

$$\kappa_{04} = \frac{n(n+1)}{(n-1)(n-2)(n-3)} \left\{ \sum_{i=1}^{n} (z_i^\beta)^4 - \frac{3(n-1)^3}{n(n+1)} \right\}$$

$$\kappa_{40} = \frac{n(n+1)}{(n-1)(n-2)(n-3)} \left\{ \sum_{i=1}^{n} (z_i^\alpha)^4 - \frac{3(n-1)^3}{n(n+1)} \right\}.$$

4.5 Independent Component Analysis

Independent component analysis (ICA) is a method that extracts the hidden components or underlying factors from multivariate data [Hyvärinen, Karhunen, and Oja, 2001]. Aspects of ICA are similar to some of the approaches we described in Chapter 2, such as PCA and factor analysis. However, ICA is different from these in that it looks for factors or components that are statistically *independent* and *non-Gaussian*. We include ICA in this chapter because it can be shown [Stone, 2004] that it is closely related to projection pursuit, where the search index being optimized measures independence and nonnormality.

Historically, the development of ICA came from what is called the **blind source separation problem** in signal processing. In these applications, one would like to separate an observed noisy mixture of signals into separate source signals. For example, say we have several signals that are emitted by some physical objects or sources. These sources could be people talking in the same room, two radio stations broadcasting at the same frequency, or different brain areas emitting electric signals. We further assume that there are several sensors located in different positions that receive these signals as a mixture of the original ones, but with slightly different weights. ICA can be used to recover those original signals.

We will use the example of two radio signals from different sources to explain the general idea of ICA [Martinez and Martinez, 2015]. If the two signals are broadcast by two different stations and using a fine time scale, then we can assume that the amplitude of one signal at a particular point in time is unrelated to the amplitude of the other signal at the same point. So, to separate the signals from the mixture, we might look for time-varying signals that are unrelated. The final assumption being that if we find such signals, then they probably arise from different physical processes. Stone [2004] points out that this last assumption, that goes in the reverse direction, is not necessarily correct in all cases.

The fact that two or more signals are unrelated can be expressed in terms of statistical independence. As we know from basic probability and statistics, if random variables (or signals) are statistically independent, then the value of one of the variables would not give us any information about the value of the other variables (or source signals) in the mixture. It is important to note that statistical independence is a much stronger requirement than lack of correlation. Two variables that are statistically independent will also be uncorrelated, but the fact that two variables are uncorrelated does not imply that they are independent. An exception to this is with Gaussian data, because in this case uncorrelated variables are also independent. ICA seeks to separate the data into a set of statistically independent or unrelated component signals or variables, which are then assumed to be from some meaningful sources or factors.

In general, we have a set of p-dimensional observations, and we are seeking a representation or transformation of the variables that reveals information that is not otherwise known. The assumption is that the transformed variables correspond to the underlying hidden components that describe the fundamental structure of the observed data. Only linear transformations are considered for independent component analysis to make it simpler to interpret the results and to make the computations faster.

Say we are given a set of observations $x_j(t)$,[2] where the index $j = 1, ..., p$ is attached to the variable and $t = 1, ..., n$. We assume that they are generated as a linear mixture of components as given here:

$$
\begin{bmatrix} x_1(t) \\ x_2(t) \\ \vdots \\ x_p(t) \end{bmatrix} = \mathbf{A} \begin{bmatrix} s_1(t) \\ s_2(t) \\ \vdots \\ s_p(t) \end{bmatrix}.
\tag{4.12}
$$

The matrix $\mathbf{A}$ is an unknown matrix containing the mixing coefficients; so it is often called the ***mixing matrix***. The independent components $s_j(t)$ are also unknown. Thus, independent component analysis consists of estimating $\mathbf{A}$ and the $s_j(t)$ using only the $x_j(t)$ that we observed.

An alternative model to Equation 4.12 and one that is more commonly seen in the ICA literature is the following

$$
\begin{bmatrix} s_1(t) \\ s_2(t) \\ \vdots \\ s_p(t) \end{bmatrix} = \mathbf{W} \begin{bmatrix} x_1(t) \\ x_2(t) \\ \vdots \\ x_p(t) \end{bmatrix}.
\tag{4.13}
$$

Here we can view the goal of ICA as one of finding the linear transformation given by the matrix $\mathbf{W}$ so the components $s_j(t)$ are as independent as possible. The matrix $\mathbf{W}$ is known as the ***separating matrix*** in the literature.

---

[2] The index t is used because of the historical connection between ICA and signal processing.

There are many methods for finding a set of independent components, but most of them are similar to projection pursuit in the two major pieces or steps. First, one has to have an objective function that measures the quantity one is trying to optimize, which in this case is statistical independence and nonnormality. Next, one requires a procedure or algorithm to find matrix **W** (or **A**) that optimizes the objective function.

In practice, many ICA algorithms adjust the matrix **W** until the entropy of a fixed function g of the variables recovered by **W** is maximized [Stone, 2002], where the function g is assumed to be the cumulative distribution function of the source variables. Another popular option for the objective function is based on mutual information and the Kullback-Leibler divergence. The interested reader should look at the review paper by Hyvärinen [1999a] for more options. Optimization algorithms include many of the classical ones, such as stochastic gradient approaches or Newton-like methods.

Hyvärinen [1999b] developed a computationally efficient algorithm for ICA based on the model in Equation 4.13. Hyvärinen starts from an information-theoretic viewpoint to derive a new type of objective function that minimizes mutual information and then uses projection pursuit to find the independent components. We do not go into details here because this algorithm has been implemented in the FastICA Toolbox. We demonstrate its use in the next example.

Example 4.5

We return to the **oronsay** data set to illustrate the results of ICA. We are using a function called **fastica**, which is part of the FastICA Toolbox, as well as the EDA Toolbox.

```
% Load oronsay data
load oronsay
% Perform the ICA.
% Note that the matrix must be transposed.
X = oronsay';
icasig = fastica(X);
```

The **fastica** function requires the input matrix to be in the form of variables (as rows) by observations (as columns); so we must transpose the input and output to match our convention. We will look at a scatterplot of the data in the space given by two of the independent components.

```
% Transpose the results to match
% observations as rows.
Xica = icasig';
% Look at the last two components in scatterplot.
ind0 = find(midden == 0);
ind1 = find(midden == 1);
ind2 = find(midden == 2);
plot(Xica(ind0,11),Xica(ind0,12),'.')
```

```
hold on
plot(Xica(ind1,11),Xica(ind1,12),'+')
plot(Xica(ind2,11),Xica(ind2,12),'*')
hold off
```

The scatterplot is shown in Figure 4.7, where some interesting non-Gaussian structure is visible.
❑

FIGURE 4.7
This is a scatterplot showing the results of applying ICA to the **oronsay** data. We have the three sampling sites displayed using different symbols.

As we stated earlier, ICA is somewhat related to the linear methods of dimensionality reduction (e.g., PCA and factor analysis) we presented in Chapter 2. We provide a brief discussion here comparing them. However, we note that ICA is not really meant to be used for dimensionality reduction, but there is nothing stopping one from extracting a smaller number of independent components $s_j(t)$ on the left-hand side of Equation 4.13. For instance, the **fastica** function we used in the previous example has options for accomplishing dimensionality reduction using ICA. We also note that there are over-complete versions of ICA that provide more than p independent components [Hyvärinen, 1999a].

Like independent component analysis, PCA looks for a set of components that are a linear transformation of the observed variables and will produce transformed variables or (principal components) that are uncorrelated with each other. One difference between the two approaches has to do with their goals. PCA seeks components that explain the maximum amount of variance, while independence and nonnormality are maximized in ICA. Thus, both of these methods have an objective that defines the interestingness of the linear transformation and then find components that optimize that function. We also know that uncorrelatedness does not imply statistical independence. So, the principal components are not guaranteed to be independent, while those in ICA are optimized for independence.

ICA is closely related to factor analysis. Hyvärinen [1999a] states that ICA can be thought of as a non-Gaussian factor analysis. One of the main differences is that dimensionality reduction is not the main objective, as it usually is with factor analysis. Another important difference relates to the rotations that are often used with factor analysis. Recall that extracted factors can be rotated to aid interpretation without changing the lack of correlation between the factors. Thus, the transformation from factor analysis is not unique. This is not so with ICA, since any rotation of the independent components would yield dependent components.

4.6 Summary and Further Reading

In this chapter, we discussed several methods for data tours that can be used to search for interesting features or structures in high-dimensional data. These include the torus winding method for the grand tour, the pseudo grand tour, the interpolation tour, and projection pursuit for EDA. The grand tour methods are dynamic, but are not typically interactive. The interpolation tour is interactive in the sense that the user can specify starting and target planes to guide the tour. Projection pursuit is not a visual tour (although it could be implemented that way); this tour seeks planes that have maximal structure as defined by some measure. Finally, independent component analysis was presented as a method that is somewhat similar to projection pursuit in the EDA context.

Some excellent papers describing the underlying mathematical foundation for tour methods and motion graphics include Wegman and Solka [2002], Hurley and Buja [1990], Buja and Asimov [1986], and Asimov and Buja [1994]. Another method for implementing a tour based on a fractal space-filling curve is described in Wegman and Solka [2002]. The grand tour combined with projection pursuit is described in Cook et al. [1995].

Many articles have been written on projection pursuit and its use in EDA and other applications. Jones and Sibson [1987] describe a steepest-ascent algorithm that starts from either principal components or random starts.

Friedman [1987] combines steepest-ascent with a stepping search to look for a region of interest. Crawford [1991] uses genetic algorithms to optimize the projection index. An approach for projection pursuit in three dimensions is developed by Nason [1995]. A description of other projection pursuit indexes can also be found in Cook, Buja, and Cabrera [1993].

Other uses for projection pursuit have been proposed. These include projection pursuit probability density estimation [Friedman, Stuetzle, and Schroeder, 1984], projection pursuit regression [Friedman and Stuetzle, 1981], robust estimation [Li and Chen, 1985], and projection pursuit for pattern recognition [Flick et al., 1990]. For a theoretical and comprehensive description of projection pursuit, the reader is directed to Huber [1985], where he discusses the important matter of sphering the data before exploring the data with projection pursuit. This invited paper with discussion also presents applications of projection pursuit to computer tomography and to the deconvolution of time series. Another paper that provides applications of projection pursuit is Jones and Sibson [1987]. Montanari and Lizzani [2001] apply projection pursuit to the variable selection problem. Bolton and Krzanowski [1999] describe the connection between projection pursuit and principal component analysis.

There are two excellent books on independent component analysis. One is a book by Stone [2004]. This is a tutorial introduction to the topic, and it includes many examples, as well as MATLAB code. Another excellent resource is a book by Hyvärinen, Karhunen, and Oja [2001]. The text is very readable and serves as a good resource for the theory and applications of ICA. Several review and tutorial articles have been published on ICA. These include Hyvärinen [1999a], Hyvärinen and Oja [2000], and Fodor [2002]. Finally, for a more statistical view of ICA, the reader can check out Hastie, Tibshirani, and Friedman [2009].

Exercises

4.1   Run the tour as in Example 4.1 and vary the number of iterations. Do you see any interesting structure along the way?

4.2   Apply the torus grand tour to the following data sets and comment on the results.

a. **environmental**

b. **oronsay**

c. **iris**

d. **posse** data sets

e. **skulls**

f. **spam**

      g. **pollen**

      h. gene expression data sets.

4.3   Run the pseudo grand tour in Example 4.2. Comment on the structures found, if any.

4.4   Apply the pseudo grand tour to the following data sets and comment on the results. Compare with the results you got with the grand tour.

      a. **environmental**

      b. **yeast**

      c. **iris**

      d. **posse** data sets

      e. **skulls**

      f. **spam**

      g. **pollen**

      h. gene expression data sets

4.5   Apply the interpolation tour of Example 4.3 to the data sets in Problem 4.4.

4.6   Repeat the interpolation tour in Example 4.3 using other target planes (Hint: 9 and 10 make an interesting one). Do you see any structure?

4.7   Apply projection pursuit EDA to the data sets in Problem 4.4. Search for several structures. Use both projection pursuit indexes. Show your results in a scatterplot and discuss them.

4.8   Repeat Example 4.4 and look for more than two best projection planes. Describe your results. Do you find planes using the moment index where the planes exhibit structure other than outliers?

4.9   Apply ICA to the data sets in Problem 4.4. Display any interesting structures in a scatterplot.

Chapter 5

Finding Clusters

We now turn our attention to the problem of finding groups or clusters in our data, which is an important objective in EDA and data mining. We present two of the basic methods in this chapter: agglomerative clustering and k-means clustering. Another method, fuzzy clustering based on estimating a finite mixture probability density function, is described in the following chapter. In addition, we discuss how nonnegative matrix factorization (first presented in Chapter 2) and probabilistic latent semantic analysis can be used to cluster document collections. We also describe an innovative method called spectral clustering based on the graph Laplacian and an approach using the minimum spanning tree. Finally, we address the important issue of assessing the quality of the resulting clusters at the end of the chapter, where we describe several statistics and plots that will aid in the analysis.

5.1 Introduction

Clustering is the process of organizing a set of data into groups in such a way that observations within a group are more similar to each other than they are to observations belonging to a different cluster. It is assumed that the data represent features that would allow one to distinguish one group from another. An important starting point in the process is choosing a way to represent the objects to be clustered. Many methods for grouping or clustering data and representation schemes can be found in various communities, such as statistics, machine learning, data mining, and computer science. We note, though, that no clustering technique is universally appropriate for finding all varieties of groupings that can be represented by multidimensional data [Jain, Murty, and Flynn, 1999]. So in the spirit of EDA, the user should try different clustering methods on a given data set to see what patterns emerge.

Clustering is also known as **unsupervised learning** in the literature. To understand clustering a little better, we will compare it to **discriminant analysis** or **supervised learning**. In supervised learning, the collection of observations has a class label associated with it. Thus, we know the true number of groups in the data, as well as the actual group membership of each data point. We use the data, along with the class labels, to create a classifier. Then, when we encounter a new, unlabeled observation, we can use the classifier to attach a label to it [Hastie, Tibshirani, and Friedman, 2009; Duda, Hart, and Stork, 2001; Webb, 2002].

However, with clustering (or unsupervised learning), we usually do not have class labels for the observations. Thus, we do not know how many groups are represented by the data, what the group membership or structure is, or even if there are any groups in the first place. As we said earlier, most clustering methods will find some desired number of groups, but what we really want is some meaningful clusters that represent the true phenomena. So, the analyst must look at the resulting groups and determine whether or not they aid in understanding the problem. Of course, nothing prevents us from using clustering methods on data that have class labels associated with the observations. As we will see in some of the examples, knowing the true clustering helps us assess the performance of the methods.

One can group the usual steps of clustering into the following [Jain and Dubes, 1988]:

1. ***Pattern representation***: This includes much of the preparation and initial work, such as choosing the number of clusters to look for, picking what measurements to use (***feature selection***), determining how many observations to process, and choosing the scaling or other transformations of the data (***feature extraction***). Some of this might be beyond the control of analysts.

2. ***Pattern proximity measure***: Many clustering methods require a measure of distance or proximity between observations and maybe between clusters. As one might suspect, different distances give rise to different partitions of the data. We discuss various distances and measures of proximity in Appendix A.

3. ***Grouping***: This is the process of partitioning the data into clusters. The grouping can be *hard*, which means that an observation either belongs to a group or not. In contrast, it can be *fuzzy*, where each data point has a degree of membership in each of the clusters. It can also be *hierarchical*, where we have a nested sequence of partitions.

4. ***Data abstraction***: This is the optional process of obtaining a simple and compact representation of the partitions. It could be a description of each cluster in words (e.g., one cluster represents lung cancer, while another one corresponds to breast cancer). It might be something quantitative such as a representative pattern, e.g., the centroid of the cluster.

5. ***Cluster assessment***: This could involve an assessment of the data to see if it contains any clusters. However, more often, it means an

examination of the output of the algorithm to determine whether or not the clusters are meaningful.

In our discussion so far, we have assumed that we know what a cluster is. However, several authors have pointed out the difficulty of formally defining such a term [Everitt, Landau, and Leese, 2001; Estivill-Castro, 2002]. Most clustering methods assume some sort of structure or model for the clusters (e.g., spherical, elliptical). Thus, they find clusters of that type, regardless of whether they are really present in the data or not.

Humans are quite adept at locating clusters in 2–D scatterplots, as we show in Figure 5.1. Bonner [1964] argued that the meaning of terms like *cluster* and *group* is in the 'eye of the beholder.' We caution the reader that it is usually easy to assign some structure or meaning to the clusters that are found just because we think there *should* be something there. However, we should keep in mind that the groups might be a result of the clustering method and that we could be *imposing* a pattern rather than *discovering* one.

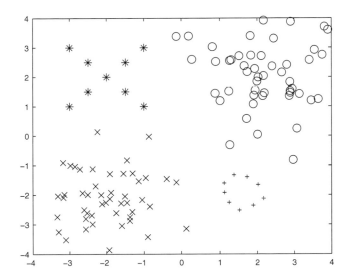

FIGURE 5.1
Here we show an example of some clusters. Keep in mind that what constitutes a cluster is based on one's definition and application.

5.2 Hierarchical Methods

One of the most common approaches to clustering is to use a hierarchical method. This seems to be popular in the areas of data mining and gene expression analysis [Hand, Mannila, and Smyth, 2001; Hastie, Tibshirani, and Friedman, 2009]. In ***hierarchical clustering***, one does not have to know the number of groups ahead of time; i.e., the data are not divided into a pre-determined number of partitions. Rather, the result consists of a hierarchy or set of nested partitions. As we point out below, there are several types of hierarchical clustering, and each one can give very different results on a given data set. Thus, there is not one recommended method, and the analyst should use several in exploring the data.

The process consists of a sequence of steps, where two groups are either merged (***agglomerative***) or divided (***divisive***) according to some optimality criterion. In their simplest and most commonly used form, each of these hierarchical methods have n observations in their own group (i.e., n total groups) at one end of the process and one group with all n data points at the other end. The difference between them is where we start the grouping process. With agglomerative clustering, we have n singleton clusters and end up with all points belonging to one group. Divisive methods are just the opposite; we start with everything in one group and keep splitting them until we have n singleton clusters.

One of the issues with hierarchical clustering is that once points are grouped together or split apart, the step cannot be undone. Another issue, of course, is how many clusters are appropriate.

We will not cover the divisive methods in this book, because they are less common and they can be computationally intensive (except in the case of binary variables, see Everitt, Landau, and Leese [2001]). However, Kaufman and Rousseeuw [1990] point out that an advantage with divisive methods is that most data sets have a small number of clusters and that structure would be revealed in the beginning of a divisive approach, whereas, with agglomerative methods, this does not happen until the end of the process.

Agglomerative clustering requires the analyst to make several choices, such as how to measure the proximity (distance) between data points and how to define the distance between two clusters. Determining what distance to use is largely driven by the type of data one has: continuous, categorical or a mixture of the two, as well as what aspect of the features one wants to emphasize. Please see Appendix A for a description of various distances, including those that are implemented in the MATLAB Statistics Toolbox.

The input required for most agglomerative clustering methods is the $n \times n$ interpoint distance matrix (as was used before in multidimensional scaling); some also require the full data set. The next step is to specify how we will determine what clusters to link at each stage of the method. The usual way is to link the two closest clusters at each stage of the process, where *closest* is defined by one of the linkage methods described below. It should be noted that using different definitions of distance and linkage can give rise to very different cluster structures. We now describe each of the linkage methods that are available in the MATLAB Statistics Toolbox.

We set up some notation before we continue with a description of the various approaches. Given a cluster r and a cluster s, the number of objects in each cluster is given by n_r and n_s. The distance between cluster r and s is denoted by $d_c(r,s)$.

Single Linkage

Single linkage is perhaps the method used most often in agglomerative clustering, and it is the default method in the MATLAB **linkage** function, which produces the hierarchical clustering. Single linkage is also called *nearest neighbor*, because the distance between two clusters is given by the smallest distance between objects, where each one is taken from one of the two groups. Thus, we have the following distance between clusters

$$d_c(r, s) = min\{d(x_{ri}, x_{sj})\} \qquad i = 1, \ldots, n_r; j = 1, \ldots, n_s,$$

where $d(x_{ri}, x_{sj})$ is the distance between observation i from group r and observation j from group s. Recall that this is the interpoint distance (e.g., Euclidean, etc.), which is the input to the clustering procedure.

Single linkage clustering suffers from a problem called *chaining*. This comes about when clusters are not well separated, and snake-like chains can form. Observations at opposite ends of the chain can be very dissimilar, but yet they end up in the same cluster. Another issue with single linkage is that it does not take the cluster structure into account [Everitt, Landau, and Leese, 2001].

Complete Linkage

Complete linkage is also called the *furthest neighbor* method, since it uses the largest distance between observations, one in each group, as the distance between the clusters. The distance between clusters is given by

$$d_c(r, s) = max\{d(x_{ri}, x_{sj})\} \qquad i = 1, \ldots, n_r; j = 1, \ldots, n_s.$$

Complete linkage is not susceptible to chaining. Additionally, the resulting clusters tend to be spherical, and it has difficulty recovering nonspherical groups. Like single linkage, complete linkage does not account for cluster structure.

Average (Unweighted and Weighted) Linkage

The *average linkage* method defines the distance between clusters as the average distance from all observations in one cluster to all points in another cluster. In other words, it is the average distance between pairs of observations, where one is from one cluster and one is from the other. Thus, we have the following distance

$$d_c(r, s) = \frac{1}{n_r n_s} \sum_{i=1}^{n_r} \sum_{j=1}^{n_s} d(x_{ri}, x_{sj}).$$

This method tends to combine clusters that have small variances, and it also tends to produce clusters with approximately equal variance. It is relatively robust and does take the cluster structure into account. Like single and complete linkage, this method takes the interpoint distances as input.

Version 5 of the Statistics Toolbox has another related type of linkage called *weighted average distance* or WPGMA. The average linkage mentioned above is *unweighted* and is also known as UPGMA.

Centroid Linkage

Another type, called *centroid linkage*, requires the raw data, as well as the distances. It measures the distance between clusters as the distance between their centroids. Their centroids are usually the mean, and these change with each cluster merge. We can write the distance between clusters, as follows

$$d_c(r, s) = d(\bar{x}_r, \bar{x}_s),$$

where $\bar{x}_r$ is the average of the observations in the r-th cluster, and $\bar{x}_s$ is defined similarly.

The distance between the centroids is usually taken to be Euclidean. The MATLAB **linkage** function for centroid linkage works only when the interpoint distances are Euclidean, and it does not require the raw data as input. A somewhat related method called *median linkage* computes the distance between clusters using weighted centroids. A problem with both centroid and median linkage is the possibility of reversals [Morgan and Ray, 1995]. This can happen when the distance between one pair of cluster centroids is less than the distance between the centroid of another pair that was merged earlier. In other words, the fusion values (distances between clusters) are not monotonically increasing. This makes the results confusing and difficult to interpret.

Ward's Method

Ward [1963] devised a method for agglomerative hierarchical clustering where the merging of two clusters is determined by the size of the incremental sum of squares. It looks at the increase in the total within-group sum of squares when clusters r and s are joined. The distance between two clusters using Ward's method is given by

$$d(r, s) = n_r n_s d_{rs}^2 \div (n_r + n_s),$$

where d_{rs}^2 is the distance between the r-th and s-th cluster as defined in the centroid linkage definition. In other words, to get each merge in the procedure, the within-cluster sum of squares is minimized over all possible partitions that can be obtained by combining two clusters from the current set of groups.

Ward's method tends to combine clusters that have a small number of observations. It also has a tendency to locate clusters that are of the same size and spherical. Due to the sum of squares criterion, it is sensitive to the presence of outliers in the data set.

Visualizing Hierarchical Clustering Using the Dendrogram

We discuss the dendrogram in more detail in Chapter 8, where we present several ways to visualize the output from cluster analysis. We briefly introduce it here, so we can use the dendrograms to present the results of this chapter to the reader.

A *dendrogram* is a tree diagram that shows the nested structure of the partitions and how the various groups are linked at each stage. The dendrogram can be shown horizontally or vertically. However, we will concentrate on the vertical version for now. There is a numerical value associated with each stage of the method where the branches (i.e., clusters) join, which usually represents the distance between the two clusters. The scale for this numerical value is shown on the vertical axis.

We show an example of a dendrogram in Figure 5.2 for a very small data set. Notice that the tree is made up of inverted U-shaped links, where the top of the U represents a fusion between two clusters. In most cases, the fusion levels will be monotonically increasing, yielding an easy to understand dendrogram.

We already discussed the problem of reversals with the centroid and median linkage methods. With reversals, these merge points can decrease, which can make the results confusing. Another problem with some of these methods is the possibility of nonuniqueness of the hierarchical clustering or dendrogram. This can happen when there are ties in the distances between its clusters. How these are handled depends on the software. Sadly, this information is often left out of the documentation. Morgan and Ray [1995] provide a detailed explanation of the inversion and nonuniqueness problems in hierarchical clustering. This will be explored further in the exercises.

Example 5.1

In this example, we use the **yeast** data to illustrate the procedure in MATLAB for obtaining agglomerative hierarchical clustering. The first step is to load the data and get all of the interpoint distances.

```
load yeast
% Get the distances. The output from this function
% is a vector of the n(n-1)/2 interpoint distances.
```

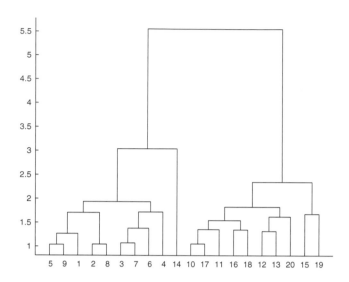

FIGURE 5.2

This is an example of a dendrogram for the two spherical clusters in Figure 5.1 (shown with **x**'s and **o**'s), where average linkage has been used to generate the hierarchy. Note that we are showing only 20 leaf nodes. See Chapter 8 for more information on what this means.

```
% The default is Euclidean distance.
Y = pdist(data);
```

The output from the **pdist** function is just the upper triangular portion of the complete $n \times n$ interpoint distance matrix. It can be converted to a full matrix using the function **squareform**. However, this is not necessary for the next step, which is to get the hierarchy of partitions. See the **help** on **pdist** for more information on the other distances that are available in MATLAB.

```
% Single linkage (the default) shows chaining.
Z = linkage(Y);
dendrogram(Z);
```

The default for the **linkage** function is single linkage. The output is a matrix **Z**, where the first two columns indicate what groups were linked and the third column contains the corresponding distance or fusion level. The dendrogram for this is shown in Figure 5.3 (top), where we can see the chaining that can happen with single linkage. Now, we show how to do the same thing using complete linkage.

```
% Complete linkage does not have the chaining.
Z = linkage(Y,'complete');
dendrogram(Z);
```

This dendrogram is given in Figure 5.3 (bottom), and we see no chaining here. Since the dendrogram shows the entire set of nested partitions, one could say the dendrogram *is* the actual clustering. However, it is useful to know how to get the grouping for any desired number of groups. MATLAB provides the **cluster** function for this purpose. One can specify the number of clusters, as we do below, among other options. See the **help** on **cluster** for other uses.

```
% To get the actual clusters - say based
% on two partitions, use the following
% syntax.
cind = cluster(Z, 'maxclust',2);
```

The output argument **cind** is an n-dimensional vector of group labels.
❑

5.3 Optimization Methods — k-Means

The methods discussed in the previous section were all hierarchical, where the output consists of a complete set of nested partitions. Another category of clustering methods consists of techniques that optimize some criterion in order to partition the observations into a *specified* or *predetermined* number of groups. These *partition* or *optimization methods* differ in the nature of the objective function, as well as the optimization algorithm used to come up with the final clustering. One of the issues that must be addressed when employing these methods (as is also the case with the hierarchical methods) is determining the number of clusters in the data set. We will discuss ways to tackle this problem in a later section. However, one of the major advantages of the optimization-based methods is that they require only the data as input (along with some other parameters), not the interpoint distances, as in hierarchical methods. Thus, these are usually more suitable when working with large data sets.

One of the most commonly used optimization-based methods is k-means clustering, which is the only one we will discuss in this book. The reader is referred to Everitt, Landau, and Leese [2001] or other books mentioned at the end of the chapter for more information on the other types of partition methods. The MATLAB Statistics Toolbox has a function that implements the k-means algorithm.

The goal of k-means clustering is to partition the data into k groups such that the within-group sum-of-squares is minimized. We start by defining the *within-class scatter matrix* given by

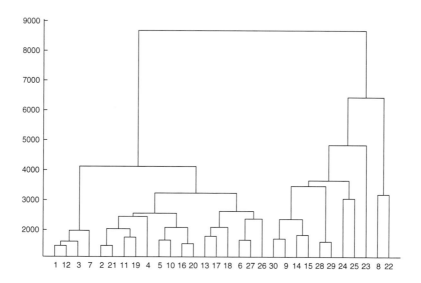

FIGURE 5.3
The first dendrogram shows the results of using Euclidean distance and single linkage on
the **yeast** data, and we can see what chaining looks like in the partitions. The second
dendrogram is what we obtain using complete linkage.

$$\mathbf{S}_W = \frac{1}{n} \sum_{j=1}^{g} \sum_{i=1}^{n} I_{ij} (\mathbf{x}_i - \bar{\mathbf{x}}_j)(\mathbf{x}_i - \bar{\mathbf{x}}_j)^T,$$

where I_{ij} is one if $\mathbf{x}_i$ belongs to group j and zero otherwise, and g is the number of groups. The criterion that is minimized in k-means is given by the sum of the diagonal elements of $\mathbf{S}_W$, (the trace of the matrix), as follows

$$\mathrm{Tr}(\mathbf{S}_W) = \sum \mathbf{S}_{W_{ii}}.$$

If we minimize the trace, then we are also minimizing the total within-group sum of squares about the group means. Everitt, Landau, and Leese [2001] show that minimizing the trace of $\mathbf{S}_W$ is equivalent to minimizing the sum of the squared Euclidean distances between individuals and their group mean. Clustering methods that minimize this criterion tend to produce clusters that have a hyperellipsoidal shape. This criterion can be affected by the scale of the variables; so standardization should be done first.

We briefly describe two procedures for obtaining clusters via k-means. The basic algorithm for k-means clustering is a two step procedure. First, we assign each observation to its closest group, usually using the Euclidean distance between the observation and the cluster centroid. The second step of the procedure is to find the new centroids using the assigned observations. These steps are alternated until there are no changes in cluster membership or until the centroids do not change. This algorithm is sometimes referred to as HMEANS [Späth, 1980] or the basic ISODATA method.

Procedure – k-Means

1. Specify the number of clusters k.
2. Determine initial cluster centroids. These can be randomly chosen or the user can specify them.
3. Calculate the distance between each observation and each cluster centroid.
4. Assign every observation to the closest cluster.
5. Calculate the centroid (i.e., the d-dimensional mean) of every cluster using the observations that were just grouped there.
6. Repeat steps 3 through 5 until no more changes result.

The k-means algorithm could lead to empty clusters; so users should be aware of this possibility. Another issue concerns the optimality of the partitions. With k-means, we are searching for partitions where the within-group sum-of-squares is a minimum. It can be shown [Webb, 2002] that in some cases the final k-means cluster assignment is not optimal, in the sense that moving a single point from one cluster to another may reduce the sum of squared errors. The following procedure called the **enhanced k-means** helps address the second problem.

Procedure – Enhanced k-means

1. Obtain a partition of k groups via k-means as described previously.
2. Take each data point $\mathbf{x}_i$ and calculate the Euclidean distance between it and every cluster centroid.
3. Here $\mathbf{x}_i$ is in the r-th cluster, n_r is the number of points in the r-th cluster, and d_{ir}^2 is the Euclidean distance between $\mathbf{x}_i$ and the centroid of cluster r. If there is a group s such that

$$\frac{n_r}{n_r - 1} d_{ir}^2 > \frac{n_s}{n_s + 1} d_{is}^2 ,$$

then move $\mathbf{x}_i$ to cluster s.

4. If there are several clusters that satisfy the above inequality, then move the $\mathbf{x}_i$ to the group that has the smallest value for

$$\frac{n_s}{n_s + 1} d_{is}^2 .$$

5. Repeat steps 2 through 4 until no more changes are made.

We note that there are many algorithms for k-means clustering described in the literature that improve the efficiency, allow clusters to be created and deleted during the process, and other improvements. See Webb [2002] and the other references at the end of the chapter for more information.

Example 5.2

For this example, we turn to a data set that is familiar to most statisticians: the iris data. These data consist of three classes of iris: *Iris setosa*, *Iris versicolor*, and *Iris virginica*. They were originally analyzed by Fisher [1936]. He was interested in developing a method for discriminating the species of iris based on their sepal length, sepal width, petal length, and petal width. The **kmeans** function in MATLAB requires the data as input, along with the desired number of groups. MATLAB also allows the user to specify a distance measure used in the minimization process. In other words, **kmeans** computes the centroid clusters differently for the different distance measures. We will use the default squared Euclidean distance.

```
load iris
% First load up the data and put into one data
% matrix.
data = [setosa; versicolor; virginica];
vars = ['Sepal Length';
        'Sepal Width ';
        'Petal Length';
        'Petal Width '];
  kmus = kmeans(data,3);
```

We illustrate the results in a scatterplot matrix shown in Figure 5.4.[1] The first plot shows the results from *k*-means, and the second shows the true groups in the data. Different symbols correspond to the different groups, and we see that *k*-means produced reasonable clusters that are not too far off from the truth. We will explore this more in a later example.
❑

The *k*-means method is dependent on the chosen initial cluster centers. MATLAB allows the user to specify different starting options, such as randomly selecting *k* data points as centers (the default), uniformly generated *p*-dimensional vectors over the range of **X**, or user-defined centers. As in many optimization problems that rely on a designated starting point, *k*-means can get stuck in a locally optimal solution. Thus, *k*-means should be performed several times with different starting points. MATLAB provides this option also as an input argument to the **kmeans** function. For another implementation of *k*-means in MATLAB, see Martinez and Martinez [2015].

5.4 Spectral Clustering

Spectral clustering has grown very popular in recent years. It can be easily implemented with linear algebra software packages such as MATLAB, and it often performs better than standard algorithms such as *k*-means and agglomerative clustering [Verma and Meila, 2003]. Many spectral clustering algorithms have been developed [von Luxburg, 2007] and applied in many fields such as image analysis, data mining, and document clustering. Most spectral clustering algorithms use the top *k* eigenvectors of a matrix that is based on the distance between the observations. We will present one of the most popular approaches by Ng, Jordan, and Weiss [2002].

As usual, we are given a set of *p*-dimensional observations $x_1, ..., x_n$, and we want to divide them into *k* groups, where data points in each group are similar to each other and different to points in other groups. We then calculate the affinity or similarity (see Appendix A) between each pair of

---

[1] See the file **Example54.m** for the MATLAB code used to construct the plots.

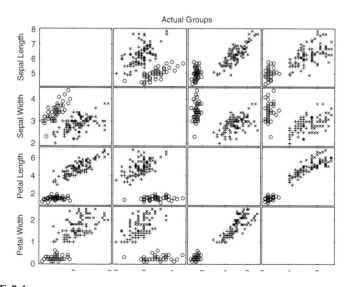

FIGURE 5.4
The first scatterplot matrix shows the results from *k*-means, while the second one shows the true groups in the data. We see that for the most part, *k*-means did a reasonable job finding the groups. (**SEE COLOR INSERT.**)

points $\mathbf{x}_i$ and $\mathbf{x}_j$, and we enter these values into the affinity matrix $\mathbf{A}$. The elements of $\mathbf{A}$ are given by

$$A_{ij} = \exp\left[-\frac{\|\mathbf{x}_i - \mathbf{x}_j\|^2}{2\sigma^2}\right] \quad \text{for } i \ne j \tag{5.1}$$

$$A_{ii} = 0 .$$

The scaling parameter σ^2 determines how fast the affinity decreases with the distance between $\mathbf{x}_i$ and $\mathbf{x}_j$. Ng, Jordan, and Weiss [2002] describe a method to choose the scaling parameter automatically.

We define the matrix $\mathbf{D}$ as the diagonal matrix whose ii-th element is the sum of the elements in the i-th row of $\mathbf{A}$:

$$D_{ii} = \sum_{j=1}^{n} A_{ij} . \tag{5.2}$$

We then construct the following matrix:

$$\mathbf{L} = \mathbf{D}^{-1/2} \mathbf{A} \mathbf{D}^{-1/2} .$$

The matrix $\mathbf{L}$ is similar to a graph Laplacian, which is the main tool used in spectral clustering. However, von Luxburg [2007] notes that there is no unique convention or definition of a graph Laplacian; so care should be taken when implementing spectral clustering algorithms.

The next step is to find the eigenvectors and eigenvalues of $\mathbf{L}$. We denote the k largest eigenvalues of $\mathbf{L}$ as $\lambda_1 \ge \lambda_2 \ge \dots \ge \lambda_k$ and their corresponding eigenvectors as $\mathbf{u}_1, \mathbf{u}_2, \dots, \mathbf{u}_k$. We use the eigenvectors to form the matrix $\mathbf{U}$ as follows

$$\mathbf{U} = \begin{bmatrix} \mathbf{u}_1 & \mathbf{u}_2 & \dots & \mathbf{u}_k \end{bmatrix} . \tag{5.3}$$

Equation 5.3 just means that the columns of $\mathbf{U}$ correspond to the eigenvectors of $\mathbf{L}$. As usual, all eigenvectors have unit length and are chosen to be orthogonal in the case of repeated eigenvalues.

Now, we form the matrix $\mathbf{Y}$ by normalizing each *row* of $\mathbf{U}$ to have unit length. Thus, the elements of $\mathbf{Y}$ are given by

$$Y_{ij} = \frac{U_{ij}}{\sqrt{\sum_j U_{ij}^2}} . \tag{5.4}$$

We are now ready to do some clustering. We can think of the n rows of **Y** as an observation in a k-dimensional space and group them using k-means clustering to get the final grouping. We assign an observation x_i to the m-th cluster if the i-th row of **Y** was assigned to the m-th cluster. The steps in the Ng, Jordan, and Weiss (NJW) algorithm that we just described are listed below.

Procedure – NJW Algorithm

1. We first form the affinity matrix **A** according to Equation 5.1.
2. Construct the matrix **D** using elements of **A** (Equation 5.2).
3. Obtain the matrix $\mathbf{L} = \mathbf{D}^{-1/2}\mathbf{A}\mathbf{D}^{-1/2}$.
4. Find the eigenvectors and eigenvalues of **L**.
5. Put the eigenvectors that correspond to the k largest eigenvalues into a matrix **U**, as shown in Equation 5.3.
6. Normalize the rows of **U** to have unit length (Equation 5.4) and call it matrix **Y**.
7. Apply k-means clustering to the rows of **Y**.
8. An observation x_i is assigned to the m-th cluster if the i-th row of **Y** was assigned to the m-th cluster.

As Ng, Jordan, and Weiss [2002] note, this approach does not make a lot of sense initially, since we could apply k-means to the data directly without going through all of the matrix calculations and manipulations. However, some analyses show [von Luxburg, 2007; Verma and Meila, 2003] that mapping the points to this k-dimensional space can produce tight clusters that are easily found using k-means.

Example 5.3

In the following example, we will explore the spectral clustering algorithm applied to bivariate data with two groups. We use the Spectral Clustering Toolbox written by Verma and Meila [2003]. The code is also included with the EDA Toolbox.

```
% We first generate some bivariate normal data.
mu1 = [2 2];
cov1 = eye(2);
mu2 = [-1 -1];
cov2 = [1 .9; .9 1];
X1 = mvnrnd(mu1,cov1,100);
X2 = mvnrnd(mu2,cov2,100);
% Plot the data showing the known clusters.
plot(X1(:,1),X1(:,2),'.',...
```

```
     X2(:,1),X2(:,2),'+');
```

The resulting plot is shown in Figure 5.5 (top), where we see two clusters that overlap slightly. The different symbols in the plot correspond to the known groups, so we can compare those with the output from spectral clustering. We now create our data matrix **X** and then cluster the data.

```
% Put the data into one matrix.
X = [X1; X2];
% Get the affinity matrix using a sigma of 1.
A = AffinitySimilarityMatrix(X',1);
% Get the cluster ID's according to the spectral
% clustering according to Ng, Jordan and Weiss and
% also using k-means.
cids = cluster_spectral_general(A,2,...
    'njw_gen','njw_kmeans');
```

Next we plot the observations in Figure 5.5 (bottom) with observations belonging to group one shown as x's and those belonging to the second group as stars.

```
% Plot the data according to the cluster IDs.
% First plot group 1 as x's.
figure
ind1 = find(cids == 1);
plot(X(ind1,1),X(ind1,2),'x');
hold on
% Now plot group 2 as stars.
ind2 = find(cids == 2);
plot(X(ind2,1),X(ind2,2),'*')
```

We see that spectral clustering provided a reasonable grouping of the data. However, as one would expect, it did make some mistakes in the area where they overlap.
❏

5.5 Document Clustering

In this section, we present two methods that are often used for clustering documents. These are called nonnegative matrix factorization and probabilistic latent semantic indexing (PLSI). These methods are described in the context of clustering collections of documents based on the term-document matrix, but they can be applied in many other situations.

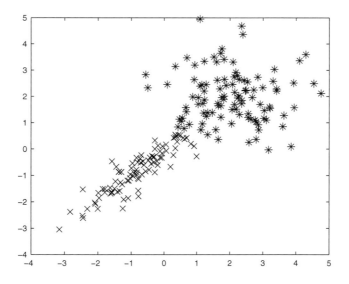

FIGURE 5.5

The top panel shows a scatterplot of the clusters that were generated for Example 5.3. The two groups are displayed using different symbols. One of the groups is plotted as points, and the other is displayed using crosses. Note that there is some overlap between the clusters.The scatterplot in the bottom panel shows the groups obtained using the spectral clustering algorithm of Ng, Jordan, and Weiss [2002]. One of the groups is shown as **x**'s, and the other group is plotted with stars. The spectral clustering did a good job, but was incorrect in the region that overlaps.

5.5.1 Nonnegative Matrix Factorization — Revisited

Let us revisit our discussions of nonnegative matrix factorization (NMF) with an eye towards clustering. Recall that the application of NMF in Chapter 2 was to data transformation. We follow Xu et al. [2003] in our development of the mathematical formalism of NMF; so the notation given below is slightly different than what we saw previously.

Let $\mathbf{X}$ be an $n \times p$ term-document matrix, where n represents the number of words in the lexicon, and p denotes the number of documents in the corpus. Recall that the ij-th element of the term-document matrix is equal to the number of times the i-th word appears in the j-th document. We could convert this to a weighted term-document matrix, where the entries are multiplied by factors that improve tasks in text analysis such as information retrieval or text data mining [Berry and Browne, 2005]. However, this step is not required to use nonnegative matrix factorization. Next, each column in the matrix $\mathbf{X}$ is normalized to have a magnitude of one.

Assuming that our document collection has k clusters, we want to factorize $\mathbf{X}$ as $\mathbf{U}\mathbf{V}^{T}$, where $\mathbf{U}$ is a nonnegative $n \times k$ matrix and $\mathbf{V}^{T}$ is a nonnegative $k \times p$ matrix that minimize

$$ J = \frac{1}{2}\|\mathbf{X} - \mathbf{U}\mathbf{V}^{T}\|, \tag{5.5} $$

where the notation $\|\cdot\|$ represents the squared sum of the elements in the matrix argument. Constrained optimization via Lagrange multipliers leads to the following update equations

$$ u_{ij} \leftarrow u_{ij}\frac{(\mathbf{X}\mathbf{V})_{ij}}{(\mathbf{U}\mathbf{V}^{T}\mathbf{V})_{ij}} $$
$$ v_{ij} \leftarrow v_{ij}\frac{(\mathbf{X}^{T}\mathbf{U})_{ij}}{(\mathbf{V}\mathbf{U}^{T}\mathbf{U})_{ij}} \ . \tag{5.6} $$

The equations given above are just a slightly different version of the multiplicative update equations that were discussed in Chapter 2. In the final step of the NMF clustering procedure, these are normalized using

$$ v_{ij} \leftarrow v_{ij}\sqrt{\sum_{i} u_{ij}^{2}} $$
$$ u_{ij} \leftarrow \frac{u_{ij}}{\sum_{i} u_{ij}^{2}} \ . \tag{5.7} $$

We may interpret the meaning of the **U** and **V** matrices within an SVD type framework. The *ij*-th element of the matrix **U** (u_{ij}) represents the degree that the *i*-th term belongs to the *j*-th cluster. Similarly, the *ij*-th element of the **V** matrix encodes the degree that the *i*-th document belongs to the *j*-th cluster. So, if the *i*-th document belongs to cluster *m*, then the v_{im} value in **V** will be large while the rest of the elements in the *i*-th row of the matrix will be small. The procedure of NMF clustering as applied to documents is summarized below and is illustrated in the next example, where we apply it to a small document data set.

Procedure – NMF Clustering

1. Construct the (possibly weighted) term-document matrix **X** from a set of documents.
2. Normalize each column of **X** to have unit norm.
3. Apply the update equations in Equation 5.6 to solve for **U** and **V**.
4. Normalize **U** and **V** as given in Equation 5.7.
5. Use the **V** matrix to cluster each document. Document d_i is assigned to cluster *m* if

$$m = \operatorname{argmax}_j \{v_{ij}\}.$$

TABLE 5.1

List of Documents Used in Example 5.4

d1	*Human* machine *interface* for Lab ABC *computer* applications
d2	A *survey* of *user* opinion of *computer system response time*
d3	The *EPS user interface* management *system*
d4	*System* and *human system* engineering testing of *EPS*
d5	Relation of *user*-perceived *response time* to error measurement
d6	The generation of random, binary, unordered *trees*
d7	The intersection *graph* of paths in *trees*
d8	*Graph minors* IV: Widths of *trees* and well-quasi-ordering
d9	*Graph minors*: A *survey*

Example 5.4

We will use a data set discussed in Deerwester et al. [1990] in order to illustrate NMF-based clustering. The documents used in this example are provided in Table 5.1. Following Deerwester, we have italicized those terms that will ultimately be used to form our term counts. In Table 5.2, we list the term counts where each column constitutes a document. Recall that this table is the term-document matrix **X**. We notice through an examination of the original document collection or by looking at the columns of the term-document matrix, that these documents should cluster into two groups. One of the groups could consist of {*d*1, *d*2, *d*3, *d*4, *d*5} , which are about computer

TABLE 5.2

Term-document Matrix Corresponding to Table 5.1

	d1	d2	d3	d4	d5	d6	d7	d8	d9
human	1	0	0	1	0	0	0	0	0
interface	1	0	1	0	0	0	0	0	0
computer	1	1	0	0	0	0	0	0	0
user	0	1	1	0	1	0	0	0	0
system	0	1	1	2	0	0	0	0	0
response	0	1	0	0	1	0	0	0	0
time	0	1	0	0	1	0	0	0	0
EPS	0	0	1	1	0	0	0	0	0
survey	0	1	0	0	0	0	0	0	1
trees	0	0	0	0	0	1	1	1	0
graph	0	0	0	0	0	0	1	1	1
minors	0	0	0	0	0	0	0	1	1

systems and interfaces. The other group of documents, $\{d6, d7, d8, d9\}$, is about graphs and trees. We will use nonnegative matrix factorization to solve for these clusters, as shown in the following MATLAB steps.

```
% First, we load the data set with the
% term-document matrix in it.
load nmfclustex
% Next, normalize each column.
[n,p] = size(nmfclustex);
for i = 1:p
    termdoc(:,i) = nmfclustex(:,i)/...
        (norm(nmfclustex(:,i)));
end
```

We now perform nonnegative matrix factorization with $k = 2$ to extract the cluster structure. Notice that NMF is formulated as $\mathbf{X} - \mathbf{U}\mathbf{V}^T$; so we use **VT** to denote $\mathbf{V}^T$ to make the following notation a little easier to connect with the procedure presented in the text.

```
[U,VT] = nnmf(termdoc,2,'algorithm','mult');
% It is easier to proceed forward with our
% calculations if we have V.
V = VT';
```

The next step is to normalize **U** and **V**, as described in Equation 5.7.

```
[nu,pu] = size(U);
[nv,pv] = size(V);
% Normalize the V entries as discussed above
for i = 1:nv
    for j = 1:pv
        V(i,j) = V(i,j) * norm(U(:,j));
    end
end
```

```
% Normalize the U entries as discussed above
for j = 1:pu
        U(:,j) = U(:,j)/(norm(U(:,j)));
end
```

We plot the rows of **V** to illustrate that the document cluster structure is easily discernible in the NMF space, as shown in Figure 5.6. One cluster is shown as asterisks, and they follow the vertical axis. The other cluster is seen by the circles that are aligned along the horizontal axis.

```
% First we set up a cell array of labels, so
% we can use these in the plot.
lab = {'d1','d2','d3','d4','d5','d6','d7','d8','d9'};
plot(V(1:5,1), V(1:5,2),'k*')
text(V(1:5,1)+.05, V(1:5,2),lab(1:5))
hold on
plot(V(6:9,1), V(6:9,2),'ko')
text(V(6:9,1), V(6:9,2)+.05,lab(6:9))
xlabel('V1')
ylabel('V2')
hold off
```

FIGURE 5.6

This scatterplot shows the cluster structure in the document collection of Table 5.1, once we have applied nonnegative matrix factorization. The first five documents are shown as asterisks along the vertical axis, and the second set of documents cluster along the horizontal axis.

Now, we extract the cluster (or axis) to which each document belongs.

```
% Here we extract the axes corresponding to
% each document.
[Y,I] = max(V,[],2);
% View the contents of I, which is the column where
% the maximum value occurred.
I'
ans = 2 2 2 2 2 1 1 1 1
```

As expected, we find that the first five documents have maximum values in column 2 (the vertical axis), while the last four documents have maximum values in column 1 (the horizontal axis).
❏

One of the interesting aspects of using nonnegative matrix factorization for clustering is that the clustering falls naturally out of the transformation. No other type of clustering methodology, such as k-means, needs to be applied to the transformed data. We simply see which axis is most relevant to each document.

This leads to another important characteristic of using nonnegative matrix factorization for document clustering. We saw in Chapter 2 how latent semantic indexing is often used to transform the term-document matrix, but the resulting space is not so easily interpreted. In the case of nonnegative matrix factorization, we can think of each axis or dimension as corresponding to a cluster of documents or topic.

It is important to note that NMF can be used in many applications besides document clustering. It can be used whenever the data are comprised of nonnegative values. Examples of some applications where the approach has proved successful are provided at the end of this chapter.

5.5.2 Probabilistic Latent Semantic Analysis

Probabilistic latent semantic analysis (PLSA or PLSI) is one approach to cast the identification of latent structure of a document collection within a statistical framework. We follow Hofmann [1999a, 1999b], the originator of PLSA, in our discussions.

The PLSA technique uses an approach known as the *aspect* (or *factor*) *model*, where we associate an unobserved class variable $Z = \{z_1, ..., z_k\}$ for each occurrence of a word $W = \{w_1, ..., w_M\}$ in a document collection $D = \{d_1, ..., d_N\}$. We can use this approach to generate our document collection within a probabilistic framework.

We first select a document d with probability $P(d)$ and then pick a latent class z with probability $P(z|d)$. The next step of the model is to generate a word with probability $P(w|z)$. As a result of this process, one obtains an observed document–word pair (d, w), where the latent class variable z is no longer needed.

We can calculate joint probabilities via the following equations

$$P(d, w) = P(d)P(w|d),\qquad(5.8)$$

where

$$P(w|d) = \sum_{z \in Z} P(w|z)P(z|d).\qquad(5.9)$$

The aspect (or factor) model given in Equations 5.8 and 5.9 is a statistical mixture model based on two assumptions. The first assumption is that the pairs (d, w) are generated independently, which corresponds to the usual bag-of-words approaches, such as latent semantic analysis. The second is that, conditional on the latent class z, words are generated independently of the specific document d. This means that we have to solve the inverse problem, where we have the documents with their word content, and we have to infer the latent variables Z and any applicable parameters.

We apply the maximum likelihood estimation principle to determine the quantities $P(d)$, $P(z|d)$, and $P(w|z)$ for our aspect model. Thus, we seek to maximize the log-likelihood function given here

$$L = \sum_{d \in D} \sum_{w \in W} n(d, w)\log[P(d, w)],\qquad(5.10)$$

where $n(d, w)$ denotes the number of times word w occurred in document d.

The typical way to solve the likelihood equation when we have latent variables is to use the *expectation maximization* (EM) approach [Hofmann, 1999a, 1999b; Dempster et al., 1977]. This is a general framework and can be applied in many problems where the variables to be estimated depend on hidden or missing information. We already saw it being used with generative topographic maps (Chapter 3), and we will see it again in the context of model-based clustering (Chapter 6).

The EM algorithm has two steps. First is the expectation or E-step, where we calculate posterior probabilities for the latent variable z, using current values of any parameters. Next is the maximization step or M-step, where the other estimates are updated using the posterior probabilities calculated in the previous E-step. These two steps are repeated until the algorithm converges; i.e., there are no more changes in the estimated parameters.

In the case of PLSI, the E-Step of the EM algorithm is given by

$$P(z|d, w) = \frac{P(z)P(d|z)P(w|z)}{\sum_{z_k} P(z_k)P(d|z_k)P(w|z_k)}, \tag{5.11}$$

which is the probability that the word w in the given document d is explained by the factor z.

The M-step equations are given by the following

$$P(w|z) = \frac{\sum_{d} n(d, w)P(z|d, w)}{\sum_{d, w_k} n(d, w_k)P(z|d, w_k)}, \tag{5.12}$$

$$P(d|z) = \frac{\sum_{w} n(d, w)P(z|d, w)}{\sum_{d_k, w} n(d_k, w)P(z|d_k, w)}, \tag{5.13}$$

$$P(z) = \frac{1}{R}\sum_{d, w} n(d, w)P(z|d, w), \tag{5.14}$$

where

$$R \equiv \sum_{d, w} n(d, w).$$

By iteratively repeating the E-step (Equation 5.11) and the M-step (Equations 5.12 through 5.14) until convergence, we arrive at a local maximum of the log-likelihood function.

One of the outcomes from this solution is the probability of a word w for each of the factors or aspects z, as given by $P(w|z)$. Thus, the factors can be represented by the ten most probable words for each z and can be ordered by their associated probability $P(w|z)$.

Hofmann [1999a] provides an example of this where he constructed the PLSA factor solution using the TDT pilot study corpus from the Linguistic Data Consortium.[2] He removed stop words using a standard word list, but he did not do any further processing. He provides some examples of factors and the words that characterize them. The top four words of one factor are *plane, airport, crash,* and *flight*. Another factor that is also somewhat related to

---

[2] Their website is **http://www.ldc.upenn.edu/**, and the collection Hofmann used was catalog number LDC98T25, 1998.

this factor has words *space, shuttle, mission,* and *astronauts.* So, we see that both of these factors or aspects are related to flight with one pertaining to planes and the other to space ships.

The reader may still be a little confused as to how we relate PLSA to LSA. One way to help understand this is to recast our aspect generative model within a matrix framework. An equivalent version of the aspect model (Equation 5.8) can be found by applying Bayes' rule to invert the $P(z|d)$. Thus, our generative model can be defined as follows

$$P(d, w) = \sum_{z \in Z} P(z)P(w|z)P(d|z).$$
(5.15)

Rewriting this in matrix notation we obtain the following joint probability model

$$\mathbf{P} = \mathbf{U\Sigma V}^T,$$

where the elements of the matrices are defined as

$$u_{ik} = P(d_i|z_k)$$
$$\Sigma_{kk} = P(z_k)$$
$$v_{jk} = P(w_j|z_k).$$

Using the above definitions, we provide the joint probability model as the matrix product

$$\mathbf{P} = \begin{bmatrix} P(d_1|z_1) & \dots & P(d_1|z_K) \\ \dots & \dots & \dots \\ P(d_N|z_1) & \dots & P(d_N|z_K) \end{bmatrix} \times \begin{bmatrix} P(z_1) & \dots & 0 \\ \dots & \dots & \dots \\ 0 & \dots & P(z_K) \end{bmatrix} \times \begin{bmatrix} P(w_1|z_1) & \dots & P(w_1|z_K) \\ \dots & \dots & \dots \\ P(w_M|z_1) & \dots & P(w_M|z_K) \end{bmatrix}^T.$$

Several things might be noted about this joint probability model. First, we see that aspects z are discarded via the multiplication. In fact, the $P(z_k)$ serve as mixing proportions or weights. When we compare this to LSA (Chapter 2), we see that these are similar to the singular values of the SVD approach. Also, the left and right eigenvectors in SVD correspond to the factors $P(w|z)$ and $P(d|z)$. We have to be careful making these connections, because there is a fundamental difference between the objective function used to obtain the best decomposition or factorization. LSA (or SVD) is based on the L_2 norm under the assumption of Gaussian noise, while the PLSA solution is found by maximizing the log-likelihood and the predictive power of the resulting model. Finally, we note that the SVD can be computed exactly, while the EM algorithm is an iterative method and is only guaranteed to find a locally-optimal solution to the log-likelihood function.

Example 5.5

PLSA has a natural role in clustering. We return to the previous example, where we used nonnegative matrix factorization to cluster a small document collection (Tables 5.1 and 5.2). We will use the PLSA methodology to solve for a cluster solution of this document collection. To do this, we utilize a very simple implementation of the PLSA algorithm as provide by Ata Kaban.[3] Her implementation of PLSA is in terms of $P(z|d)$ rather than $P(d|z)$, as in our discussions above. This allows her formulation to be very compact. The function is included in the EDA Toolbox, and the interested reader can look at her function for the details on her approach. We now load the data and then call the PLSA function requesting two factors or classes.

```
load nmfclustex
% Next we call the PLSA function with 2 topics and
% 100 iterations of the algorithm.
[V,S] = PLSA(nmfclustex,2,100);
```

TABLE 5.3

Elements of Matrix **V**

Words	$P(w|z_1)$	$P(w|z_2)$
human	0.0625	0.0768
interface	0.0623	0.0771
computer	0.0000	0.1537
user	0.0621	0.1543
system	0.1876	0.0769
response	0.0000	0.1537
time	0.0000	0.1537
EPS	0.1251	0.0000
survey	0.0000	0.1537
trees	0.1876	0.0000
graph	0.1876	0.0000
minors	0.1251	0.0000

The elements of the matrix **V** contain the probability of each word for the given factor, $P(w|z)$. Therefore, in this example the **V** matrix has dimensions 12×2. Because of this slightly different formulation, the elements of the matrix **S** correspond to the probability of the latent class z given each document. So, **S** has two rows and nine columns. We provide the elements of matrix **V** in Table 5.3. We note that the top five words for latent class (or topic) $z = 1$ are *system, EPS, trees, graph*, and *minors*. The top five terms for $z = 2$ are *computer, user, response, time*, and *survey*. Now, we examine the **S** matrix, which is given in Table 5.4. The entries in the **S** matrix reflect the probability

---

TABLE 5.4

Elements of Matrix **S**

Class	d1	d2	d3	d4	d5	d6	d7	d8	d9
$z = 1$	0	0	0.997	1	0	1	1	1	0.776
$z = 2$	1	1	0.003	0	1	0	0	0	0.333

of the class z given the document, and we see that the first class or topic includes documents $d3$, $d4$, $d6$, $d7$, $d8$, and $d9$, while the second one has documents $d1$, $d2$, and $d5$. The first group of documents includes the EPS and the graph tree articles, and the second one has the remaining articles on system response time. The reader will be asked in the exercises to repeat this for a three class solution.

❏

5.6 Minimum Spanning Trees and Clustering

In this section, we describe a clustering method that first converts a data set to a weighted graph and then finds a minimum spanning tree (MST). We can divide the data into groups based on the longest edges in this spanning tree. We first provide some definitions used in graph theory and then describe the steps in minimum spanning tree clustering.

5.6.1 Definitions

The usual definition of a *graph* $G = (V, E)$ consists of two sets denoted by V and E. The set V is a collection of objects known as *vertices* or *nodes*. For example, these can be such things as cities, networked computers, and even observations in a general data set, as we will see later. The set E contains the *edges* (or *links*) of the graph, which are pairs of vertices specifying a link.

These edges can be *directed* or *undirected*. The vertex pairs are unordered, if we have an undirected graph. If the edges have a direction associated with them, then we traverse the graph (or edges) in the given direction. Using u and v to represent nodes in the graph, we denote an edge in an undirected graph as an unordered pair $\{u, v\}$. For a directed graph, we can write them as $u \rightarrow v$ or an ordered pair (u, v), specifying a direction from vertex u to vertex v. We are concerned only with undirected graphs in this text.

An edge with the same endpoints (or nodes) is called a *one-node loop*. By definition, a graph $G = (V, E)$ has no one-node loops in the set of edges E.

As an example, the graph in the top panel of Figure 5.7 can be represented by the following sets V and E:

- $V = \{A, B, C, D, E, F, G, H\}$
- $E = \{\{A, B\}, \{B, C\}, \{B, E\}, \{E, G\}, \{E, F\}, \{F, H\}, \{F, D\}\}$

It is the usual practice to also denote the number of vertices in the graph as V and the number of edges as E. However, the meaning (set or cardinality) should be clear from the context. The number of edges in an undirected graph satisfy

$$0 \le E \le \binom{V}{2}.$$

A graph $G' = (V', E')$ is a ***subgraph*** of G, if $V' \subseteq V$ and $E' \subseteq E$.

A ***path*** is a sequence of edges in a graph specifying a way to travel from an ***origin*** node to a ***destination*** node. Each successive pair of edges in the sequence shares one vertex. The origin and destination vertices are called the ***endpoints*** of the path. We usually specify a path as an ordered list of directed edges, even in an undirected graph. In this case, the direction specifies the order in which vertices and edges are traversed for the path. If the origin and destination nodes are the same, then we have a ***closed path*** or a ***cycle***. In a closed path, the only vertices that can be repeated are the endpoints. An undirected graph is said to be ***acyclic***, if no subgraph has a cycle.

A ***walk*** is also a sequence of edges traversing a graph from one node to another. However, a walk allows for the possibility of revisiting vertices in the graph. There is no requirement that a walk has to repeat nodes, so any path is also a walk, by definition.

If there is a walk between any two vertices, then the undirected graph is ***connected***. A disconnected graph consists of ***components***. This is the set of maximal connected subgraphs. Two vertices will belong to the same component if and only if there is a path connecting them.

A ***tree*** is a connected acyclic graph. Another name for acyclic graphs is a ***forest***. Alternatively, a tree is one component of a forest. A ***spanning tree*** of a graph G is a subgraph that (1) is a tree and (2) contains every vertex in V. Note that a graph G will have a spanning tree if and only if it is connected. A ***spanning forest*** of G is the set of spanning trees, one for each connected component. Figure 5.7 contains two graphs. The top one is a tree, and the lower one is not a tree.

In addition to the sets of vertices V and edges E, our graph might also have a function w: $E \to R^+$ that assigns a weight $w(e)$ to each edge. A ***minimum spanning tree*** of a graph G is the spanning tree T that minimizes

$$w(T) = \sum_{e \in T} w(e).$$

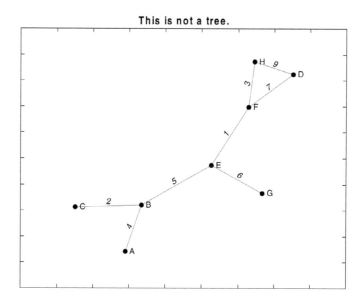

FIGURE 5.7

The graph at the top is a tree. The one shown in the bottom plot is not a tree because vertices D, F, and H form a cycle (also sometimes called a loop).

If the weights are distinct for all edges, then there will be a unique minimum spanning tree [Erickson, 2015]. If a graph has no weights, then any spanning tree is a minimum spanning tree.

5.6.2 Minimum Spanning Tree Clustering

Gower and Ross [1969] and Zahn [1971] were among the first to illustrate the relationship between hierarchical clustering and minimum spanning trees. We do not go into any details on algorithms for finding a minimum spanning tree because we are more interested in how they can be used for clustering. Two popular approaches were developed by Kruskal [1956] who made the connection between minimum spanning trees and the traveling salesman problem in operations research. The second algorithm was proposed by Prim [1957]. These are the algorithms implemented in the function **minspantree**, which is part of the base MATLAB software.

This approach to clustering proceeds by first forming the graph using our n data points as vertices. The edge weights are obtained using the interpoint distances. As we will see in the next example, one way to specify a graph in MATLAB is to use an *adjacency matrix*. An adjacency matrix is a square symmetric matrix where the rows and columns correspond to nodes in the graph. A nonzero entry of one represents an edge between node i and j in an unweighted graph. In MATLAB, an adjacency matrix can have a nonzero entry greater than one, indicating the edge weight. Once we have a graph, we can find a minimum spanning tree. Next, we remove the k largest edges from this tree. This will divide the graph into components or subgraphs. Each of these subgraphs will constitute a cluster.

Procedure – Minimum Spanning Tree Clustering

1. Designate the nodes (set V) as the n observations in the data set.
2. Find the edge weights by forming the interpoint distance matrix using an appropriate measure. This is the *weight matrix* (or adjacency matrix in MATLAB).
3. Form the minimum spanning tree.
4. Remove the k edges with the largest weights (or longest edges) from the tree.
5. This will produce $k + 1$ subgraphs, each of which is considered to be a group or cluster.
6. Assign cluster IDs according to the subgraph membership.

Cutting the minimum spanning tree based on the longest edges is similar to hierarchical clustering [Zahn 1971]. Recall from a previous discussion that we get a dendrogram as a result of this approach. We cut the dendrogram to get the clusters, and a reasonable choice is where the links are large.

Example 5.6

We generate a data set with a known cluster structure to illustrate minimum spanning tree clustering.

```
% First, we generate some data with some clusters.
x = randn(100,2);
x(26:50,:) = x(26:50,:) + 4;
x(51:75,:) = x(51:75,:) - 4;
x(76:100,1) = x(76:100,1) + 4;
x(76:100,2) = x(76:100,2) - 4;
labs = [zeros(1,25), ...
     ones(1,25), ...
     2*ones(1,25), ...
     3*ones(1,25)];
```

We show the data in a scatterplot in Figure 5.8.

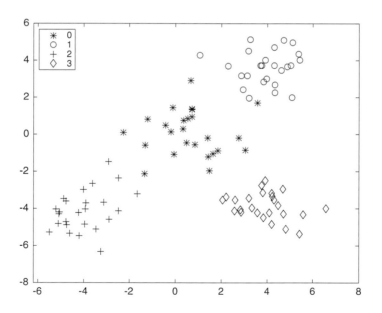

FIGURE 5.8

These are the data we generated in Example 5.6. There are four groups or clusters.

```
% Construct a grouped scatterplot.
gscatter(x(:,1),x(:,2),labs,...
     ['r','g','b','k'],...
```

```
['*','o','+','d'])
```

The next step is to construct a graph using the data points as nodes. The edge weights are obtained using the distance between the observations.

```
% Next, we have to get the distances between
% the observations. This will be the edge weights.
% Use Euclidean distance and put it in matrix form.
D = pdist(x,'euclidean');
DS = squareform(D);
```

The base MATLAB software includes a function called **graph** that has several options for creating graphs. For instance, one could specify the edges using node pairs or an adjacency matrix. See the documentation on **graph** for more options. We use the interpoint distance matrix **D** found above as the adjacency matrix and build a graph.

```
% We get a graph object here.
Gx = graph(DS);
```

We can easily view the graph using the **plot** function. See Figure 5.9 (top).

```
% Let's take a look at the graph.
plot(Gx,'XData',x(:,1),'YData',...
    x(:,2),'EdgeAlpha',.01)
```

The next step is to obtain a minimum spanning tree, and we do that using the following code.

```
% We are now ready to get a
% minimum spanning tree.
Gxtree = minspantree(Gx);
% Plot the tree using this code.
plot(Gxtree,'XData',x(:,1),'YData',x(:,2))
```

The minimum spanning tree is shown in Figure 5.9 (bottom). We can also look at the spanning tree in the command window by typing **Gxtree** at the prompt. This is what we have for the tree:

```
Gxtree =

graph with properties:

Edges: [99x2 table]
Nodes: [100x0 table]
```

Our clustering approach with minimum spanning trees is based on removing the largest edges from the tree. So, we first have to sort the edges.

```
% Extract the edges from the tree and sort.
junk = Gxtree.Edges;
[sortwt,index] = sort(junk.Weight,'descend');
```

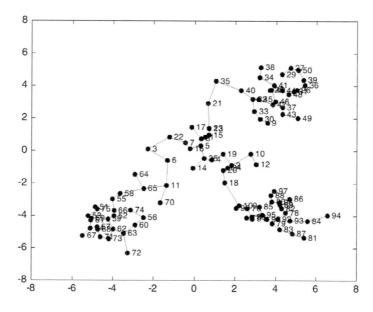

FIGURE 5.9

The top plot shows the graph we obtained using the interpoint distance matrix as our edge weights. The plot in the lower panel is a minimum spanning tree.

```
% Display the 8 largest edges.
junk(index(1:8),:)

ans =

      EndNodes        Weight

     _____      _____


      18     100      1.5847
       6      11      1.5386
      21      23      1.5297
      21      35      1.4267
      35      40      1.3628
       3      22      1.2772
      63      72      1.2254
      58      65      1.2116
```

We want to remove weights with the highest values, and we should look for a natural break in them. Given that the largest three weights are all approximately 1.5, it seems reasonable to delete the first three edges. Deleting more will produce different cluster solutions. The function **rmedge** deletes edges from the tree.

```
% Remove the edges of the tree with
% the highest weights.
Gxtree = rmedge(Gxtree,index(1:3));
```

Now, we get the observations belonging to each of the connected subgraphs. These are the clusters.

```
% Get the connected components.
binx = conncomp(Gxtree);
% Show in a scatterplot.
gscatter(x(:,1),x(:,2),binx,...
    ['r','g','b','k'],...
    ['*','o','+','d'])
legend off
```

The clusters (or subgraphs) we found are shown in Figure 5.10. The clusters we get are good ones, except for a few incorrect points near the boundaries. ❏

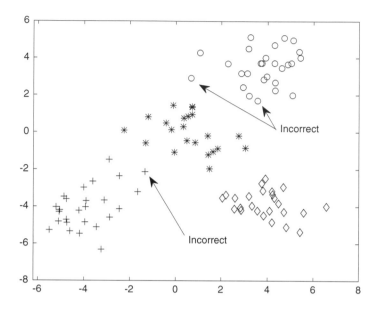

FIGURE 5.10

We removed the edges in the minimum spanning tree with the largest weights. This created disconnected components in the graph, each of which corresponds to a cluster. The clusters found with this approach are shown in the lower panel. We see that there are some mistakes when we compare these groups with the true ones in Figure 5.8.

5.7 Evaluating the Clusters

In this section, we turn our attention to understanding more about the quality of our cluster results and to estimating the 'correct' number of groups in our data. We present the following measures that can be used to address both of these issues. The first is the Rand index that can be used to compare two different groupings of the same data set. Next, we discuss the cophenetic correlation coefficient that provides a way to compare a set of nested partitions from hierarchical clustering with a distance matrix or with another set of nested partitions. We also cover a method due to Mojena [1977] for determining the number of groups in hierarchical clustering based on the fusion levels. We illustrate the silhouette plot and silhouette statistic that can be used to help decide how many groups are present. We discuss a novel method called the gap statistic [Tibshirani, Walther, and Hastie, 2001] that seems to be successful at estimating the number of clusters. An added benefit of the gap statistic approach is that it addresses the issue of whether or not there are any clusters at all. We conclude the section with descriptions of several cluster validity indices that are provided with the MATLAB Statistics Toolbox.

5.7.1 Rand Index

Say we have two partitions of the same data set called G_1 and G_2, with g_1 groups and g_2 groups, respectively. This can be represented in a $g_1 \times g_2$ matching matrix **N** with elements n_{ij}, where n_{ij} is the number of observations in group i of partition G_1 that are also in group j of partition G_2. Note that the number of groups in each partition do not have to be equal, and that the classifications can be obtained through any method.

The ***Rand index*** [Rand, 1971] was developed to help analysts answer four questions about their cluster results. These are:

- How well does a method retrieve natural clusters?
- How sensitive is a method to perturbation of the data?
- How sensitive is a method to missing data?
- Given two methods, do they produce different results when applied to the same data?

In this book, we are more interested in the last question.

The motivation for the Rand index follows three assumptions. First, clustering is considered to be discrete in that every point is assigned to a specific cluster. Second, it is important to define clusters with respect to the observations they *do not* contain, as well as by the points that they *do* contain. Finally, all data points are equally important in determining the cluster structure.

Since the cluster labels are arbitrary, the Rand index looks at pairs of points and how they are partitioned in groupings G_1 and G_2. There are two ways that data points x_i and x_j can be grouped similarly (i.e., the groupings agree):

- x_i and x_j are put in the same cluster in G_1 and G_2.
- x_i and x_j are in different clusters in G_1 and in different clusters in G_2.

There are also two ways that x_i and x_j can be grouped differently:

- x_i and x_j are in the same cluster in G_1 and different clusters in G_2.
- x_i and x_j are in different clusters in G_1 and the same cluster in G_2.

The Rand index calculates the proportion of the total n choose 2 objects that agree between the two groupings. It is given by the following

$$RI = \frac{nC2 + \sum_{i=1}^{g_1}\sum_{j=1}^{g_2}n_{ij}^2 - \frac{1}{2}\sum_{i=1}^{g_1}\left[\sum_{j=1}^{g_2}n_{ij}\right]^2 - \frac{1}{2}\sum_{j=1}^{g_2}\left[\sum_{i=1}^{g_1}n_{ij}\right]^2}{nC2},$$

where nCk denotes the binomial coefficient or the number of ways to choose k objects from a set of n objects. The Rand index is a measure of similarity between the groupings, and it ranges from zero when the two groupings are not similar to a value of one when the groupings are exactly the same.

Fowlkes and Mallows [1983] developed their own index for the case $g_1 = g_2$, which will be presented in the exercises. They point out that the Rand index increases as the number of clusters increase, and the possible range of values is very narrow. Hubert and Arabie [1985] developed an ***adjusted Rand index*** that addresses these issues. This is given by $RI_A = N/D$, where

$$N = \sum_{i=1}^{g_1}\sum_{j=1}^{g_2}n_{ij}C2 - \sum_{i=1}^{g_1}n_{i\cdot}C2\sum_{j=1}^{g_2}n_{\cdot j}C2 \div nC2,$$

$$D = \left[\sum_{i=1}^{g_1}n_{i\cdot}C2 + \sum_{j=1}^{g_2}n_{\cdot j}C2\right] \div 2 - \sum_{i=1}^{g_1}n_{i\cdot}C2\sum_{j=1}^{g_2}n_{\cdot j}C2 \div nC2,$$

and

$$n_{\cdot j} = \sum_{i=1}^{g_1}n_{ij} \qquad n_{i\cdot} = \sum_{j=1}^{g_2}n_{ij}.$$

The binomial coefficient $mC2$ is defined as 0 when $m = 0$ or $m = 1$. The adjusted Rand index provides a standardized measure such that its expected value is zero when the partitions are selected at random and one when the partitions match completely.

Example 5.7

We return to the **iris** data of the previous example to illustrate the Rand index, by comparing the k-means results with the true class labels. First we get some of the needed information to construct the matching matrix.

```
% Get some of the preliminary information.
% You can load the data using: load example52
ukmus = unique(kmus); ulabs = unique(labs);
n1 = length(ukmus); n2 = length(ulabs);
n = length(kmus);
```

Now we find the matrix **N**, noting that it is not necessarily square (the number of partitions do not have to be equal), and it is usually not symmetric.

```
% Now find the matching matrix N
N = zeros(n1,n2);
I = 0;
for i = ukmus(:)'
    I = I + 1;
    J = 0;
    for j = ulabs(:)'
        J = J + 1;
        indI = find(kmus == i);
        indJ = find(labs == j);
        N(I,J) = length(intersect(indI,indJ));
    end
end
nc2 = nchoosek(n,2);
nidot = sum(N);
njdot = sum(N');
ntot = sum(sum(N.^2));
num = nc2+ntot-0.5*sum(nidot.^2)-0.5*sum(njdot.^2);
ri = num/nc2;
```

The resulting Rand index has a value of 0.8797, which indicates good agreement between the classifications. We provide a function called **randind** that implements this code for any two partitions **P1** and **P2**. We also implement the adjusted Rand index in the function **adjrand**. It is used in the following manner.

```
% Now use the adjusted Rand index function.
ari = adjrand(kmus,labs);
```

This yields 0.7302, indicating an agreement above what is expected by chance alone.
❏

5.7.2 Cophenetic Correlation

In some applications, we might be interested in comparing the output from two hierarchical partitions. We can use the *cophenetic correlation coefficient* for this purpose. Perhaps the most common use of this measure is in comparing hierarchical clustering results with the proximity data (e.g., interpoint distances) that were used to obtain the partitions. For example, as discussed before, the various hierarchical methods impose a certain structure on the data, and we might want to know if this is distorting the original relationships between the points as specified by their proximities.

We start with the cophenetic matrix, **H**. The *ij*-th element of **H** contains the fusion value where object *i* and *j* were first clustered together. We only need the upper triangular entries of this matrix (i.e., those elements above the diagonal). Say we want to compare this partition to the interpoint distances. Then, the cophenetic correlation is given by the product moment correlation between the values (upper triangular only) in **H** and their corresponding entries in the interpoint distance matrix. This has the same properties as the product moment correlation coefficient. Values close to one indicate a higher degree of correlation between the fusion levels and the distances. To compare two hierarchical clusters, one would compare the upper triangular elements of the cophenetic matrix for each one.

Example 5.8

The cophenetic correlation coefficient is primarily used to assess the results of a hierarchical clustering method by comparing the fusion level of observations with their distance. MATLAB provides a function called **cophenet** that calculates the desired measure. We return to the **yeast** data in Example 5.1 to determine the cophenetic coefficient for single linkage and complete linkage.

```
load yeast
% Get the Euclidean distances.
Y = pdist(data);
% Single linkage output.
Zs = linkage(Y);
% Now get the cophenetic coefficient.
scoph = cophenet(Zs,Y);
```

The cophenetic coefficient is 0.9243, indicating good agreement between the distances and the hierarchical clustering. Now we apply the same procedure to the complete linkage.

```
% Now do the same thing for the complete linkage.
Zc = linkage(Y,'complete');
ccoph = cophenet(Zc,Y);
```

The coefficient in this case is 0.8592, showing less correlation. The **cophenet** function only does the comparison between the clustering and the distances and not the comparison between two hierarchical structures.
❏

5.7.3 Upper Tail Rule

The upper tail rule was developed by Mojena [1977] as a way of determining the number of groups in hierarchical clustering. It uses the relative sizes of the fusion levels in the hierarchy. Let the fusion levels $\alpha_0, \alpha_1, ..., \alpha_{n-1}$

correspond to the stages in the hierarchy with $n, n-1, \ldots, 1$ clusters. We also denote the average and standard deviation of the j previous fusion levels by $\bar{\alpha}$ and s_α. To apply this rule, we estimate the number of groups as the first level at which we have

$$\alpha_{j+1} > \bar{\alpha} + cs_\alpha, \tag{5.16}$$

where c is a constant. Mojena suggests a value of c between 2.75 and 3.50, but Milligan and Cooper [1985] offer a value of 1.25 based on their study of simulated data sets. One could also look at a plot of the values

$$\frac{(\alpha_{j+1} - \bar{\alpha})}{s_\alpha} \tag{5.17}$$

against the number of clusters j. A break in the plot is an indication of the number of clusters. Given the dependence on the value of c in Equation 5.16, we recommend the graphical approach of Equation 5.17.

Example 5.9

We now show how to implement the graphical Mojena procedure in a way that makes it similar to the elbow plots of previous applications. We turn to the **lungB** data set for this example, and we use the standardized Euclidean distance, where each coordinate in the sum of squares is inversely weighted by its sample variance.

```
load lungB
% Get the distances and the linkage.
% Use the standardized Euclidean distance.
Y = pdist(lungB','seuclidean');
Z = linkage(Y,'complete');
% Plot dendrogram with fewer leaf nodes.
dendrogram(Z,15);
```

The dendrogram is shown in Figure 5.11 (top). We are going to flip the **Z** matrix to make it easier to work with, and we will find the values in Equation 5.17 for a maximum of 10 clusters.

```
nc = 10;
% Flip the Z matrix - makes it easier.
Zf = flipud(Z);
% Now get the vectors of means
% and standard deviations
for i = 1:nc
    abar(i) = mean(Zf(i:end,3));
    astd(i) = std(Zf(i:end,3));
end
```

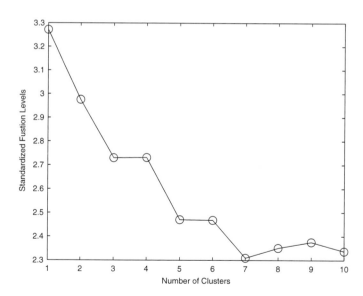

FIGURE 5.11

In the top panel, we show the dendrogram (with 15 leaf nodes) for the **lungB** data set, where standardized Euclidean distance is used with complete linkage. The second panel is the plot of the standardized fusion levels. The elbow in the curve indicates that three clusters is reasonable. However, some other elbows at 5 and 7 might provide interesting clusters, too.

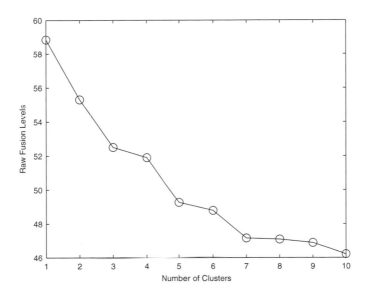

FIGURE 5.12
This is a plot of the raw fusion levels for Example 5.8. Again, we see an elbow at three clusters.

```
% Get the y values for plotting.
yv = (Zf(1:nc,3) - abar(:))./astd(:);
xv = 1:nc;
plot(xv,yv,'-o')
```

This plot is shown in Figure 5.11 (bottom), where the elbow in the curve seems to indicate that three clusters could be chosen for this application. We could also plot the raw fusion values in a similar plot, as follows.

```
% We can also plot just the fusion levels
% and look for the elbow.
plot(1:nc,Zf(1:nc,3),'o-')
```

This plot is given in Figure 5.12, and we see again that three seems to be a reasonable estimate for the number of clusters. We provide a function called **mojenaplot** to construct the plot, given the output from **linkage**.
❑

5.7.4 Silhouette Plot

Kaufman and Rousseeuw [1990] present the *silhouette statistic* as a way of estimating the number of groups in a data set. Given observation i, we denote the average dissimilarity to all other points in its own cluster as a_i. For any other cluster c, we let $\bar{d}(i, c)$ represent the average dissimilarity of i to all objects in cluster c. Finally, we let b_i denote the minimum of these average dissimilarities $\bar{d}(i, c)$. The **silhouette width** for the i-th observation is

$$sw_i = \frac{(b_i - a_i)}{\max(a_i, b_i)}. \qquad (5.18)$$

We can find the **average silhouette width** by averaging the sw_i over all observations:

$$\overline{sw} = \frac{1}{n} \sum_{i=1}^{n} sw_i.$$

Observations with a large silhouette width are well clustered, but those with small values tend to be those that are scattered between clusters. The silhouette width sw_i in Equation 5.18 ranges from -1 to 1. If an observation has a value close to 1, then the data point is closer to its own cluster than a neighboring one. If it has a silhouette width close to -1, then it is not very well-clustered. A silhouette width close to zero indicates that the observation could just as well belong to its current cluster or one that is near to it.

Kaufman and Rousseeuw use the average silhouette width to estimate the number of clusters in the data set by using the partition with two or more clusters that yields the largest average silhouette width. They state that an average silhouette width greater than 0.5 indicates a reasonable partition of the data, and a value of less than 0.2 would indicate that the data do not exhibit cluster structure.

There is also a nice graphical display called a **silhouette plot**, which is illustrated in the next example. This type of plot displays the silhouette values for each cluster, ranking them in decreasing order. This allows the analyst to rapidly visualize and assess the cluster structure.

Example 5.10

MATLAB provides a function in the Statistics Toolbox called **silhouette** that will construct the silhouette plot and also returns the silhouette values, if desired. We illustrate its functionality using the **iris** data and k-means, where we choose both $k = 3$ and $k = 4$. In this example, we will use **replicates** (i.e., repeating the k-means procedure five times) and the **display** option that summarizes information about the replicates.

```
load iris
data = [setosa; versicolor; virginica];
% Get a k-means clustering using 3 clusters,
% and 5 replicates. We also ask MATLAB to
% display the final results for each replicate.
kmus3 = kmeans(data,3,...
```

```
'replicates',5,'display','final');
```

When we run this code, the following is echoed to the command window. Since this procedure uses a random selection of starting points, you might see different results when implementing this code.

```
5 iterations, total sum of distances = 78.8514
4 iterations, total sum of distances = 78.8514
7 iterations, total sum of distances = 78.8514
6 iterations, total sum of distances = 78.8514
8 iterations, total sum of distances = 142.754
```

From this, we see that local solutions do exist, but the final result from MATLAB will be the one corresponding to the lowest value of our objective function, which is 78.8514. Now we repeat this for four clusters and get the corresponding silhouette plots.

```
% Get a k-means clustering using 4 clusters.
kmus4 = kmeans(data,4,...
    'replicates',5,'display','final');
% Get the silhouette plots and the values.
[sil3, h3] = silhouette(data, kmus3);
[sil4, h4] = silhouette(data, kmus4);
```

The silhouette plots for both $k = 3$ and $k = 4$ are shown in Figure 5.13. We see that we have mostly large silhouette values in the case of three clusters, and all are positive. On the other hand, four clusters yield some with negative values and some with small (but positive) silhouette indexes. To get a one-number summary describing each clustering, we find the average of the silhouette values.

```
mean(sil3)
mean(sil4)
```

The three cluster solution has an average silhouette value of 0.7357, and the four cluster solution has an average of 0.6714. This indicates that the grouping into three clusters using k-means is better than the one with four groups.

❏

5.7.5 Gap Statistic

Tibshirani et al. [2001] define a technique called the *gap statistic method* for estimating the number of clusters in a data set. This methodology applies to any technique used for grouping data. The clustering method could be k-means, hierarchical, etc. This technique compares the within-cluster dispersion with what one might expect given a reference null distribution (i.e., no clusters). Fridlyand and Dudoit [2001] note a type of gap test used by

FIGURE 5.13

Here we have the silhouette plots for $k = 3$ and $k = 4$ clusters using the **iris** data. The top one indicates large values for cluster 2 and a few large values for clusters 1 and 3. In the second plot with 4 clusters, we see that there are some negative values in cluster 4, and clusters 1, 3, and 4 have many low silhouette values. These plots indicate that 3 clusters are a better fit to the data.

Bock [1985] in cluster analysis to test the null hypothesis of a homogeneous population versus a heterogeneous one. However, they are defined somewhat differently.

We start with a set of k clusters $C_1, ..., C_k$ from some clustering method, where the r-th group has n_r observations. We denote the sum of the pairwise distances for all points in cluster r as

$$D_r = \sum_{i,j \in C_r} d_{ij}.$$

We now define W_k as

$$W_k = \sum_{r=1}^{k} \frac{1}{2n_r} D_r. \tag{5.19}$$

Tibshirani et al. [2001] note that the factor 2 in Equation 5.19 makes this the same as the pooled within-cluster sum of squares around the cluster means, if the distance used is squared Euclidean distance.

The gap statistic method compares the standardized graph of the within-dispersion index $\log(W_k)$, $k = 1,...,K$ with its expectation under a reference null distribution. The null hypothesis of a single cluster is rejected in favor of a model with k groups if there is strong evidence for this based on the gap statistic (see Equation 5.20). Their estimate for the number of groups is the value of k where $\log(W_k)$ is the furthest below this reference curve. The reference curve shows the expected values of $\log(W_k)$ and is found using random sampling.

Tibshirani et al. [2001] show that there are two possible choices for the reference distribution. These are the uniform distribution over the range of observed values for a given variable or a uniform distribution over a box aligned with the principal components of the data set. The second method is preferred when one wants to ensure that the shape of the distribution is accounted for. We first discuss how to generate data from these distributions, then we go on to describe the entire gap statistic procedure.

Generating Data from Reference Distributions

1. **Gap-Uniform**: For each of the i dimensions (or variables), we generate n one-dimensional variates uniformly distributed over the range of x_i^{min} to x_i^{max}, where x_i represents the i-th variable or the i-th column of X.

2. **Gap-PC**: We assume that the data matrix X has been column-centered. We then compute the singular value decomposition

$$X = UDV^T.$$

Next we transform X using

$$X' = XV.$$

We generate a matrix of random variates Z' as in the gap-uniform case, using the range of the columns of X' instead. We transpose back using

$$Z = Z'V^T.$$

The basic gap statistic procedure is to simulate B data sets according to either of the null distributions and then apply the clustering method to each of them. We can calculate the same index $\log(W_k^*)$ for each simulated data set. The estimated expected value would be their average, and the estimated gap statistic is given by

$$gap(k) = \frac{1}{B}\sum_b \log(W_{k,b}^*) - \log(W_k).$$

Tibshirani et al. [2001] did not offer insights as to what value of B to use, but Fridlyand and Dudoit [2001] use $B = 10$ in their work. We use the gap statistic to decide on the number of clusters as outlined in the following procedure.

Procedure – Gap Statistic Method

1. Cluster the given data set to obtain partitions $k = 1, 2, \dots , K$, using any desired clustering method.
2. For each partition with k clusters, calculate the observed $\log(W_k)$.
3. Generate a random sample of size n using either the gap-uniform or the gap-PC procedure. Call this sample X^*.
4. Cluster the random sample X^* using the same clustering method as in step 1.
5. Find the within-dispersion measures for this sample, call them $\log(W_{k,b}^*)$.
6. Repeat steps 3 through 5 for a total of B times. This yields a set of measures $\log(W_{k,b}^*)$, $k = 1, \dots, K$ and $b = 1, \dots, B$.
7. Find the average of these values, using

$$\overline{W}_k = \frac{1}{B}\sum_b \log(W_{k,b}^*),$$

and their standard deviation

$$sd_k = \sqrt{\frac{1}{B} \sum_b [\log(W_{k,b}^*) - \overline{W}]^2}.$$

8. Calculate the estimated gap statistic

$$gap(k) = \overline{W}_k - \log(W_k) . \tag{5.20}$$

9. Define

$$s_k = sd_k\sqrt{1 + 1/B},$$

and choose the number of clusters as the smallest k such that

$$gap(k) \geq gap(k+1) - s_{k+1}.$$

Example 5.11

We show how to implement the gap statistic method for a uniform null reference distribution using the **lungB** data set. We use agglomerative clustering (with complete linkage) in this application rather than k-means, mostly because we can get up to K clusters without performing the clustering again for a different k. Note also that we standardized the columns of the data matrix.

```
load lungB
% Take the transpose, because the
% columns are the observations.
X = lungB';
[n,p] = size(X);
% Standardize the columns.
for i = 1:p
    X(:,i) = X(:,i)/std(X(:,i));
end
```

We now find the observed $\log(W_k)$ for a maximum of $K = 10$ clusters.

```
% Test for a maximum of 10 clusters.
K = 10;
Y = pdist(X,'euclidean');
Z = linkage(Y,'complete');
% First get the observed log(W_k).
% We will use the squared Euclidean distance
% for the gap statistic.
```

```
% Get the one for 1 cluster first.
W(1) = sum(pdist(X).^2)/(2*n);
for k = 2:K
    % Find the index for k.
    inds = cluster(Z,k);
    for r = 1:k
        indr = find(inds==r);
        nr = length(indr);
        % Find squared Euclidean distances.
        ynr = pdist(X(indr,:)).^2;
        D(r) = sum(ynr)/(2*nr);
    end
    W(k) = sum(D);
end
```

We now repeat the same procedure B times, except that we use the uniform reference distribution as our data.

```
% Now find the estimated expected
% values.
B = 10;
% Find the range of columns of X for gap-uniform
minX = min(X);
maxX = max(X);
Wb = zeros(B,K);
% Now do this for the bootstrap.
Xb = zeros(n,p);
for b = 1:B
    % Generate according to the gap-uniform method.
    % Find the min values and max values.
    for j = 1:p
        Xb(:,j) = unifrnd(minX(j),maxX(j),n,1);
    end
    Yb = pdist(Xb,'euclidean');
    Zb = linkage(Yb,'complete');
    % First get the observed log(W_k)
    % We will use the squared Euclidean distance.
    % Get the one for 1 cluster first.
    Wb(b,1) = sum(pdist(Xb).^2)/(2*n);
    for k = 2:K
        % Find the index for k.
        inds = cluster(Zb,k);
        for r = 1:k
            indr = find(inds==r);
            nr = length(indr);
            % Find squared Euclidean distances.
            ynr = pdist(Xb(indr,:)).^2;
```

```
                 D(r) = sum(ynr)/(2*nr);
           end
           Wb(b,k) = sum(D);
     end
end
```

The matrix **Wb** contains our $\log(W_k^*)$ values, one set for each row. The following code gets the gap statistics, as well as the observed and expected $\log(W_k)$.

```
% Find the mean and standard deviation
Wobs = log(W);
muWb = mean(log(Wb));
sdk = (B-1)*std(log(Wb))/B;
gap = muWb - Wobs;
% Find the weighted version.
sk = sdk*sqrt(1 + 1/B);
gapsk = gap - sk;
% Find the lowest one that is larger:
ineq = gap(1:9) - gapsk(2:10);
ind = find(ineq > 0);
khat = ind(1);
```

The estimated number of clusters is two, which corresponds to the correct number of cancer types. This can be compared to the results of Example 5.9, where the Mojena graphical rule indicated three clusters. In any event, the results of the gap statistic method seem to indicate that there is evidence to reject the hypothesis that there is only one group. We plot the gap curve in Figure 5.14, where we can see the strong maximum at $k = 2$.
❏

The gap statistic method relies on B random samples from the reference null distribution to estimate the expected value of $\log(W_k)$, each of which is clustered using the same procedure that was used to get the observed ones. The data analyst should keep in mind that this can be computationally intensive if the sample size is large and a clustering method such as agglomerative clustering is used.

5.7.6 Cluster Validity Indices

The silhouette and gap statistics are examples of approaches we might use to validate our clusters. These are known as *cluster validity indices*. We can use these to explore how well the cluster solution fits the data and to provide estimates of the number of groups. These indices are used when we do not have the correct classifications available for external validation; i.e., when the data are unlabeled.

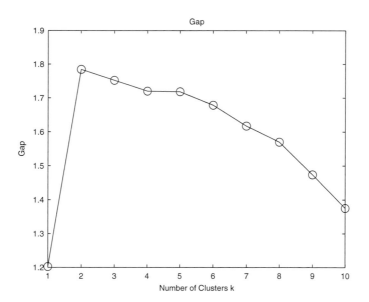

FIGURE 5.14
The upper panel shows the estimated expected and observed values of the $\log(W_k)$. The lower plot is the gap statistic curve, where we see a clear maximum at $k = 2$ clusters.

Most of the cluster validity indices measure two aspects of the partition: compactness and separation of the clusters. We could think of *compactness* as cluster cohesion or how similar data points are within each of the clusters. *Separation* is related to the distance between clusters. The measures are then combined using ratios or summations.

Arbelaitz et al. [2013] published an extensive study of thirty cluster validity indices that include the silhouette statistic and others. They found that the silhouette was among those that performed the best. In the remainder of this section, we describe additional superior cluster validity indices, as found by Arbelaitz et al. These include the Dunn and its variants, the Calinski-Harabasz (CH), and the Davies-Bouldin (DB). We follow the notation of Halkidi et al. [2001] in our discussion.

Dunn Index

The *Dunn index* was developed by Dunn [1973], and several variations of the index have been proposed in the literature [Pal and Biswas, 1997; Arbelaitz et al., 2013]. It measures the compactness or cohesion of a cluster as the maximum distance between all points within the cluster. The separation between two clusters is measured as the minimum distance between two data points, where one is from each of the clusters.

The Dunn index for a given partition of K clusters is defined below.

$$Dunn(K) = \min_i \left[\min_j \left\{ \frac{d(c_i, c_j)}{\max_k \{ \mathrm{diam}(c_k) \}} \right\} \right], \qquad (5.21)$$

for $i = 1, ..., K$, $j = i + 1, ..., K$, and $k = 1, ..., K$. The numerator $d(c_i, c_j)$ is the distance between the i-th and j-th clusters

$$d(c_i, c_j) = \min_{\mathbf{x} \in c_i, \mathbf{y} \in c_j} \{ d(\mathbf{x}, \mathbf{y}) \}.$$

The denominator $\mathrm{diam}(c_k)$ is the diameter of the k-th cluster given by

$$\mathrm{diam}(c_k) = \max_{\mathbf{x}, \mathbf{y} \in c_k}.$$

One can view $\mathrm{diam}(c_k)$ as a measure of the variation or dispersion of the observations in the k-th group.

If we have compact clusters that are separated well, then the Dunn index should be large. Cohesive clusters result in small diameters (denominator), and well-separated ones yield a large minimum distance between clusters (numerator). Thus, we seek cluster solutions corresponding to large values of the Dunn index.

Looking at Equation 5.21, we can see that quantifying the compactness of a cluster in this manner is sensitive to the presence of noisy data because it is based on the maximum interpoint distance in a given cluster (denominator). Pal and Biswas [1997] developed modifications to the Dunn index using different graph-based methods to quantify the cohesion or diameter of the clusters. These are known as the *generalized Dunn index*.

We illustrate the Dunn index based on the minimum spanning tree (MST). We first create a connected graph G_i from the data in the i-th cluster using their interpoint distances as the edge weights. Next, we find an MST for graph G_i. The largest edge weight in the MST is used as the diameter of the cluster, which is $\text{diam}(c_k)$ in Equation 5.21. Other variations of the generalized Dunn index find the neighborhood graphs G_i in different ways, such as a relative neighborhood graph (RNG) and a Gabriel graph. Other neighborhood graphs can also be used [Ilc, 2012].

A *relative neighborhood graph* was defined by Toussaint [1980]. Two data points are relative neighbors if they are "at least as close to each other as they are to any other point." Mathematically, we say that two points are relatively close if the distance between the i-th and j-th observations satisfy

$$d(\mathbf{x}_i, \mathbf{x}_j) \leq \max[d(\mathbf{x}_i, \mathbf{x}_k), d(\mathbf{x}_j, \mathbf{x}_k)],$$

for $k = 1, \ldots, n$, $k \neq i, j$. We then construct the relative neighborhood graph by putting an edge between points if and only if they are relative neighbors.

Two data points $\mathbf{x}_i$ and $\mathbf{x}_j$ are connected in a *Gabriel graph* if the squared distance between them is less than the sum of the squared distance between each of them and any other observation $\mathbf{x}_k$. Mathematically, we can state this as the following

$$d^2(\mathbf{x}_i, \mathbf{x}_j) < d^2(\mathbf{x}_i, \mathbf{x}_k) + d^2(\mathbf{x}_j, \mathbf{x}_k),$$

for all $k \neq i, j$.

A Gabriel graph is easiest to understand and visualize in 2–D. We first connect the i-th and j-th data points with a straight line, and a circle is constructed with this line as a diameter. Thus, the two data points lie on the circle. If no other observations lie within the circle, then we construct an edge between $\mathbf{x}_i$ and $\mathbf{x}_j$ in the Gabriel graph.

We can use the silhouette statistic, the gap statistic, the Dunn index, and other cluster validity indices we cover later to estimate the number of clusters in the data set by following this general process.

Procedure – Number of Clusters using a Cluster Validity Index

  1. Establish a maximum number of clusters K_{max}.

2. Determine the type of clustering (e.g., hierarchical, *k*-means, etc.), the associated parameters (e.g., distance, linkage, etc.), and a cluster validity index (e.g., silhouette, gap, Dunn, etc.).

3. Set $K = 2$, $K = 2, \ldots, K_{max}$.

4. Cluster the data and obtain the cluster IDs.

5. Calculate the validity index or statistic using the data and the cluster IDs.

6. Increment K and repeat from Step 4 until $K = K_{max}$.

7. The estimated number of groups in the data set corresponds to the number of clusters yielding either the maximum or minimum index from Step 5. Whether one uses the maximum or minimum value depends on the validity index.

We illustrate this procedure in the next example.

Example 5.12

We use a data set from Jain and Law [2005] to illustrate the above procedure with the Dunn index and its variants.

```
% First, load the data.
load jain
% Get the data from the matrix.
X = jain(:,1:2);
% View in scatterplot
gscatter(X(:,1),X(:,2),...
    jain(:,3),...
    'rg','o*')
```

The scatterplot of the data is shown in Figure 5.15, where we easily see two clusters. However, it might be difficult for a cluster method like *k*-means to find the right number of clusters. There is a package for calculating both the original and generalized Dunn index on MATLAB Central [Ilc, 2013]. We include it with the EDA Toolbox and use it below. We use *k*-means as our clustering method, and partition the data into $K = 2, \ldots, 10$ groups. We evaluate each clustering using the original Dunn index and the one based on RNG.

```
% Loop over values of k - number of clusters.
for k = 2:10
    idx = kmeans(X,k,...
        'replicates',5);
    % Save the cluster IDs.
    IDX(:,k) = idx(:);
    DunnI(k) = indexDN(X,idx(:),'euclidean');
    DunnIrng(k) = indexDNg(X,idx(:),'rng');
```

end

The best estimate of the number of clusters corresponds to the maximum value of the Dunn index. We plot the values in Figure 5.16.

```
% Plot the values of the indices.
subplot(2,1,1),plot(DunnI,'o-')
ylabel('Dunn Index')
subplot(2,1,2),plot(DunnIrng,'o-')
ylabel('Dunn Index - RNG')
```

We see that the original Dunn index has higher variation than the index using the RNG. This is likely due to the sensitivity the original Dunn index has to the estimated cluster diameters. There is a clear maximum at three clusters with the generalized Dunn index, and we use that partition in the scatterplot shown in the bottom of Figure 5.15.
❑

Calinski-Harabasz Index

Milligan and Cooper [1985] conducted one of the first comparison studies of cluster validation indices, and they found that one of the best is the *Calinski-Harabasz* [1974], at least for the data sets they used. As it turns out, Arbelaitz et al. [2013] came to the same conclusion. This index is a ratio-type estimator and is also known as the *variance ratio criterion*. The Calinski-Harabasz (CH) measures the compactness of a cluster as the average distance from points in the cluster to its centroid. Cluster separation is based on the distance from the cluster centers to the overall center of the data.

We can define the CH index for K groups as follows

$$CH(K) = \frac{n-K}{K-1} \times \frac{\sum_{k=1}^{K} n_k \|\bar{\mathbf{x}}_k - \bar{\mathbf{X}}\|^2}{\sum_{k=1}^{K} \sum_{\mathbf{x} \in C_k} \|\mathbf{x} - \bar{\mathbf{x}}_k\|^2}, \tag{5.22}$$

where n_k is the number of observations in the k-th cluster, and $\bar{\mathbf{x}}_k$ is its centroid. The overall mean of the data is denoted by $\bar{\mathbf{X}}$, and $\|\mathbf{x} - \mathbf{y}\|^2$ is the L^2 norm between two vectors $\mathbf{x}$ and $\mathbf{y}$.

The numerator of the second factor in Equation 5.22 is related to the overall between-cluster variance, and the denominator corresponds to the within-cluster variance. This index is similar in form to the Dunn index, with compactness and separation measured differently. Well-separated clusters would have large between-cluster variance (numerator), and clusters with good cohesion would have small values for the within-cluster variance

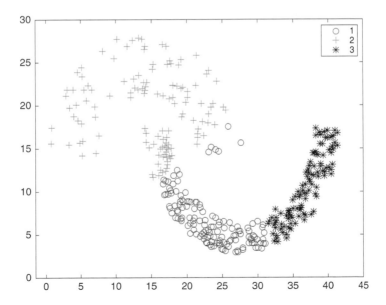

FIGURE 5.15
The top plot is a scatterplot of the **jain** data. This is a synthetic data set with two clusters. These clusters are easily seen by the human eye, but not so easily found with unsupervised learning methods. We applied two versions of the Dunn index to k-means partitions and found that the partition with three groups was the best fit. This grouping is shown in the bottom panel. See Figure 5.16 for the values of the Dunn index.

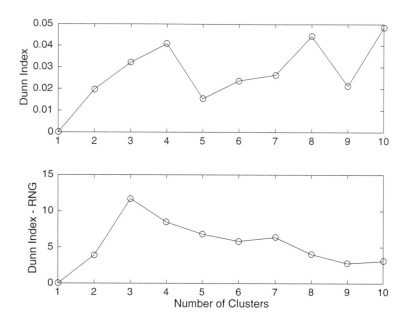

FIGURE 5.16

The top line plot shows the values for the original Dunn index as a function of the number of clusters used in k-means clustering (Euclidean distance). We see that there is considerable variation in the index, which is likely due to the way cluster diameters are estimated. The bottom plot is the generalized Dunn index using a relative neighborhood graph (RNG). There is a clear maximum at three clusters.

(denominator). Thus, we look for cluster solutions that produce high values of the CH index.

Davies-Bouldin Index

The **Davies-Bouldin** (DB) **index** [1979] is also based on a ratio of quantities that measure cohesion and separation. Like the CH index, it estimates cohesion as the distance from the points in the cluster to its centroid. The separation of the clusters is measured using the distance between cluster centroids. The DB index can be thought of as the average similarity of each cluster and the one it is closest or most similar to.

The Davies-Bouldin index is defined as

$$DB(K) = \frac{1}{K} \sum_{i=1}^{K} \max_{i \neq j} \left\{ \frac{\bar{d}_i + \bar{d}_j}{d_{i,j}} \right\}, \tag{5.23}$$

where $\bar{d}_i$ is the average distance of points in the i-th cluster to its centroid, and $d_{i,j}$ is the Euclidean distance between the i-th and j-th cluster centroids.

Because we take the maximum in Equation 5.23, we can think of this as the worst case for a cluster solution. Thus, a good cluster solution for a data set would correspond to the smallest value of the Davies-Bouldin index.

Example 5.13

We return to the data set from the previous example to illustrate the CH and DB cluster validity indices. First, make sure the data set is loaded.

```
% First, load the data.
load jain
% Get the data from the matrix.
X = jain(:,1:2);
```

The MATLAB Statistics Toolbox has a function for estimating and evaluating cluster solutions called **evalclusters**. There are several options for this function. For instance, we could have the **evalclusters** function partition the data using some values of K, a given cluster method, and validity index. It then returns the criterion values and the optimal K. Alternatively, we could find cluster solutions and obtain the desired validity index for each one using the **evalclusters** function. This is what we do next. We first get the cluster solutions using k-means.

```
% Loop over values of k - number of clusters.
for k = 1:10
    idx = kmeans(X,k,...
        'replicate',5);
    % Save the cluster IDs.
    IDX(:,k) = idx(:);
end
```

This produces an $n \times K$ matrix of cluster solutions. This is a convenient way to apply different cluster validity indices to the same groupings of the data, which is the approach we use below.

```
% Get the Calinski-Harabasz index for each solution.
CHcvi = evalclusters(X,IDX,'CalinskiHarabasz');
% Get the Davies-Bouldin index for each solution.
DBcvi = evalclusters(X,IDX,'DaviesBouldin');
```

We plot the cluster indices in Figure 5.17. The optimal number of clusters for the CH index is 9, and the DB index indicates 8 groups. We can find this by entering the output from **evalclusters** at the command line. This is what we get.

```
% Optimal K is at the maximum of the CH index.
CHcvi =

    CalinskiHarabaszEvaluation with properties:
```

```
    NumObservations: 373
         InspectedK: [1 2 3 4 5 6 7 8 9 10]
    CriterionValues: [1x10 double]
           OptimalK: 9
 % Optimal K is at the minimum of the DB index.
 DBcvi =

    DaviesBouldinEvaluation with properties:

    NumObservations: 373
         InspectedK: [1 2 3 4 5 6 7 8 9 10]
    CriterionValues: [1x10 double]
           OptimalK: 8
```

We plot the indices in Figure 5.17. It is also interesting to see the clusters we get from the optimal solutions. These are shown in Figure 5.18.
❏

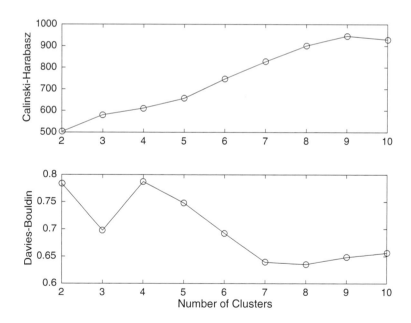

FIGURE 5.17
Here are line plots for the Calinski-Harabasz and Davies-Bouldin cluster validity indices. The optimal number of clusters of the CH index is the maximum, and the optimal number from the DB index corresponds to the number of groups with the minimum value. Thus, the CH indicates 9 groups is optimal, and the DB index suggests 8 groups. See Figure 5.18 for scatterplots showing these groupings.

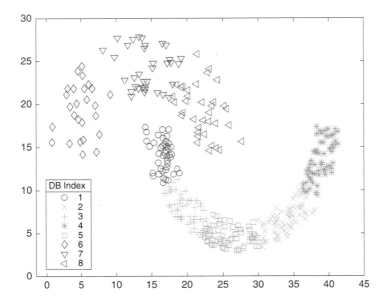

FIGURE 5.18

K-means clustering used to cluster the `jain` data. The top panel shows 9 groups (optimal *K* from CH index), and the bottom panel shows 8 groups (optimal *K* from the DB index).

5.8 Summary and Further Reading

Clustering is a technique that has been used in many diverse areas, such as biology, psychiatry, archaeology, geology, marketing, and others [Everitt, Landau, and Leese, 2001]. Because of this, clustering has come to be called different things, such as unsupervised learning [Duda and Hart, 1973; Jain, Murty, and Flynn, 1999], numerical taxonomy [Sneath and Sokal, 1973], and vector quantization [Hastie, Tibshirani, and Friedman, 2009].

In this chapter, we presented examples of the two most commonly used types of clustering: hierarchical methods and optimization-based methods. Hierarchical methods yield an entire sequence of nested partitions. On the other hand, optimization or partition methods, like k-means, group the data into k nonoverlapping data sets. Hierarchical methods use the $n(n-1)/2$ interpoint distances as inputs (note that some also require the data), while optimization methods just require the data, making them suitable for large data sets. In addition, we touched upon modern spectral-based methods, as well as nonnegative matrix factorization and probabilistic latent semantic analysis, which are suitable for document clustering.

In the next chapter, we describe another grouping technique called model-based clustering based on estimating finite mixture probability density functions. Note that some of the methods discussed in Chapter 3, such as self-organizing maps, generative topographic maps, and multidimensional scaling, can be considered a type of clustering, where the clusters are sometimes assessed visually. Since clustering methods (in most cases) will yield a grouping of the data, it is important to perform some validation or assessment of the cluster output. To that end, we presented several of these methods in this chapter.

We feel we should mention that, for the most part, we discussed clustering *methods* in this chapter. This can be contrasted with clustering *algorithms*, which are the underlying computational steps to achieve each of the clustering structures. For example, for any given hierarchical or optimization based method, many different algorithms are available that will achieve the desired result.

This is not meant to be an exhaustive treatment of the subject and much is left out; so we offer pointers to some additional resources. An excellent book to consult is Everitt, Landau, and Leese [2001], which has been updated to include some of the latest developments in clustering based on the classification likelihood and also on neural networks. It is relatively nonmathematical, and it includes many examples for the student or practitioner. For a good summary and discussion of methods for estimating the number of clusters, as well as for other clustering information, we recommend Gordon [1999]. Most books that focus strictly on clustering are Kaufman and Rousseeuw [1990], Jain and Dubes [1988], Späth [1980],

Hartigan [1975], and Anderberg [1973]. Other books on statistical pattern recognition usually include coverage of unsupervised learning or clustering. An excellent book like this is Webb [2002]. It is very readable and the author includes many applications and step-by-step algorithms. Of course, one of the seminal books in pattern recognition is Duda and Hart [1973], which has recently been updated to a newer edition [Duda, Hart, and Stork, 2001]. For more of the neural network point of view, one could consult Ripley [1996] or Hastie, Tibshirani, and Friedman [2009].

Some survey papers on clustering are available. For an overview from the machine learning point of view see Jain, Murty, and Flynn [1999]. The goal in their paper is to provide useful advice on clustering to a broad community of users. Another summary paper written by a *Panel on Discriminant Analysis, Classification, and Clustering* [1989] describes methodological and theoretical aspects of these topics. A presentation of clustering from an EDA point of view is Dubes and Jain [1980]. An early review paper on grouping is by Cormack [1971], where he provides a summary of distances, clustering techniques, and their various limitations. An interesting position paper on clustering algorithms from a data mining point of view is Estivill-Castro [2002].

Many books and survey papers have been written on text data mining and document clustering in recent years. Two excellent survey books on these methods were edited by Berry [2003] and Berry and Castellanos [2007].

For a survey and analysis of procedures for estimating the number of clusters, see Milligan and Cooper [1985]. One of the successful ones in their study uses a criterion based on the within-group sum-of-squares objective function, which was developed by Krzanowski and Lai [1988]. Roeder [1994] proposes a graphical technique for this purpose. Arbelaitz et al. [2013] describe a cluster validity index called COP. It is a ratio index with cohesion measured using the distance of points within a cluster to its centroid. The cluster separation is measured as the furthest neighbor distance. Tibshirani et al. [2001] develop an innovative method called prediction strength for validating clusters and assessing the number of groups, based on cross-validation. Bailey and Dubes [1982] develop something called cluster validity profiles that quantify the interaction between a cluster and its environment in terms of its compactness and isolation, thus providing more information than a single index would. One should apply several cluster methods, along with the appropriate cluster assessment and validation methods with each data set, to search for interesting and informative groups.

In his paper on PLSI, Hofmann [1999b] provides a generalization of the EM algorithm and calls it *tempered EM* (TEM). The basic idea behind TEM is to introduce a parameter that corresponds to a pseudo-temperature. This can be adjusted to control the optimization process and avoids overfitting in PLSI. This in turn leads to a method that will generalize and will work better with new data.

Exercises

5.1   Get the Euclidean distances for the **iris** data. Apply centroid linkage and construct the dendrogram. Do you get inversions? Do the same thing with Mahalanobis distance. Do you get similar results? Try some of the other distances and linkages. Do you still get inversions?

5.2   Apply single linkage hierarchical clustering to the following data sets. Do you get chaining? Explore some of the other types of distance/linkage combinations and compare results.

a. **geyser**

b. **singer**

c. **skulls**

d. **spam**

e. **sparrow**

f. **oronsay**

g. gene expression data sets

5.3   Construct the dendrogram for the partitions found in Problem 5.2. Is there evidence of groups or clusters in each of the data sets?

5.4   The inconsistency coefficient can be used to determine the number of clusters in hierarchical clustering. This compares the length of a link in the hierarchy with the average length of neighboring links. If the merge is consistent with those around it, then it will have a low inconsistency coefficient. A higher inconsistency coefficient indicates the merge is inconsistent and thus indicative of clusters. One of the arguments to the **cluster** function can be a threshold for the inconsistency coefficient corresponding to the **cutoff** argument. The cutoff point of the dendrogram occurs where links are greater than this value. The MATLAB Statistics Toolbox has a separate function called **inconsistent** that returns information in a matrix, where the last column contains the inconsistency coefficients. Generate some bivariate data containing two well-separated clusters. Apply a suitable hierarchical clustering method and construct the dendrogram. Obtain the output from the **inconsistent** function and use these values to get a threshold for the **cutoff** argument in **cluster**. Knowing that there are only two clusters, does the inconsistency coefficient give the correct result?

5.5   Apply the inconsistent threshold to get partitions for the hierarchical clustering in Problem 5.2. Where possible, construct scatterplots or a scatterplot matrix (using **plotmatrix** or **gplotmatrix**) of the resulting groups. Use different colors and/or symbols for the groups. Discuss your results.

5.6   Do a **help** on the **cophenet** function. Write your own MATLAB function to calculate the cophenetic coefficient comparing two dendrograms.

5.7   Generate 2–D uniform random variables. Apply the gap statistic procedure and plot the expected gap statistic and observed value for each k. Compare the curves. What is the estimated number of clusters?

5.8   Now generate bivariate normal random variables with two well-separated clusters. Apply the gap statistic procedure and plot the expected gap statistic and observed value for each k. What is the estimated number of clusters?

5.9   Write a MATLAB function to implement the gap-PC approach.

5.10   Apply the gap statistic procedure to the cluster results from Problem 5.2. Is there evidence for more than one group? How many clusters are present according to the gap statistic? Compare with your results in Problem 5.5.

5.11   Apply the upper tail rule and **mojenaplot** to the data and groupings found in Problem 5.2. How many clusters are present?

5.12   In what situation would the Rand index be zero? Use **adjrand** and **randind** functions to verify your answer.

5.13   Using the cluster results from one of the data sets in Problem 5.2, use the **adjrand** function with the input arguments the same. The value should be one. Do the same thing using **randind**.

5.14   Apply k-means and the agglomerative clustering method of your choice to the **oronsay** data set (both classifications), using the correct number of known groups. Use the silhouette plot and average silhouette values. Repeat this process varying the value for k. Discuss your results.

5.15   Using the same data and partitions in Problem 5.14, use the Rand index and adjusted Rand index to compare the estimated partitions with the true ones.

5.16   Repeat Example 5.8 using different distances and linkages. Discuss your results.

5.17   Repeat Example 5.11 using gap-PC method. Are the results different?

5.18   Hartigan [1985] also developed an index to estimate the number of clusters. This is given by

$$hart_k = \left[\frac{tr(\mathbf{S}_{W_k})}{tr(\mathbf{S}_{W_{k+1}})} - 1 \right] (n - k - 1).$$

The estimated number of clusters is the smallest value of k, where

$$1 \le k \le 10.$$

Implement this in MATLAB, apply it to the data used in Example 5.11, and compare with previous results.

5.19　The Fowlkes-Mallows [1983] index can be calculated using the matching matrix $\mathbf{N}$ (which in this case must be $g \times g$, where $g = g_1 = g_2$) as

$$B_g = T_g \div \sqrt{P_g Q_g},$$

where

$$T_g = \sum_{i=1}^{g} \sum_{j=1}^{g} n_{ij}^2 - n$$

$$P_g = \sum_{i=1}^{g} n_{i\cdot}^2 - n$$

$$Q_g = \sum_{j=1}^{g} n_{\cdot j}^2 - n$$

Implement this in a MATLAB function. Use the Fowlkes-Mallows index with the **oronsay** data, as in Problem 5.14 and compare with previous results.

5.20　In this chapter, we used the simple implementation of PLSA. The reader will now explore a more traditional implementation. First, download the PLSA code from this website

http://www.robots.ox.ac.uk/~vgg/software/

Unzip the Sivic [2010] PLSA code to a directory and apply it to the same data set **nmfclustex** from this chapter. Use the function **pLSA_EM** in the package. Compare the answers produced by this code with what was produced in the previous example.

5.21　Repeat Example 5.5 using three classes and assess the results.

5.22　Apply nonnegative matrix factorization clustering to the following data sets and evaluate your results.

　　a. **iris**

　　b. **oronsay** (both classifications)

5.23　Apply spectral clustering to the following data sets and evaluate your results using suitable clustering indexes.

　　a. **iris**

　　b. **leukemia**

　　c. **oronsay** (both classifications)

5.24　Apply MST clustering to the following data sets and evaluate your results using suitable clustering indexes.

　　a. **iris**

  b. **leukemia**

  c. **oronsay** (both classifications)

5.25 Generate a data set as we did in Example 5.6. Create a MST for your data. Explore various cluster solutions by removing different sets of edges from the tree.

Chapter 6

Model-Based Clustering

In this chapter, we present a method for clustering that is based on a finite mixture probability model. We first provide an overview of a comprehensive procedure for model-based clustering, so the reader has a framework for the methodology. We then describe the constituent techniques used in model-based clustering, such as finite mixtures, the EM algorithm, and model-based agglomerative clustering. We then put it all together and include more discussion on its use in EDA, clustering, probability density estimation, and discriminant analysis. Finally, we show how to use a GUI tool to generate random samples based on the finite mixture models presented in the chapter.

6.1 Overview of Model-Based Clustering

In Chapter 5, we discussed two main types of clustering—hierarchical and partition-based (k-means). The hierarchical method we discussed in detail was agglomerative clustering, where two groups are merged at each step, such that some criterion is optimized. We will also use agglomerative clustering in this chapter, where the objective function is now based on optimizing the classification likelihood function.

We mentioned several issues with the methods in Chapter 5. One problem is that the number of clusters must be specified in k-means or chosen later in agglomerative hierarchical clustering. We presented several ways to handle this problem (e.g., the gap statistic, Mojena's upper tail rule). The model-based clustering framework is another approach to address the issue of choosing the number of groups represented by the data. Another problem we mentioned is that many clustering methods are heuristic and impose a certain structure on the clusters. In other words, using Ward's method tends to produce same size, spherical clusters (as does k-means clustering). Additionally, some of the techniques are sensitive to outliers (e.g., Ward's method), and the statistical properties are generally unknown.

The model-based clustering methodology is based on probability models, such as the finite mixture model for probability densities. The idea of using probability models for clustering has been around for many years. See Bock [1996] for a survey of cluster analysis in a probabilistic and inferential framework. In particular, finite mixture models have been proposed for cluster analysis by Edwards and Cavalli-Sforza [1965], Day [1969], Wolfe [1970], Scott and Symons [1971], and Binder [1978]. Later researchers recognized that this approach can be used to address some of the issues in cluster analysis we just discussed [Fraley and Raftery, 2002; Everitt, Landau, and Leese, 2001; McLachlan and Peel, 2000; McLachlan and Basford, 1988; Banfield and Raftery, 1993].

The finite mixture approach assumes that the probability density function can be modeled as the sum of weighted component densities. As we will see shortly, when we use finite mixtures for cluster analysis, the clustering problem becomes one of estimating the parameters of the assumed mixture model, i.e., probability density estimation. Each component density corresponds to a cluster and posterior probabilities are used to determine cluster membership.

The most commonly used method for estimating the parameters of a finite mixture probability density is the Expectation-Maximization (EM) algorithm, which is based on maximum likelihood estimation [Dempster, Laird, and Rubin, 1977]. In order to apply the finite mixture–EM methodology, several issues must be addressed:

1. We must specify the number of component densities (or groups).[1]
2. The EM algorithm is iterative; so we need initial values of the parameters to get started.
3. We have to assume some form (e.g., multivariate normal, *t*-distribution, etc.) for the component densities.

The model-based clustering framework provides a principled way to tackle all of these problems.

Let's start with the second problem: initializing the EM algorithm. We use model-based agglomerative clustering for this purpose. Model-based agglomerative clustering [Murtagh and Raftery, 1984; Banfield and Raftery, 1993] uses the same general ideas of hierarchical agglomerative clustering, where all observations start out in a single group, and two clusters are merged at each step. However, in the model-based case, two clusters are merged such that the classification likelihood is maximized.[2] Recall that we get a complete set of nested partitions with hierarchical clustering. From this, we can obtain initial estimates of our component parameters based on a given partition from the hierarchical clustering.

This brings us to issue three, which is the form of the component densities. Banfield and Raftery [1993] devise a general framework for multivariate

---

[1] This would be similar to specifying the *k* in *k*-means clustering.
[2] The classification likelihood is similar to the mixture likelihood, except that each observation is allowed to belong to only one component density. See Section 6.4 for a discussion.

normal mixtures based on constraining the component covariance matrices. Imposing these constraints governs certain geometric properties of the clusters, such as orientation, volume, and shape. In model-based clustering, we assume that the finite mixture is composed of multivariate normal densities, where constraints on the component covariance matrices yield different models.

So, why is this called *model*-based clustering? This really pertains to issues (1) and (2), together, which are specifying the number of components and initializing the EM algorithm. We can consider the number of groups, combined with the form of the component densities, as producing different statistical models for the data. Determining the final model is accomplished by using the Bayesian Information Criterion (BIC), which is an approximation to Bayes factors [Schwarz, 1978; Kass and Raftery, 1995]. The model with the optimal BIC value is chosen as the 'best' model. The reader should be aware that we also use the word *model* to represent the type of constraints and geometric properties of the covariance matrices. Hopefully, the meaning of the word *model* will be clear from the context.

The steps of the model-based clustering framework are illustrated in Figure 6.1. First, we choose our constraints on the component covariance matrices (i.e., the model), and then we apply agglomerative model-based clustering. This provides an initial partition of our data for any given number of groups and any desired model. We use this partition to get initial estimates of our component density parameters for use in the EM algorithm. Once we converge to an estimate using the EM, we calculate the BIC, which will be defined later in the chapter. We continue to do this for various models (i.e., number of groups and forms of the covariance matrices). The final model that we choose is the one that produces the highest BIC. We now present more information and detail on the constituent parts of model-based clustering.

FIGURE 6.1
This shows the flow of the steps in the model-based clustering procedure.

6.2 Finite Mixtures

In this section, we provide more in-depth information regarding finite mixtures, as well as the different forms of the constrained covariance matrices used in model-based clustering. However, first we investigate an example of a *univariate* finite mixture probability density function to facilitate understanding of the multivariate approach.

Example 6.1

In this example, we show how to construct a probability density function that is the weighted sum of two univariate normal densities. First, we set up the various component parameters. We have equal mixing proportions or weights. The means are given by –2 and 2, and the corresponding variances are 0.25 and 1.44. Mathematically, we have

$$f(x; \pi_k, \mu_k, \sigma_k^2) = \sum_{k=1}^{2} \pi_k N(x; \mu_k, \sigma_k^2)$$

$$= 0.5 \times N(x; -2, 0.25) + 0.5 \times N(x; 2, 1.44)$$

where $N(x; \mu, \sigma^2)$ denotes a univariate normal probability density function with mean μ and variance σ^2. The following MATLAB code assigns the component density parameters.

```
% First the mixing coefficients will be equal.
pie1 = 0.5;
pie2 = 0.5;
% Let the means be -2 and 2.
mu1 = -2;
mu2 = 2;
```

The MATLAB function **normpdf** that we use below requires the standard deviation instead of the variance.

```
% The standard deviation of the first term is 0.5
% and the second one is 1.2.
sigma1 = 0.5;
sigma2 = 1.2;
```

Now we need to get a domain over which to evaluate the density, making sure that we have enough points to encompass the interesting parts of this density.

```
% Now generate a domain over which to evaluate the
% function.
```

```
x = linspace(-6, 6);
```

We use the Statistics Toolbox function **normpdf** to evaluate the component probability density functions. To get the finite mixture density, we need to weight these according to their mixing proportions and then add them together.

```
% The following is a Statistics Toolbox function.
y1 = normpdf(x,mu1,sigma1);
y2 = normpdf(x,mu2,sigma2);
% Now weight and add to get the final curve.
y = pie1*y1 + pie2*y2;
% Plot the final function.
plot(x,y)
xlabel('x'), ylabel('Probability Density Function')
title('Univariate Finite Mixture - Two Terms')
```

The plot is shown in Figure 6.2. The two terms are clearly visible, but this is not always the case. In the cluster analysis context, we would speculate that one group is located at –2 and another at 2.
❑

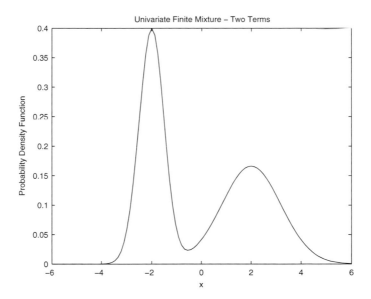

FIGURE 6.2
This shows the univariate probability density function as described in Example 6.1.

6.2.1 Multivariate Finite Mixtures

The *finite mixture* approach to probability density estimation (and cluster analysis) can be used for both univariate and multivariate data. In what follows, we will concern ourselves with the multivariate case only.

Finite mixtures encompass a family of probability density functions that are a weighted sum of component densities. The form of the density is given by

$$f(\mathbf{x}; \pi_k, \theta_k) = \sum_{k=1}^{c} \pi_k \, g_k(\mathbf{x}; \theta_k). \tag{6.1}$$

The *component density* is denoted by $g_k(\mathbf{x}; \theta_k)$ with associated parameters represented by θ_k. Note that θ_k is used to denote any type and number of parameters. The *weights* are given by π_k, with the constraint that they are nonnegative and sum to one. These weights are also called the *mixing proportions* or *mixing coefficients*.

Say we want to use the finite mixture in Equation 6.1 as a model for the distribution that generated our data. Then, to estimate the density, we must know the number of components c, and we need to have a form for the function g_k.

The component densities can be any *bona fide* probability density, but one of the most commonly used ones is the multivariate normal. This yields the following equation for a multivariate Gaussian finite mixture

$$f(\mathbf{x}; \pi_k, \mu_k, \Sigma_k) = \sum_{k=1}^{c} \pi_k \, N(\mathbf{x}; \mu_k, \Sigma_k), \tag{6.2}$$

where $N(\mathbf{x}; \mu_k, \Sigma_k)$ represents a multivariate normal probability density function given by

$$N(\mathbf{x}; \mu_k, \Sigma_k) = \frac{\exp\left\{ -\frac{1}{2}(\mathbf{x}_i - \mu_k)^T \Sigma_k^{-1} (\mathbf{x}_i - \mu_k) \right\}}{(2\pi)^{p/2} \sqrt{|\Sigma_k|}},$$

where $|\cdot|$ represents the determinant of a matrix. It is apparent from Equation 6.2 that this model has parameters π_k, μ_k (each one is a p-dimensional vector of means), and Σ_k (each is a $p \times p$ covariance matrix).

Now that we have a form for the component densities, we know what we have to estimate using our data. We need to estimate the weights π_k, the p-dimensional means for each term, and the covariance matrices. Before we describe how to do this using the EM algorithm, we first look at the form of the multivariate normal densities in a little more detail.

6.2.2 Component Models — Constraining the Covariances

Banfield and Raftery [1993] and Celeux and Govaert [1995] provide the following eigenvalue decomposition of the k-th covariance matrix

$$\Sigma_k = \lambda_k \mathbf{D}_k \mathbf{A}_k \mathbf{D}_k^T. \tag{6.3}$$

The factors in Equation 6.3 are given by the following:

- The volume of the k-th cluster or component density is governed by the λ_k, which is proportional to the volume of the standard deviation ellipsoid. We note that the volume is different from the size of a cluster. The *size* is the number of observations falling into the cluster, while the *volume* is the amount of space encompassed by the cluster.
- $\mathbf{D}_k$ is a matrix with columns corresponding to the eigenvectors of Σ_k. It determines the *orientation* of the cluster.
- $\mathbf{A}_k$ is a diagonal matrix. $\mathbf{A}_k$ contains the normalized eigenvalues of Σ_k along the diagonal. By convention, they are arranged in decreasing order. This matrix is associated with the *shape* of the distribution.

We now describe the various models in more detail, using the notation and classification of Celeux and Govaert. The eigenvalue decomposition given in Equation 6.3 produces a total of 14 models, by keeping factors λ_k, $\mathbf{A}_k$, and $\mathbf{D}_k$ constant across terms and/or restricting the form of the covariance matrices (e.g., restrict it to a diagonal matrix). There are three main families of models: the spherical family, the diagonal family, and the general family.

Celeux and Govaert provide covariance matrix update equations based on these models for use in the EM algorithm. Some of these have a closed form, and others must be solved in an iterative manner. We concern ourselves only with the nine models that have a closed form update for the covariance matrix. Thus, what we present in this text is a subset of the available models. These are summarized in Table 6.1.

Spherical Family

The models in this family are characterized by diagonal matrices, where each diagonal element of the covariance matrix Σ_k has the same value. Thus, the distribution is spherical; i.e., each variable of the component density has the

TABLE 6.1

Description of Multivariate Normal Mixture Models with Closed Form Solution to Covariance Matrix Update Equation in EM Algorithm

Model #	Covariance	Distribution	Description
1	$\Sigma_k = \lambda \mathbf{I}$	Family: Spherical Volume: Fixed Shape: Fixed Orientation: NA	• Diagonal covariance matrices • Diagonal elements are equal • Covariance matrices are equal • $\mathbf{I}$ is a $p \times p$ identity matrix
2	$\Sigma_k = \lambda_k \mathbf{I}$	Family: Spherical Volume: Variable Shape: Fixed Orientation: NA	• Diagonal covariance matrices • Diagonal elements are equal • Covariance matrices may vary • $\mathbf{I}$ is a $p \times p$ identity matrix
3	$\Sigma_k = \lambda \mathbf{B}$	Family: Diagonal Volume: Fixed Shape: Fixed Orientation: Axes	• Diagonal covariance matrices • Diagonal elements may be unequal • Covariance matrices are equal • $\mathbf{B}$ is a diagonal matrix
4	$\Sigma_k = \lambda \mathbf{B}_k$	Family: Diagonal Volume: Fixed Shape: Variable Orientation: Axes	• Diagonal covariance matrices • Diagonal elements may be unequal • Covariance matrices may vary among components • $\mathbf{B}$ is a diagonal matrix
5	$\Sigma_k = \lambda_k \mathbf{B}_k$	Family: Diagonal Volume: Variable Shape: Variable Orientation: Axes	• Diagonal covariance matrices • Diagonal elements may be unequal • Covariance matrices may vary among components • $\mathbf{B}$ is a diagonal matrix
6	$\Sigma_k = \lambda \mathbf{D} \mathbf{A} \mathbf{D}^T$	Family: General Volume: Fixed Shape: Fixed Orientation: Fixed	• Covariance matrices can have nonzero off-diagonal elements • Covariance matrices are equal
7	$\Sigma_k = \lambda \mathbf{D}_k \mathbf{A} \mathbf{D}_k^T$	Family: General Volume: Fixed Shape: Fixed Orientation: Variable	• Covariance matrices can have nonzero off-diagonal elements • Covariance matrices may vary among components
8	$\Sigma_k = \lambda \mathbf{D}_k \mathbf{A}_k \mathbf{D}_k^T$	Family: General Volume: Fixed Shape: Variable Orientation: Variable	• Covariance matrices can have nonzero off-diagonal elements • Covariance matrices may vary among components
9	$\Sigma_k = \lambda_k \mathbf{D}_k \mathbf{A}_k \mathbf{D}_k^T$	Family: General Volume: Variable Shape: Variable Orientation: Variable	• Covariance matrices can have nonzero off-diagonal elements • Covariance matrices may vary among components

same variance. We have two closed-form cases in this family, each corresponding to a fixed spherical shape.

The first has equal volume across all components, so the covariance is of the form

$$\Sigma_k = \lambda \mathbf{DAD}^T = \lambda \mathbf{I},$$

where $\mathbf{I}$ is the $p \times p$ identity matrix. The other model allows the volumes to vary. In this case, Equation 6.3 becomes

$$\Sigma_k = \lambda_k \mathbf{DAD}^T = \lambda_k \mathbf{I}.$$

Celeux and Govaert note that these models are invariant under any isometric transformation.

Example 6.2

We will generate a data set according to model 2 in Table 6.1. Here we have the spherical model, where the covariance matrices are allowed to vary among the component densities. The data set has $n = 250$ data points in 3–D, and we choose to have $c = 2$ component densities. The parameters for the multivariate normal components are given by

$$\mu_1 = [2, 2, 2]^T \qquad \mu_2 = [-2, -2, -2]^T$$
$$\Sigma_1 = \mathbf{I} \qquad \Sigma_2 = 2\mathbf{I}$$
$$\pi_1 = 0.7 \qquad \pi_2 = 0.3$$
$$\lambda_1 = 1 \qquad \lambda_2 = 2$$

We use the **genmix** GUI to generate the sample. See the last section of this chapter for information on how to use this tool. Once we generate the data, we can display it in a scatterplot matrix, as in Figure 6.3. Note that the second component centered at $[-2, -2, -2]^T$ has fewer points, but seems to have a larger volume.
❑

Diagonal Family

The models in this family are also diagonal, but now the elements along the diagonal of the covariance matrix are allowed to be different. The covariances are of the form

$$\Sigma = \lambda \mathbf{B} = \lambda \mathbf{DAD}^T,$$

FIGURE 6.3
This shows a scatterplot matrix of the data set randomly generated according to model 2, which is a member of the spherical family.

where

$$\mathbf{B} = \mathrm{diag}(b_1, \ldots, b_p),$$

and $|\mathbf{B}| = 1$. The matrix $\mathbf{B}$ determines the shape of the cluster.

The cluster shapes arising in this family of models are elliptical, because the variance in each of the dimensions is allowed to be different. Celeux and Govaert mention that the models in the diagonal family are invariant under any scaling of the variables. However, invariance does not hold for these models under all linear transformations.

We include three of the models in this family. The first is where each component covariance matrix is equal, with the same volume and shape:

$$\Sigma_k = \lambda \mathbf{B}.$$

Next we have one where the covariance matrices are allowed to vary in shape, but not in volume:

$$\Sigma_k = \lambda \mathbf{B}_k.$$

Finally, we allow both volume and shape to vary between the components or clusters:

$$\Sigma_k = \lambda_k \mathbf{B}_k.$$

Example 6.3

To illustrate an example from the diagonal family, we again use **genmix** to generate random variables according to the following model. We will use $n = 250$, $p = 2$, and $c = 2$. The means, weights, and covariance are given by

$$\mu_1 = [2, 2]^T \qquad \mu_2 = [-2, -2]^T$$

$$\mathbf{B}_1 = \begin{bmatrix} 3 & 0 \\ 0 & \frac{1}{3} \end{bmatrix} \qquad \mathbf{B}_2 = \begin{bmatrix} \frac{1}{2} & 0 \\ 0 & 2 \end{bmatrix}$$

$$\pi_1 = 0.7 \qquad \pi_2 = 0.3$$

$$\lambda_1 = 1 \qquad \lambda_2 = 1$$

This is model 4, which is equal volume, but different shapes. A scatterplot matrix of a random sample generated according to this distribution is shown in Figure 6.4.
❏

General Family

This family includes the more general cases for each cluster or component density. The covariances are no longer constrained to be diagonal. In other words, the off-diagonal terms of the matrices can be nonzero. The models in the general family are invariant under any linear transformation of the data.

We include four models from this family. As usual, the first one has all covariances constrained to be equal. This means that the clusters have fixed shape, volume, and orientation:

$$\Sigma_k = \lambda \mathbf{D} \mathbf{A} \mathbf{D}^T.$$

Next we allow the orientation to change, but keep the volume and shape fixed:

$$\Sigma_k = \lambda \mathbf{D}_k \mathbf{A} \mathbf{D}_k^T.$$

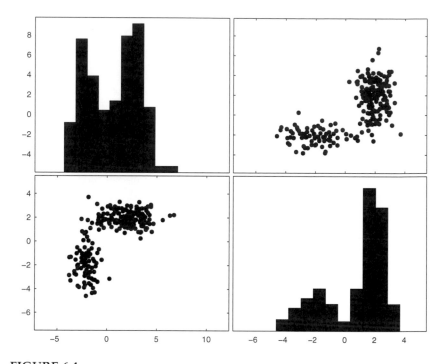

FIGURE 6.4
These data were generated according to a distribution that corresponds to model 4 of the diagonal family.

Then we allow both shape and orientation to vary, while keeping the volume fixed:

$$\Sigma_k = \lambda \mathbf{D}_k \mathbf{A}_k \mathbf{D}_k^T .$$

Finally, we have the unconstrained version, where nothing is fixed, so shape, volume and orientation are allowed to vary:

$$\Sigma_k = \lambda_k \mathbf{D}_k \mathbf{A}_k \mathbf{D}_k^T .$$

The unconstrained version is the one typically used in finite mixture models with multivariate normal components.

Example 6.4
The general family of models requires a full covariance matrix for each component density. We illustrate this using the following model ($n = 250$, $p = 2$, and $c = 2$):

$$\mu_1 = [2, 2]^T \qquad \mu_2 = [-2, -2]^T$$

$$\mathbf{A}_1 = \begin{bmatrix} 3 & 0 \\ 0 & \frac{1}{3} \end{bmatrix} \qquad \mathbf{A}_2 = \begin{bmatrix} \frac{1}{2} & 0 \\ 0 & 2 \end{bmatrix}$$

$$\mathbf{D}_1 = \begin{bmatrix} \cos(\pi \div 8) & -\sin(\pi \div 8) \\ \sin(\pi \div 8) & \cos(\pi \div 8) \end{bmatrix} \qquad \mathbf{D}_2 = \begin{bmatrix} \cos(6\pi \div 8) & -\sin(6\pi \div 8) \\ -\sin(6\pi \div 8) & \cos(6\pi \div 8) \end{bmatrix}$$

$$\pi_1 = 0.7 \qquad\qquad \pi_2 = 0.3$$

$$\lambda_1 = 1 \qquad\qquad \lambda_2 = 1$$

Note that this is model 8 that corresponds to fixed volume, variable shape, and orientation. We have to multiply these together as in Equation 6.3 to get the covariance matrices for use in the **genmix** tool. These are given below, rounded to the fourth decimal place.

$$\Sigma_1 = \begin{bmatrix} 2.6095 & 0.9428 \\ 0.9428 & 0.7239 \end{bmatrix} \qquad \Sigma_2 = \begin{bmatrix} 1.6667 & -1.3333 \\ -1.3333 & 1.6667 \end{bmatrix}.$$

A scatterplot matrix showing a data set generated from this model is shown in Figure 6.5.
❏

Now that we know more about the different models we might have for multivariate normal finite mixtures, we need to look at how we can use our data to get estimates of the parameters. The EM algorithm is used for this purpose.

6.3 Expectation-Maximization Algorithm

The problem of estimating the parameters in a finite mixture has been studied for many years. The approach we present here is called the *Expectation-Maximization* (EM) method. This is a general method for optimizing likelihood functions and is useful in situations where data might be missing or simpler optimization methods fail. The seminal paper on this method is by Dempster, Laird, and Rubin [1977], where they formalize the EM algorithm and establish its properties. Redner and Walker [1984] apply it to finite mixture probability density estimation. The EM methodology is now

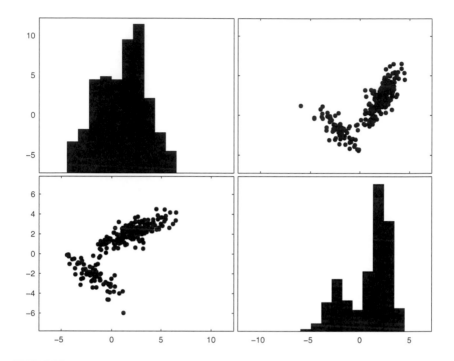

FIGURE 6.5
This shows a data set generated according to model 8 of the general family. These two components have equal volumes, but different shapes and orientations.

a standard tool for statisticians and is used in many applications besides finite mixture estimation.

We wish to estimate the parameters:

$$\theta = \pi_1, \dots, \pi_{c-1}, \mu_1, \dots, \mu_c, \Sigma_1, \dots, \Sigma_c.$$

Using the maximum likelihood approach, we maximize the log-likelihood given by

$$L(\theta \mid \mathbf{x}_1, \dots, \mathbf{x}_n) = \sum_{i=1}^{n} \ln \left[\sum_{k=1}^{c} \pi_k \, N(\mathbf{x}_i; \mu_k, \Sigma_k) \right]. \qquad (6.4)$$

We assume that the components exist in a fixed proportion in the mixture, given by the π_k. Thus, it makes sense to calculate the probability that a particular point $\mathbf{x}_i$ belongs to one of the component densities. It is this component membership (or cluster membership) that is unknown, and the reason why we need to use something like the EM algorithm to maximize

Equation 6.4. We can write the *posterior probability* that an observation $\mathbf{x}_i$ belongs to component k as

$$\hat{\tau}_{ik}(\mathbf{x}_i) = \frac{\hat{\pi}_k N(\mathbf{x}_i; \hat{\mu}_k, \hat{\Sigma}_k)}{\hat{f}(\mathbf{x}_i; \hat{\pi}_k, \hat{\mu}_k, \hat{\Sigma}_k)}; \qquad k = 1, ..., c; \; i = 1, ..., n, \qquad (6.5)$$

where

$$\hat{f}(\mathbf{x}_i; \hat{\pi}_k, \hat{\mu}_k, \hat{\Sigma}_k) = \sum_{k=1}^{c} \hat{\pi}_k N(\mathbf{x}_i; \hat{\mu}_k, \hat{\Sigma}_k).$$

For those readers not familiar with the notation, the *hat* or caret above the parameters denotes an estimate. So, the posterior probability in Equation 6.5 is really an estimate based on estimates of the parameters.

Recall from calculus, that to maximize the function given in Equation 6.4, we must find the first partial derivatives with respect to the parameters in θ and then set them equal to zero. These are called the likelihood equations, and they are not provided here. Instead, we provide the solution [Everitt and Hand, 1981] to the likelihood equations as follows

$$\hat{\pi}_k = \frac{1}{n} \sum_{i=1}^{n} \hat{\tau}_{ik} \qquad (6.6)$$

$$\hat{\mu}_k = \frac{1}{n} \sum_{i=1}^{n} \frac{\hat{\tau}_{ik} \mathbf{x}_i}{\hat{\pi}_k} \qquad (6.7)$$

$$\hat{\Sigma}_k = \frac{1}{n} \sum_{i=1}^{n} \frac{\hat{\tau}_{ik}(\mathbf{x}_i - \hat{\mu}_k)(\mathbf{x}_i - \hat{\mu}_k)^T}{\hat{\pi}_k}. \qquad (6.8)$$

The update for the covariance matrix given in Equation 6.8 is for the unconstrained case described in the previous section. This is model 9 in Table 6.1. We do not provide the update equations for the other models in this text. However, they are implemented in the model-based clustering function that comes with the EDA Toolbox. Please see Celeux and Govaert [1995] for a complete description of the update equations for all models.

Because the posterior probability (Equation 6.5) is unknown, we need to solve these equations in an iterative manner. It is a two step process, consisting of an E-Step and an M-Step, as outlined below. These two steps are repeated until the estimated values converge.

E-Step

We calculate the posterior probability that the i-th observation belongs to the k-th component, given the current values of the parameters. This is given by Equation 6.5.

M-Step

Update the parameter estimates using the estimated posterior probability and Equations 6.6 through 6.8 (or use the covariance update equation for the specified model).

Note that the E-Step allows us to weight the component parameter updates in the M-Step according to the probability that the observation belongs to that component density.

Hopefully, it is obvious to the reader now why we need to have a way to initialize the EM algorithm, in addition to knowing how many terms or components are in our finite mixture. It is known that the likelihood surface typically has many modes, and the EM algorithm may even diverge, depending on the starting point. However, the EM can provide improved estimates of our parameters if the starting point is a good one [Fraley and Raftery, 2002]. The discovery that partitions based on model-based agglomerative clustering provide a good starting point for the EM algorithm was first discussed in Dasgupta and Raftery [1998].

Procedure – Estimating Finite Mixtures

1. Determine the number of terms or component densities c in the mixture.
2. Determine an initial guess at the component parameters. These are the mixing coefficients, means, and covariance matrices for each multivariate normal density.
3. For each data point x_i, calculate the posterior probability using Equation 6.5 and the current values of the parameters. This is the E-Step.
4. Update the mixing coefficients, the means, and the covariance matrices for the individual components using Equations 6.6 through 6.8. This is the M-Step. Note that in the model-based clustering approach, we use the appropriate covariance update for the desired model (replacing Equation 6.8).
5. Repeat steps 3 through 4 until the estimates converge.

Typically, step 5 is implemented by continuing the iteration until the changes in the estimates at each iteration are less than some pre-set tolerance. Alternatively, one could keep iterating until the likelihood function converges. Note that with the EM algorithm, we use the entire data set to simultaneously update the parameter estimates at each step.

Example 6.5

Since the MATLAB code can be rather complicated for the EM algorithm, we do not provide the code in this example. Instead, we show how to use a function called **mbcfinmix** that returns the weights, means, and covariances given initial starting values. The general syntax for this is

```
[pies,mus,vars]=...
     mbcfinmix(X,muin,varin,wtsin,model);
```

The input argument **model** is the number from Table 6.1. We will use the data generated in Example 6.4. We saved the data to the workspace as variable **X**. The following commands set up some starting values for the EM algorithm.

```
% Need to get initial values of the parameters.
piesin = [0.5, 0.5];
% The musin argument is a matrix of means,
% where each column is a p-D mean.
musin = [ones(2,1), -1*ones(2,1)];
% The varin argument is a 3-D array, where
% each page corresponds to one of the
% covariance matrices.
varin(:,:,1) = 2*eye(2);
varin(:,:,2) = eye(2);
```

Note that our initial values are sensible, but they do need adjusting. We call the EM algorithm with the following:

```
% Now call the function.
[pie,mu,vars]=mbcfinmix(X,musin,varin,piesin,8);
```

The final estimates (rounded to four decimal places) are shown below.

```
pie = 0.7188    0.2812
mu =
     2.1528   -1.7680
     2.1680   -2.2400
vars(:,:,1) =
     2.9582    1.1405
     1.1405    0.7920
vars(:,:,2) =
     1.9860   -1.3329
    -1.3329    1.4193
```

We see that the estimates are reasonable ones. We now show how to construct a surface based on the estimated finite mixture. The resulting plot is shown in Figure 6.6.

```
% Now create a surface for the density.
% Get a domain over which to evaluate the function.
x1 = linspace(-7,7,50);
x2 = linspace(-7,5,50);
[X1,X2] = meshgrid(x1,x2);
% The X1 and X2 are matrices. We need to
% put them as columns into another one.
dom = [X1(:), X2(:)];
% Now get the multivariate normal pdf at
% the domain values - for each component.
% The following function is from the
% Statistics Toolbox.
Y1 = mvnpdf(dom,mu(:,1)',vars(:,:,1));
Y2 = mvnpdf(dom,mu(:,2)',vars(:,:,2));
% Weight them and add to get the final function.
y = pie(1)*Y1 + pie(2)*Y2;
% Need to reshape the Y vector to get a matrix
% for plotting as a surface plot.
[m,n] = size(X1);
Y = reshape(y,m,n);
surf(X1,X2,Y)
axis([-7 7 -7 5 0 0.12])
xlabel('X_1'),ylabel('X_2')
zlabel('PDF')
```

We compare this to the scatterplot of the random sample shown in Figure 6.5. We see that the surface matches the density of the data.
❑

6.4 Hierarchical Agglomerative Model-Based Clustering

We now describe another major component of the model-based clustering process, one that enables us to find initial values of our parameters for any given number of groups. *Agglomerative model-based clustering* works in a similar manner to the hierarchical methods discussed in the previous chapter, except that we do not have any notion of distance. Instead, we work with the *classification likelihood* as our objective function. This is given by

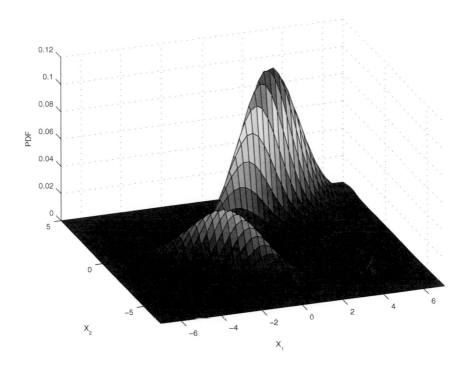

FIGURE 6.6
In Example 6.5, we estimated the density using the data from Figure 6.5 and the EM algorithm. This surface plot represents the estimated function. Compare to the scatterplot of the data in Figure 6.5.

$$l_{CL}(\theta_k, \gamma_i; \mathbf{x}_i) = \prod_{i=1}^{n} f_{\gamma_i}(\mathbf{x}_i; \theta_{\gamma_i}) , \tag{6.9}$$

where γ_i is a label indicating a classification for the i-th observation. We have $\gamma_i = k$, if $\mathbf{x}_i$ belongs to the k-th component. In the mixture approach, the number of observations in each component has a multinomial distribution with sample size of n, and probability parameters given by $\pi_1, ..., \pi_c$.

Model-based agglomerative clustering is a way to approximately maximize the classification likelihood. We start with singleton clusters consisting of one point. The two clusters producing the largest increase in the classification likelihood are merged at each step. This process continues until all observations are in one group. Note that the form of the objective function is adjusted as in Fraley [1998] to handle singleton clusters.

In theory, we could use any of the nine models in Table 6.1, but previous research indicates that the unconstrained model (number 9) with agglomerative model-based clustering yields reasonable initial partitions for the EM algorithm and any model in Table 6.1. Thus, our MATLAB implementation of agglomerative model-based clustering includes just the unconstrained model. Fraley [1998] provides efficient algorithms for the four basic models (1, 2, 6, 9) and shows how the techniques developed for these can be extended to the other models.

Example 6.6

The function for agglomerative model-based clustering is called **agmbclust**. It takes the data matrix **X** and returns the linkage matrix **Z**, which is in the same form as the output from MATLAB's **linkage** function. Thus, the output from **agmbclust** can be used with other MATLAB functions that expect this matrix. We will use the familiar **iris** data set for this example.

```
% Load up the data and put into a data matrix.
load iris
X = [setosa; versicolor; virginica];
% Then call the function for agglomerative MBC.
Z = agmbclust(X);
```

We can show the results in a dendrogram, as in Figure 6.7 (top). Two groups are obvious, but three groups might also be reasonable. We use the silhouette function from the previous chapter to assess the partition for three groups.

```
% We can apply the silhouette procedure for this
% after we find a partition. Use 3 groups.
cind = cluster(Z,3);
[S,H] = silhouette(X,cind);
```

The silhouette plot is shown in Figure 6.7 (bottom). We see one really good cluster (number three). The others have small values and even negative ones. The average silhouette value is 0.7349.
❑

6.5 Model-Based Clustering

The *model-based clustering* framework consists of three major pieces:

1. Initialization of the EM algorithm using partitions from model-based agglomerative clustering.
2. Maximum likelihood estimation of the parameters via the EM algorithm.
3. Choosing the model and number of clusters according to the BIC approximation of Bayes factors.

FIGURE 6.7

The top figure is the dendrogram derived from agglomerative model-based clustering of the **iris** data. We partition into three groups and then get their silhouette values. The corresponding silhouette plot is shown in the bottom panel.

We have already discussed the first two of these in detail; so we now turn our attention to Bayes factors and the BIC.

The Bayesian approach to model selection started with Jeffreys [1935, 1961]. Jeffreys developed a framework for calculating the evidence in favor of a null hypothesis (or model) using the Bayes factor, which is the posterior odds of one hypothesis when the prior probabilities of the two hypotheses are equal [Kass and Raftery, 1995].

Let's start with a simple two model case. We have our data $\mathbf{X}$, which we assume to have arisen according to either model M_1 or M_2. Thus, we have probability densities $p(\mathbf{X}|M_1)$ or $p(\mathbf{X}|M_2)$ and prior probabilities $p(M_1)$ and $p(M_2)$. From Bayes Theorem, we obtain the posterior probability of hypothesis M_g given data $\mathbf{X}$,

$$p(M_g|\mathbf{X}) = \frac{p(M_g)p(\mathbf{X}|M_g)}{p(M_1)p(\mathbf{X}|M_1) + p(M_2)p(\mathbf{X}|M_2)}, \qquad (6.10)$$

for $g = 1, 2$. If we take the ratio of the two probabilities in Equation 6.10 for each of the models, then we get

$$\frac{p(M_1|\mathbf{X})}{p(M_2|\mathbf{X})} = \frac{p(\mathbf{X}|M_1)}{p(\mathbf{X}|M_2)} \times \frac{p(M_1)}{p(M_2)}. \qquad (6.11)$$

The *Bayes factor* is the first factor in Equation 6.11:

$$B_{12} = \frac{p(\mathbf{X}|M_1)}{p(\mathbf{X}|M_2)}.$$

If any of the models contain unknown parameters, then the densities $p(\mathbf{X}|M_g)$ are obtained by integrating over the parameters. In this case, the resulting quantity $p(\mathbf{X}|M_g)$ is called the *integrated likelihood* of model M_g. This is given by

$$p(\mathbf{X}|M_g) = \int p(\mathbf{X}|\theta_g, M_g)p(\theta_g|M_g)d\theta_g \ , \qquad (6.12)$$

for $g = 1, ..., G$.

The natural thing to do would be to choose the model that is most likely, given the data. If the prior probabilities are equal, then this simply means that we choose the model with the highest integrated likelihood $p(\mathbf{X}|M_g)$. This procedure is relevant for model-based clustering, because it can be applied when there are more than two models, as well as being a Bayesian solution [Fraley and Raftery, 1998, 2002; Dasgupta and Raftery, 1998].

One of the problems with using Equation 6.12 is the need for the prior probabilities $p(\theta_g|M_g)$. For models that satisfy certain regularity conditions, the logarithm of the integrated likelihood can be approximated by the **Bayesian Information Criterion** (BIC), given by

$$p(\mathbf{X} \mid M_g) \approx BIC_g = 2\log p(\mathbf{X} \mid \hat{\theta}_g, M_g) - m_g\log(n),\qquad(6.13)$$

where m_g is the number of independent parameters that must be estimated for model M_g. It is well known that finite mixture models fail to satisfy the regularity conditions to make Equation 6.13 valid. However, previous applications and studies show that the use of the BIC in model-based clustering produces reasonable results [Fraley and Raftery, 1998; Dasgupta and Raftery, 1998].

Now that we have all of the pieces of model-based clustering, we offer the following procedure that puts it all together.

Procedure – Model-Based Clustering

1. Using the unconstrained model, apply the agglomerative model-based clustering procedure to the data. This provides a partition of the data for any given number of clusters.
2. Choose a model, M (see Table 6.1).
3. Choose a number of clusters or component densities, c.
4. Find a partition with c groups using the results of the agglomerative model-based clustering (step 1).
5. Using this partition, find the mixing coefficients, means, and covariances for each cluster. The covariances are constrained according to the model chosen in step 2.
6. Using the chosen c (step 3) and the initial values (step 5), apply the EM algorithm to obtain the final estimates.
7. Calculate the value of the BIC for this value of c and M:

$$BIC_g = 2L_M(\mathbf{X}, \hat{\theta}) - m_g\log(n),$$

   where L_M is the log likelihood, given the data, the model M, and the estimated parameters $\hat{\theta}$ (Equation 6.13).
8. Go to step 3 to choose another value of c.
9. Go to step 2 to choose another model M.
10. Choose the best configuration (number of clusters c and form of the covariance matrices) that corresponds to the highest BIC.

We see from this procedure that model-based clustering assumes various models for the covariance matrices and numbers of clusters, performs the cluster analysis, and then chooses the most likely clustering. So, it is an exhaustive search over the space of models that are both interesting and available. If we were interested in looking for 1 to C groups using all nine of the models, then we would have to perform the procedure $C \times 9$ times. As one might expect, this can be computationally intensive.

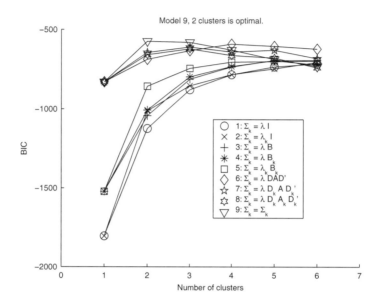

FIGURE 6.8
This shows the BIC curves for the model-based clustering of the **iris** data. We see that the highest BIC corresponds to model 9 and 2 groups.

Example 6.7

We are going to use the **iris** data for this example, as well. The function called **mbclust** invokes the entire model-based clustering procedure: agglomerative model-based clustering, finite mixture estimation via the EM algorithm, and evaluating the BIC. The outputs from this function are: (1) a matrix of BIC values, where each row corresponds to a model, (2) a structure that has fields representing the parameters (**pies, mus,** and **vars**) for the best model, (3) a structure that contains information about *all* of the models (see the next example), (4) the matrix **z** representing the agglomerative model-based clustering, and (5) a vector of group labels assigned using the best model.

```
load iris
data = [setosa;versicolor;virginica];
% Call the model-based clustering procedure with a
% maximum of 6 clusters.
[bics,bestmodel,allmodel,Z,clabs] = ...
```

```
mbclust(data,6);
```

We can plot the BIC curves using **plotbic**, as follows

```
% Display the BIC curves.
plotbic(bics)
```

This is given in Figure 6.8. We see that the best model yielding the highest BIC value is for two clusters. We know that there are three groups in this data set, but two of the clusters tend to overlap and are hard to distinguish. So, it is not surprising that two groups were found using model-based clustering. As stated earlier, the EM algorithm can diverge so the covariance matrix can become singular. In these cases, the EM algorithm is stopped for that model; so you might see some incomplete BIC curves.
❑

The model-based clustering procedure we just outlined takes us as far as the estimation of a finite mixture probability density. Recall, that in this framework for cluster analysis, each component of the density provides a template for a cluster. Once we have the best model (number of clusters and form of the covariance matrices), we use it to group data based on their posterior probability. It makes sense to cluster an observation based on its highest posterior probability:

$$\{\,\text{cluster } j \mid \hat{\tau}_{ij}^{*} = \max_{k}\hat{\tau}_{ik}\,\}.$$

In other words, we find the posterior probability that the *i*-th observation belongs to each of the component densities. We then say that it belongs to the cluster with the highest posterior probability. One benefit of using this approach is that we can use the quantity $1 - \hat{\tau}_{ij}^{*}$ as a measure of the uncertainty in the classification [Bensmail et al., 1997].

Example 6.8
Returning to the results of the previous example, we show how to use some of the other outputs from **mbclust**. We know that the **iris** data has three groups. First, let's see how we can extract the model for three groups and model 9. We can do this using the following syntax to reference it:

```
allmodel(9).clus(3)
```

The variable **allmodel** is a structure with one field called **clus**. It (**allmodel**) has nine records, one for each model. The field **clus** is another structure. It has **maxclus** records and three fields: **pies**, **mus**, and **vars**. To obtain the group labels according to this model, we use the **mixclass** function. This requires the data and the finite mixture parameters for the model.

```
% Extract the parameter information:
pies = allmodel(9).clus(3).pies;
mus = allmodel(9).clus(3).mus;
vars = allmodel(9).clus(3).vars;
```

Now we can group the observations.

```
% The mixclass function returns group labels
% as well as a measure of the uncertainty in the
% classification.
[clabs3,unc] = mixclass(data,pies,mus,vars);
```

We can assess the results, as before, using the **silhouette** function.

```
% As before, we can use the silhouette procedure
% to assess the cluster results.
[S, H] = silhouette(data,clabs3);
title('Iris Data - Model-Based Clustering')
```

The average silhouette value is 0.6503, and the silhouette plot is shown in Figure 6.9. The reader is asked in the exercises to explore the results of the two cluster case with the best model in a similar manner.
❏

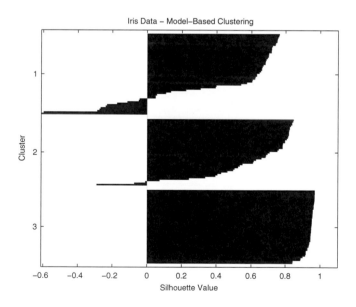

FIGURE 6.9
This is the silhouette plot showing the results of the model-based clustering, model 9, three clusters.

6.6 MBC for Density Estimation and Discriminant Analysis

6.6.1 Introduction to Pattern Recognition

Our previous discussions in this chapter focused on the application of model-based clustering to unsupervised learning or clustering. Within this section, we are going to extend these discussions and show how model-based clustering can be used to estimate probability density functions as finite mixtures and their applications within a supervised learning framework. First, we provide some background information on this topic, which is also known as *discriminant analysis* or *pattern recognition*.

Examples where pattern recognition techniques can be used are numerous and arise in disciplines such as medicine, computer vision, robotics, manufacturing, finance, and many others. Some of these include the following [Martinez and Martinez, 2015]:

- A doctor diagnoses a patient's illness based on the symptoms and test results.
- A geologist determines whether a seismic signal may represent an impending earthquake.
- A loan manager at a bank must decide whether a customer is a good credit risk based on their income, past credit history, and other variables.

In all of these applications, the human can be assisted by computational pattern recognition methods that analyze the data and recommend answers.

Pattern recognition methods for *supervised learning* situations are covered in this section. With supervised learning, we have a data set where we know the classification of each observation. Thus, each case or data point has a class label associated with it. We construct a classifier based on this data set, which is then used to classify future observations. Many of the data sets that we discussed in Chapter 1 fall into this category (e.g., the **oronsay**, **yeast**, and **leukemia** data sets).

Figure 6.10 illustrates the major steps of statistical pattern recognition. The first step in pattern recognition is to select *features* that will be used to distinguish between the classes. As the reader might expect, the choice of features is perhaps the most important part of the process. Building accurate classifiers is much easier with features that allow one to readily distinguish between classes.

Once our features are selected, we obtain a sample of these features for the different classes. This means that we find objects that belong to the classes of interest and then measure the features. Thus, each observation has a class label attached to it. Now that we have data that are known to belong to the different classes, we can use this information to create a classifier. This classifier will take a new set of feature measurements as inputs and then output the estimated class membership. Using model-based clustering as a way to create such a classifier is the topic of this section.

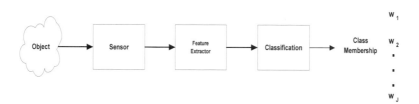

FIGURE 6.10
This shows a schematic diagram of the major steps for statistical pattern recognition [Duda and Hart, 1973; Martinez and Martinez, 2015]. Here, a sensor represents the measuring device.

6.6.2 Bayes Decision Theory

The Bayes approach to pattern classification is a fundamental technique, and we recommend it as the starting point for most pattern recognition applications. If this method proves to be inadequate, then more complicated techniques may be used (e.g., neural networks, classification trees, support vector machines). Bayes decision theory poses the classification problem in terms of probabilities. Therefore, all of the probabilities must be known or estimated from the data.

We start off by fixing some notation. Let the class membership be denoted by ω_j, $j = 1, ..., J$ for a total of J classes. For example, with the **oronsay** data, we have $J = 3$ classes:

$$\omega_1 = midden$$
$$\omega_2 = beach$$
$$\omega_3 = dune.$$

The features we are using for classification are denoted by the p-dimensional vector **x**. With the **oronsay** data, we have twelve measurements representing the size of the sand particles; so $p = 12$. In the supervised learning situation, each of the observed feature vectors will also have a class label attached to it.

Our goal is to use the data to create a decision rule or classifier that will take a feature vector **x** whose class membership is unknown and return the class it most likely belongs to. A logical way to achieve this is to assign a class label to this feature vector using the class corresponding to the highest ***posterior probability***. This probability is given by

$$P(\omega_j|\mathbf{x}); \qquad j = 1, ..., J. \tag{6.14}$$

Equation 6.14 represents the probability that the case belongs to the j-th class given the observed feature vector **x**. To use this rule, we would evaluate all of the J posterior probabilities, and the class corresponding to the highest probability would be the one chosen.

We find the posterior probabilities using Bayes theorem:

$$P(\omega_j|\mathbf{x}) = \frac{P(\omega_j)P(\mathbf{x}|\omega_j)}{P(\mathbf{x})}, \tag{6.15}$$

where

$$P(\mathbf{x}) = \sum_{j=1}^{J} P(\omega_j)P(\mathbf{x}|\omega_j). \tag{6.16}$$

We see from equation 6.15 that we must know the ***prior probability*** that it would be in class j given by

$$P(\omega_j); \qquad j = 1, ..., J, \tag{6.17}$$

and the ***class-conditional probability***

$$P(\mathbf{x}|\omega_j); \qquad j = 1, ..., J. \tag{6.18}$$

The class-conditional probability in Equation 6.18 represents the probability distribution of the features for each class. The prior probability in Equation 6.17 represents our initial degree of belief that an observed set of features is a case from the j-th class. The process of estimating these probabilities is how we build the classifier.

The prior probabilities can either be inferred from prior knowledge of the application, estimated from the data, or assumed to be equal. Using our **oronsay** data as an example, we could estimate the prior probabilities using the proportion of each class in our sample. We had 226 observed feature vectors (samples of sand), with 77 from the midden, 110 from the beach, and

39 from the dunes. Thus, we have $n = 226$, $n_1 = 77$, $n_2 = 110$, and $n_3 = 39$, and our estimated prior probabilities would be

$$\hat{P}(\omega_1) = \frac{n_1}{n} = \frac{77}{226} \approx 0.34$$

$$\hat{P}(\omega_2) = \frac{n_2}{n} = \frac{110}{226} \approx 0.49$$

$$\hat{P}(\omega_3) = \frac{n_3}{n} = \frac{39}{226} \approx 0.17.$$

In some applications, we might use equal priors when we believe each class is equally likely.

Now that we have our prior probabilities, $\hat{P}(\omega_j)$, we turn our attention to the class-conditional probabilities $P(\mathbf{x}|\omega_j)$. We can use various density estimation techniques to obtain these probabilities. In essence, we take all of the observed feature vectors that are known to come from class ω_j and estimate the density using only those cases. We will eventually use model-based clustering to obtain a probability density estimate for each class. The reader is referred to Martinez and Martinez [2015] and Scott [2015] for details and options on other approaches.

Once we have estimates of the prior probabilities (Equation 6.17) and the class-conditional probabilities (Equation 6.18), we can use Bayes theorem (Equation 6.15) to obtain the posterior probabilities. Bayes decision rule is based on these posterior probabilities. Bayes decision rule says the following:

BAYES DECISION RULE:
Given a feature vector $\mathbf{x}$, assign it to class ω_j if

$$P(\omega_j|\mathbf{x}) > P(\omega_i|\mathbf{x}); \qquad i = 1, ..., J; \ i \neq j. \tag{6.19}$$

In essence, this rule means that we will classify an observation $\mathbf{x}$ as belonging to the class that has the highest posterior probability.

We can use an equivalent rule by recognizing that the denominator of the posterior probability (see Equation 6.15) is simply a normalization factor and is the same for all classes. So, we can use the following alternative decision rule:

$$P(\mathbf{x}|\omega_j)P(\omega_j) > P(\mathbf{x}|\omega_i)P(\omega_i); \qquad i = 1, ..., J; \ i \neq j. \tag{6.20}$$

Note that if we have equal priors for each class, then our decision rule of Equation 6.20 is based only on the class-conditional probabilities. The decision rule partitions the feature space into J decision regions $\Omega_1, \Omega_2, ..., \Omega_J$. If $\mathbf{x}$ is in region Ω_j, then we will say it belongs to class ω_j.

6.6.3 Estimating Probability Densities with MBC

Let us explore the use of model-based clustering to obtain an estimate of the class-conditional probability density function (Equation 6.18) for each class. The idea is to estimate these probability densities using the mixture models obtained via the model-based clustering procedure. We can evaluate the efficacy of these estimates using maximum likelihood or the BIC, as was done within our model-based clustering discussions.

From the development of model-based clustering, we know that it clusters observations using an approach where we first estimate a probability density function using a finite mixture model and then use each of the components in the mixture as a template for a cluster. However, in a supervised learning context, we could obtain a probability density estimate from model-based clustering for each class and then use Bayes decision rule. We will explore this procedure in the next example.

Example 6.9

In this example, we use two-class data that are generated using a function called **gmdistribution** from the Statistics Toolbox. This allows us to generate data for each class known to come from a finite mixture and also to illustrate the use of this relatively new capability in the MATLAB toolbox. One can also use the **genmix** GUI that is in the EDA Toolbox for this purpose. (The **genmix** GUI is discussed in the next section.) First, we are going to generate the observations for class one.

```
% Set up a matrix of means.
MU1 = [3 3;-3 -3];
% Now we set the covariance matrices and put
% them into a 3-D array
SIGMA1 = cat(3,[1,0;0,1],[.3,0;0,.3]);
% Here we set the mixing proportions with the
% same number of observations in each class.
p1 = ones(1,2)/2;
% This Statistics Toolbox function
% creates a mixture object.
obj1 = gmdistribution(MU1,SIGMA1,p1);
% Here we draw a random sample of size 1000 from
% this mixture model
X1 = random(obj1, 1000);
```

Now, we do the same thing to generate the data that belong to the second class.

```
MU2 = [3 -3;-3 3];
SIGMA2 = cat(3,[1,.75;.75,1],[1,0;0,1]);
p2 = ones(1,2)/2;
obj2 = gmdistribution(MU2,SIGMA2,p2);
```

```
X2 = random(obj2, 1000);
```

A scatterplot of these data is shown in Figure 6.11 (top), where we have the data from class one shown as crosses and class two data shown as points. The next step in building our classifier is to estimate the class-conditional probabilities for each of our classes using the MBC procedure.

```
% Now we are going to estimate the class
% conditional probability densities for class 1
% using our model-based clustering function.
[bics1,modelout1,model1,Z1,clabs1] = mbclust(X1,5);
```

We get the following results from the function:

```
Maximum BIC is -5803.689.
Model number 2. Number of clusters is 2.
```

Recall that the number of clusters is the same as the number of terms in the mixture model, which was correctly determined by MBC. The estimated means for each term in the mixture are

```
-3.0086     3.0406
-2.9877     2.9692
```

Each column represents one of the 2–D means, and we see that our estimates are close to the true values. The covariances of each term are given here.

```
0.27 0
0     0.27

0.97 0
0     0.97
```

These estimated covariance matrices seem reasonable also. We now do the same thing for the second group of observations.

```
% We repeat this process for the second class.
[bics2,modelout2,model2,Z2,clabs2] = mbclust(X2,5);
```

We do not show them here, but the finite mixture estimate returned for the second class matches the known parameters (i.e., 2–D means, covariances, and mixing coefficients). We use the **gmdistribution** function to create a finite mixture object to use in Bayes decision rule. This object will return values of the density based on the finite mixtures.

```
% The following creates MATLAB objects that
% represent the actual density function, so we can
% use them in Bayes decision rule.
obj1fit = gmdistribution(modelout1.mus',...
      modelout1.vars,modelout1.pies);
obj2fit = gmdistribution(modelout2.mus',...
      modelout2.vars,modelout2.pies);
```

The following MATLAB code will create decision boundaries that we get from applying the Bayes decision rule using our estimated densities.

```
% Here is the code to plot the discriminant region.
% First get the grid to evaluate it.
[xx,yy] = meshgrid(-6:.1:6,-6:.1:6);
dom = [xx(:),yy(:)];
% Evaluate each class conditional probability
% at the grid locations, using the pdf function
% in the Statistics Toolbox.
ccp1 = pdf(obj1fit,dom);
ccp2 = pdf(obj2fit,dom);
% Reshape into the grid.
ccp1 = reshape(ccp1,size(xx));
ccp2 = reshape(ccp2,size(xx));
```

Since we have equal prior probabilities for each class, our decision rule is very simple. We only have to compare the two estimated class-conditional probabilities just obtained. Areas where the value of **ccp1** is greater than **ccp2** will be classified as class one, and all other areas will be classified as class two.

```
% Our decision rule is very simple.
% We just compare the two class-conditional
% probabilities.
ind = ccp1 > ccp2;
figure
surf(xx,yy,+ind)
colormap([.55 .55 .55; .85 .85 .85])
view([0 90])
axis tight
shading interp
```

The discriminant boundaries are shown in the bottom panel of Figure 6.11.
❑

We illustrated the use of the **gmdistribution** function in the previous example. This function can also be used to find clusters using the finite mixture model, but the user must specify the number of clusters to look for. It does not implement the model-based clustering approach we presented in this chapter. The general syntax for finding k clusters using this function is

```
fmfit = gmdistribution.fit(X,k);
```

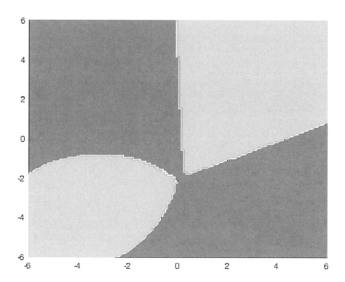

FIGURE 6.11

The top panel displays a scatterplot showing the two-class data set we generated in Example 6.9. Class one data are shown as crosses, and the class two data are shown as points. The data for each class were generated according to a two-component mixture model. The bottom panel shows the decision region for our classifier that we get using Bayes decision rule and our estimated densities from model-based clustering.

6.7 Generating Random Variables from a Mixture Model

One might often be interested in generating data from one of the finite mixture models discussed in this chapter. Of course, we could always do this manually using a multivariate normal random number generator (e.g., **mvnrnd** in the MATLAB Statistics Toolbox), but this can be tedious to use in the case of mixtures. So, we include a useful GUI tool called **genmix** that generates data according to a finite mixture.

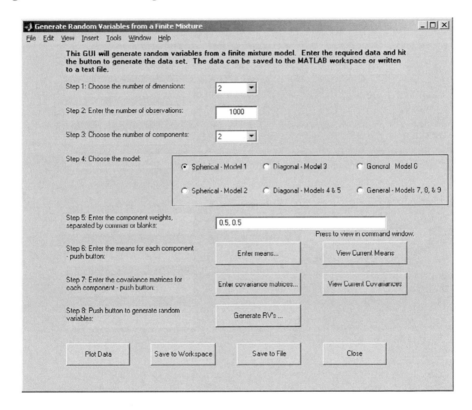

FIGURE 6.12

This shows the **genmix** GUI. The steps that must be taken to enter the parameter information and generate the data are shown on the left side of the window.

The GUI is invoked by typing **genmix** at the command line. A screen shot is shown in Figure 6.12. The steps for entering the required information are listed on the left-side of the GUI window. We outline them below and briefly describe how they work.

FIGURE 6.13

This shows the pop-up window for entering 2–D means and for the two components or terms in the finite mixture.

Step 1: Choose the number of dimensions.

This is a pop-up menu. Simply select the number of dimensions for the data.

Step 2: Enter the number of observations.

Type the total number of points in the data set. This is the value for n in Equation 6.4.

Step 3: Choose the number of components.

This is the number of terms or component densities in the mixture. This is the value for c in Equations 6.2 and 6.4. Note that one can use this GUI to generate a finite mixture with only one component, which can be useful in other applications.

Step 4: Choose the model.

Select the model for the data. The model numbers correspond to those described in Table 6.1. The type of covariance information you are required to enter depends on the model you have selected here.

Step 5: Enter the component weights, separated by commas or blanks.

Enter the corresponding weights (π_k) for each term. These must be separated by commas or spaces, and they must sum to 1.

Step 6: Enter the means for each component-push button.

Click on the button **Enter means...** to bring up a window to enter the p-dimensional means, as shown in Figure 6.13. There will be a different number of text boxes in the window, depending on the number of components selected in **Step 3**. Note that you must have the right number of values in each text box. In other words, if you have dimensionality $p = 3$ (**Step 1**), then each mean requires 3 values. If you need to check on the means that were used, then you can click on the **View Current Means** button. The means will be displayed in the MATLAB command window.

Step 7: Enter the covariance matrices for each component - push button.

Click on the button **Enter covariance matrices...** to activate a pop-up window. You will get a different window, depending on the chosen model (**Step 4**). See Figure 6.14 for examples of the three types of covariance matrix input windows. As with the means, you can push the **View Current Covariances** button to view the covariance matrices in the MATLAB command window.

Step 8: Push the button to generate random variables.

After all of the variables have been entered, push the button labeled **Generate RVs...** to generate the data set.

Once the variables have been generated, you have several options. The data can be saved to the workspace using the button **Save to Workspace**. When this is activated, a pop-up window appears, and you can enter a variable name in the text box. The data can also be saved to a text file for use in other software applications by clicking on the button **Save to File**. This brings up the usual window for saving files. An added feature to verify the output is the **Plot Data** button. The generated data are displayed in a scatterplot matrix (using the **plotmatrix** function) when this is pushed.

6.8 Summary and Further Reading

In this chapter, we have described another approach to clustering that is based on estimating a finite mixture probability density function. Each component of the density function represents one of the groups, and we can use the posterior probability that an observation belongs to the component density to assign group labels. The model-based clustering methodology consists of three main parts: (1) model-based agglomerative clustering to obtain initial partitions of the data, (2) the EM algorithm for maximum likelihood estimation of the finite mixture parameters, and (3) the BIC for choosing the best model.

Several excellent books are available on finite mixtures. Everitt and Hand [1981] is a short monograph that includes many applications for researchers. McLachlan and Basford [1988] is more theoretical, as is Titterington, Smith, and Makov [1985]. McLachlan and Peel [2000] contains a more up-to-date

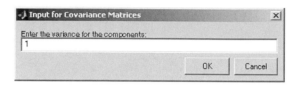

(a) This shows the pop-up window(s) for the spherical family of models. The only value that must be entered for each covariance matrix is the volume λ.

(b) This shows the pop-up window(s) for the diagonal family of models. One needs to enter the volume λ (first text box) and the diagonal elements of the matrix **B** (second text box).

(c) When the general family is selected, one of these pop-up windows will appear for each unique covariance matrix. Each text box corresponds to a *row* of the covariance matrix.

FIGURE 6.14

Here we have examples of the covariance matrix inputs for the three families of models.

treatment of the subject. They also discuss the application of mixture models to large databases, which is of interest to practitioners in EDA and data mining. Most books on finite mixtures also include information on the EM algorithm, but they also offer other ways to estimate the density parameters. McLachlan and Krishnan [1997] is one book that provides a complete account of the EM, including theory, methodology, and applications. Martinez and Martinez [2015] includes a description of the univariate and multivariate EM algorithm for finite mixtures, as well as a way to visualize the estimation process.

The EM algorithm described in this chapter is sometimes called the *iterative EM*. This is because it operates in batch mode; all data must be available for use in the EM update equations at each step. This imposes a high computational burden and storage load when dealing with massive data sets. A *recursive* form of the EM algorithm exists and can be used in an online manner. See Martinez and Martinez [2015] for more information on this algorithm and its use in MATLAB.

Celeux and Govaert [1995] present the complete set of models and associated procedures for obtaining the parameters in the EM algorithm when the covariances are constrained (multivariate normal mixtures). They also address the restricted case, where mixing coefficients are equal. We looked only at the unrestricted case, where these are allowed to vary.

There are many papers on the model-based clustering methodology, including interesting applications. Some examples include minefield detection [Dasgupta and Raftery, 1998], detecting faults in textiles [Campbell et al., 1997, 1999], classification of gamma rays in astronomy [Mukherjee et. al, 1998], and analyzing gene expression data [Yeung et al., 2001]. One of the issues with agglomerative model-based clustering is the computational load with massive data sets. Posse [2001] proposes a starting partition based on minimal spanning trees, and Solka and Martinez [2004] describe a method for initializing the agglomerative model-based clustering using adaptive mixtures. Further extensions of model-based clustering for large data sets can be found in Wehrens et al. [2004] and Fraley, Raftery, and Wehrens [2005]. Two review papers on model-based clustering are Fraley and Raftery [1998, 2002].

Other methods for choosing the number of terms in finite mixtures have been offered. We provide just a few references here. Banfield and Raftery [1993] use an approximation to the integrated likelihood that is based on the classification likelihood. This is called the AWE, but the BIC seems to perform better. An entropy criterion called the NEC (normalized entropy criterion) has been described by Biernacki and Govaert [1997] and Biernacki et al. [1999]. Bensmail et al. [1997] provide an approach that uses exact Bayesian inference via Gibbs sampling. Chapter 6 of McLachlan and Peel [2000] provides a comprehensive treatment of this issue.

Additional software is available for model-based clustering. One such package is MCLUST, which can be downloaded at

http://www.stat.washington.edu/mclust/

MCLUST is compatible with S-Plus and R.[3] Fraley and Raftery [1998, 2002, 2003] provide an overview of the MCLUST software. McLachlan et al. [1999] wrote a software package called EMMIX that fits mixtures of normals and *t*-distributions. The website for EMMIX is

http://www.maths.uq.edu.au/~gjm/emmix/emmix.html

---

[3] See Appendix B for information on the R language.

Exercises

6.1   Apply model-based agglomerative clustering to the **oronsay** data set. Use the gap statistic to choose the number of clusters. Evaluate the partition using the silhouette plot and the Rand index.

6.2   Use the model-based clustering on the **oronsay** data. Apply the gap method to find the number of clusters for one of the models. Compare the gap estimate of the number of clusters to what one gets from the BIC.

6.3   Repeat the procedure in Example 6.1 for component densities that have closer means. How does this affect the resulting density function?

6.4   Construct another univariate finite mixture density, as in Example 6.1, using a mixture of univariate exponentials.

6.5   Cluster *size* and *volume* are not directly related. Explain how they are different. In the finite mixture model, what parameter is related to cluster size?

6.6   Write a MATLAB function that will return the normal probability density function for the univariate case:

$$f(x;\mu, \sigma) = \frac{1}{\sigma\sqrt{2\pi}} \exp\left\{ -\frac{(x-\mu)^2}{2\sigma^2} \right\}.$$

Use this in place of **normpdf** and repeat Example 6.1.

6.7   Apply model-based agglomerative clustering to the following data sets. Display in a dendrogram. If necessary, first reduce the dimensionality using a procedure from Chapters 2 and 3.

a. **skulls**

b. **sparrow**

c. **oronsay** (both classifications)

d. BPM data sets

e. gene expression data sets

6.8   Generate data according to the remaining six models of Table 6.1 that are not demonstrated in this text. Display the results using **plotmatrix**.

6.9   Return to the **iris** data in Example 6.6. Try two and four clusters from the agglomerative model-based clustering. Assess the results using the silhouette procedure.

6.10   The NEC is given below. Implement this in MATLAB and apply it to the models in Example 6.7. This means that instead of a matrix of BIC values, you will have a matrix of NEC values (i.e., one value for each model and number of clusters). Are the results different with respect to the number of clusters chosen for each model? For $1 \le k \le K$ we have

$$L(K) = C(K) + E(K)$$

$$C(K) = \sum_{k=1}^{K} \sum_{i=1}^{n} \hat{\tau}_{ik} \ln[\hat{\pi}_k f(\mathbf{x}_i; \hat{\theta}_k)]$$

$$E(K) = -\sum_{k=1}^{K} \sum_{i=1}^{n} \hat{\tau}_{ik} \ln \hat{\tau}_{ik} \ge 0$$

This is a decomposition of the log-likelihood $L(K)$ into a classification log-likelihood term $C(K)$ and the entropy $E(K)$ of the classification matrix with terms given by $\hat{\tau}_{ik}$. The NEC is given by

$$NEC(K) = \frac{E(K)}{L(K) - L(1)} \qquad K > 1$$

$$NEC(1) = 1$$

One chooses the K that corresponds to the minimum value of the NEC.

6.11 Apply the NEC to the clustering found in Problem 6.2. How does the NEC estimate of the number of clusters compare with the gap estimate?

6.12 Apply other types of agglomerative clustering (e.g., different distances and linkage) to the `iris` data. Assess the results using three clusters and the `silhouette` function. Compare with the model-based agglomerative clustering results in Example 6.6. Also visually compare your results via the dendrogram.

6.13 Generate random variables according to the different models in Table 6.1. Use $n = 300$, $p = 2$, $c = 2$. Apply the model-based clustering procedure. Is the correct model (number of terms and form of covariance matrix) recovered?

6.14 Repeat Example 6.5 for a larger sample size. Are the parameter estimates closer to the true ones?

6.15 Repeat Example 6.8 using the best model, but the two cluster case. Compare and discuss your results.

6.16 Generate bivariate random samples according to the nine models. Apply model-based clustering and analyze your results.

6.17 Apply the full model-based clustering procedure to the following data sets. If necessary, first reduce the dimensionality using one of the procedures from Chapters 2 and 3. Assess your results using methods from Chapter 5.

a. **skulls**

     b. **sparrow**

     c. **oronsay** (both classifications)

     d. BPM data sets

     e. gene expression data sets

6.18   Use an appropriate method to reduce the dimensionality of the data sets listed below to 2–D. Next, use the model-based clustering method to estimate class-conditional probabilities and build classifiers. Create and display the discriminant regions as shown in Example 6.9. Finally, construct 2-D scatterplots with different colors and symbols for each class. Compare the scatterplot with the discriminant regions and comment on the results.

     a. **iris**

     b. **leukemia**

     c. **oronsay** (both classifications)

     d. BPM data sets (pick two topics)

Chapter 7

Smoothing Scatterplots

In many applications, we might make distributional and model assumptions about the underlying process that generated the data, so we can use a *parametric* approach in our analysis. The parametric approach offers many benefits (known sampling distributions, lower computational burden, etc.), but it can be dangerous if the assumed model is incorrect. At the other end of the data-analytic spectrum, we have the *nonparametric* approach, where one does not make any formal assumptions about the underlying structure or process. When our goal is to summarize the relationship between variables, then *smoothing methods* are a bridge between these two approaches. Smoothing methods make the weak assumption that the relationship between variables can be represented by a smooth curve or surface. In this chapter, we cover several methods for scatterplot smoothing, as well as some ways to assess the results.

7.1 Introduction

In most, if not all, scientific experiments and analyses, the researcher must summarize, interpret and/or visualize the data to gain insights, to search for patterns, and to make inferences. For example, with some gene expression data, we might be interested in the distribution of the gene expression values for a particular patient or experimental condition, as this might indicate what genes are most active. We could use a probability density estimation procedure for this purpose. Another situation that often arises is one in which we have a response y and a predictor x, and we want to understand and model the relationship between the y and x variables.

The main goal of smoothing from an EDA point of view is to obtain some insights into how data are related to one another and to search for patterns or structure. This idea has been around for many years, especially in the area of smoothing time series data, where data are measured at equally spaced points in time. Readers might be familiar with some of these methods, such as moving averages, exponential smoothing, and other specialized filtering techniques using polynomials.

In general, the procedures described in this chapter obtain the smooths by constructing fits in a local manner. In the case of loess, we perform local regression in a moving neighborhood around the target values. With splines, the curves are constructed using piecewise polynomials over disjoint regions.

The main loess procedures for 2–D and multivariate data are presented first, followed by a robust version to use when outliers are present in the data. We then proceed to a discussion of residuals and diagnostic plots that can be used to assess the results of the smoothing. This is followed by a description of smoothing splines, which is another commonly used method. The next section includes a discussion of some ways to choose the parameter that controls the degree of smoothing.

Cleveland and McGill [1984] developed some extensions of the loess scatterplot smoothing to look for relationships in the bivariate distribution of *x* and *y* called pairs of middle smoothings and polar smoothings, and these are covered next. Smoothing via loess and other techniques is available in the MATLAB Curve Fitting toolbox; so we finish the chapter with a brief section that describes the functions provided in this optional toolbox.

7.2 Loess

Loess (also called *lowess* in earlier works) is a locally weighted regression procedure for fitting a regression curve (or surface) by smoothing the dependent variable as a function of the independent variable. The framework for loess is similar to what is commonly used in regression. We have n measurements of the dependent variable y_i, and associated with each y_i is a corresponding value of the independent variable x_i. For now, we assume that our dimensionality is $p = 1$; the case of multivariate predictor variables $\mathbf{x}$ is covered next.

We assume that the data are generated by

$$y_i = f(x_i) + \varepsilon_i \, ,$$

where the ε_i are independent normal random variables with mean zero and variance σ^2. In the classical regression (or parametric) framework, we would assume that the function f belongs to a class of parametric functions, such as polynomials. With loess or local fitting, we assume that f is a smooth function of the independent variables, and our goal is to estimate this function f. The estimate is denoted as $\hat{y}$, a plot of which can be added to our scatterplot for EDA purposes or it can be used to make predictions of our response variable.

The point denoted by the ordered pair $(x_0, \hat{y}_0)$ is called the smoothed point at a point in our domain x_0, and $\hat{y}_0$ is the corresponding fitted value.

The curve obtained from a loess model is governed by two parameters: α and λ. The parameter α is a smoothing parameter that governs the size of the neighborhood; it represents the proportion of points that are included in the local fit. We restrict our attention to values of α between zero and one, where high values for α yield smoother curves. Cleveland [1993] addresses the case where α is greater than one. The second parameter λ determines the degree of the local regression. Usually, a first or second degree polynomial is used, so $\lambda = 1$ or $\lambda = 2$, but higher degree polynomials can be used. Some of the earlier work in local regression used $\lambda = 0$, so a constant function was fit in each neighborhood [Cleveland and Loader, 1996].

The general idea behind loess is the following. To get a value of the curve $\hat{y}_0$ at a given point x_0, we first determine a local neighborhood of x_0 based on α. All points in this neighborhood are weighted according to their distance from x_0, with points closer to x_0 receiving larger weight. The estimate $\hat{y}_0$ at x_0 is obtained by fitting a linear or quadratic polynomial using the weighted points in the neighborhood. This is repeated for a uniform grid of points x in the domain to get the desired curve.

The reason for using a weight function is that only points close to x_0 will contribute to the regression, since they should provide a better indication of the relationship between the variables in the neighborhood of x_0. The weight function W should have the following properties:

1. $W(x) > 0$ for $|x| < 1$
2. $W(-x) = W(x)$
3. $W(x)$ is a nonincreasing function for $x \geq 0$
4. $W(x) = 0$ for $|x| \geq 1$.

The basic idea is, for each point x_0 where we want to get a smoothed value $\hat{y}_0$, we define weights using the weight function W. We center the function W at x_0 and scale it so that W first becomes zero at the k-th nearest neighbor of x_0. As we will see shortly, the value for k is governed by the parameter α.

We use the tri-cube weight function in our implementation of loess. Thus, the weight $w_i(x_0)$ at x_0 for the i-th data point x_i is given by the following

$$w_i(x_0) = W\left(\frac{|x_0 - x_i|}{\Delta_k(x_0)}\right), \qquad (7.1)$$

with

$$W(u) = \begin{cases} \left(1 - u^3\right)^3; & 0 \le u < 1 \\ 0; & \text{otherwise.} \end{cases} \qquad (7.2)$$

The denominator $\Delta_k(x_0)$ is defined as the distance from x_0 to the k-th nearest neighbor of x_0, where k is the greatest integer less than or equal to $\alpha \times n$. We denote the neighborhood of x_0 as $N(x_0)$. The tri-cube weight function is illustrated in Figure 7.1.

FIGURE 7.1
This is the tri-cube weight function.

In step 6 of the loess procedure outlined below, one can fit either a straight line to the weighted points (x_i , y_i), for x_i in the neighborhood $N(x_0)$, or a quadratic polynomial can be used (in our implementation). If a line is used as the local model, then $\lambda = 1$. The values of β_0 and β_1 are found such that the following is minimized

$$\sum_{i=1}^{k} w_i(x_0)(y_i - \beta_0 - \beta_1 x_i)^2 , \qquad (7.3)$$

for (x_i , y_i), with x_i in $N(x_0)$. Letting $\hat{\beta}_0$ and $\hat{\beta}_1$ be the values that minimize Equation 7.3, the loess fit at x_0 is given by

$$\hat{y}(x_0) = \hat{\beta}_0 + \hat{\beta}_1 x_0. \tag{7.4}$$

When $\lambda = 2$, then we fit a quadratic polynomial using weighted least-squares, again using only those points in $N(x_0)$. In this case, we find the values for the β_i that minimize

$$\sum_{i=1}^{k} w_i(x_0)\left(y_i - \beta_0 - \beta_1 x_i - \beta_2 x_i^2\right)^2. \tag{7.5}$$

Similar to the linear case, if $\hat{\beta}_0$, $\hat{\beta}_1$, and $\hat{\beta}_2$ minimize Equation 7.5, then the loess fit at x_0 is

$$\hat{y}(x_0) = \hat{\beta}_0 + \hat{\beta}_1 x_0 + \hat{\beta}_2 x_0^2. \tag{7.6}$$

For more information on weighted least squares see Draper and Smith [1981].
We describe below the steps for obtaining a loess curve, which is illustrated in Figure 7.2. Using a set of generated data, we show the loess fit for a given point x_0. The top panel shows the linear fit in the neighborhood of x_0, and the bottom panel shows the quadratic fit. The open circle on the respective curves is the smoothed value at that point.

Procedure – Loess Curve Construction

1. Let x_i denote a set of n values for a predictor variable and let y_i represent the corresponding response.
2. Choose a value for α such that $0 < \alpha < 1$. Let $k = \lfloor \alpha \times n \rfloor$, where k is the greatest integer less than or equal to $\alpha \times n$.
3. For each x_0 where we want to obtain an estimate of the smooth $\hat{y}_0$, find the k points x_i in the data set that are closest to x_0. These x_i comprise a neighborhood of x_0, and this set is denoted by $N(x_0)$.
4. Compute the distance of the x_i in $N(x_0)$ that is furthest away from x_0 using

$$\Delta_k(x_0) = \max_{x_i \in N_0}|x_0 - x_i|.$$

5. Assign a weight to each point (x_i, y_i), x_i in $N(x_0)$, using the tri-cube weight function (Equations 7.1 and 7.2).
6. Obtain the value $\hat{y}_0$ of the curve at the point x_0 for the given λ using a weighted least squares fit of the points x_i in the neighborhood $N(x_0)$. (See Equations 7.3 through 7.6.)

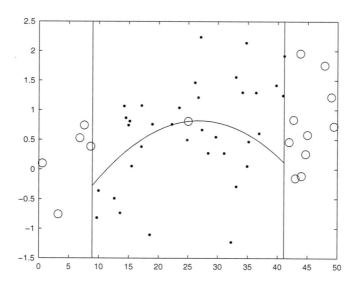

FIGURE 7.2

The top panel shows the local fit at a point $x_0 = 25$, with $\lambda = 1$ and $\alpha = 0.5$. The vertical lines indicate the limits of the neighborhood. The second panel shows the local fit at x_0, where $\lambda = 2$ and $\alpha = 0.7$. The point $(x_0, \hat{y}_0)$ is the open circle on the curves.

7. Repeat steps 3 through 6 for all x_0 of interest.

We illustrate the loess procedure for univariate data in Example 7.1 using a well-known data set discussed in the loess literature [Cleveland and Devlin, 1988; Cleveland and McGill, 1984; Cleveland, 1993]. This data set contains 111 measurements of four variables, representing ozone and other meteorological data. They were collected during May 1 and September 30, 1973 at various sites in the New York City region. The goal is to describe the relationship between ozone (PPB) and the meteorological variables (solar radiation measured in Langleys, temperature in degrees Fahrenheit, and wind speed in MPH) so one might predict ozone concentrations.

Example 7.1
We illustrate the univariate loess procedure using the ozone concentration as our response variable (y) and the temperature as our predictor variable (x). The next few lines of MATLAB code load the data set and display the scatterplot shown in Figure 7.3.

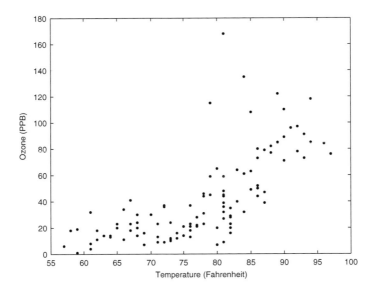

FIGURE 7.3
This is the scatterplot for ozone as it depends on temperature. We see that, in general, the ozone increases as temperature increases.

```
load environmental
% This has all four variables. We will
% use the ozone as the response and
% temperature as the predictor.
```

```
% First do a scatterplot
plot(Temperature,Ozone,'.')
xlabel('Temperature (Fahrenheit)')
ylabel('Ozone (PPB)')
```

We see in the scatterplot that the ozone tends to increase when the temperature increases, but the appropriate type of relationship for these data is not clear. We show how to use the loess procedure to find the estimate of the ozone level for a given temperature of 78 degrees Fahrenheit. First, we find some of the parameters.

```
n = length(Temperature); % Number of points
% Find the estimate at this point:
x0 = 78;
% Set up the other constants:
alpha = 2/3;
lambda = 1;
k = floor(alpha*n);
```

Next, we find the neighborhood at $x_0 = 78$.

```
% First step is to get the neighborhood.
dist = abs(x0 - Temperature);
% Find the closest distances.
[sdist,ind] = sort(dist);
% Get the points in the neighborhood.
Nx = Temperature(ind(1:k));
Ny = Ozone(ind(1:k));
% Maximum distance of neighborhood:
delxo = sdist(k);
```

We now (temporarily) delete all of the points outside the neighborhood and use the remaining points as input to the tri-cube weight function.

```
% Delete the ones outside the neighborhood.
sdist((k+1):n) = [];
% These are the arguments to the weight function.
u = sdist/delxo;
% Get the weights for all points in the neighborhood.
w = (1 - u.^3).^3;
```

The next section of code prepares the matrix for use in the weighted least squares regression (see Draper and Smith [1981]). In other words, we have the values for the x_i, but we need a matrix where the first column contains ones, the second contains the x_i, and the third contains x_i^2 (in the case of $\lambda = 2$).

```
% Now using only those points in the neighborhood,
% do a weighted least squares fit of degree lambda.
% We will follow the procedure in 'polyfit'.
```

```
x = Nx(:); y = Ny(:); w = w(:);
% Get the weight matrix
W = diag(w);
% Get the right matrix form: 1, x, x^2.
A = vander(x);
A(:,1:length(x)-lambda-1) = [];
V = A'*W*A;
Y = A'*W*y;
[Q,R] = qr(V,0);
p = R\(Q'*Y);
% The following is to fit MATLAB's convention
% for polynomials
p = p';
```

Now that we have the polynomial for the fit, we can use the MATLAB function **polyval** to get the value of the loess smooth at 78 degrees.

```
% This is the polynomial model for the local fit.
% To get the value at that point, use polyval.
yhat0 = polyval(p,x0);
```

We obtain a value of 33.76 PPB at 78 degrees. The following lines of code produce a loess smooth over the range of temperatures and superimposes the curve on the scatterplot. The **loess** function is included with the EDA Toolbox.

```
% Now call the loess function and plot the result.
% Get a domain over which to evaluate the curve.
X0 = linspace(min(Temperature),max(Temperature),50);
yhat = loess(Temperature,Ozone,X0,alpha,lambda);
% Plot the results.
plot(Temperature,Ozone,'.',X0,yhat)
xlabel('Temp (Fahrenheit)'),ylabel('Ozone (PPB)')
```

The resulting scatterplot with loess smooth is shown in Figure 7.4. It should be noted that we could use the loess function to get the estimated value of ozone at 78 degrees only, using

```
yhat0 = loess(Temperature,Ozone,78,alpha,lambda);
```

❑

Some readers might wonder where the word *loess* comes from. In geology, *loess* is defined as a deposit of fine clay or silt in river valleys. If one takes a vertical cross-section of the earth in such a place, then a *loess* would appear as curved strata running through the cross-section. This is similar to what one sees in the plot of the loess smooth in a scatterplot.

We now turn our attention to the multivariate case, where our predictor variables x have $p > 1$ dimensions. The procedure is essentially the same, but the weight function is defined in a slightly different way. We require a

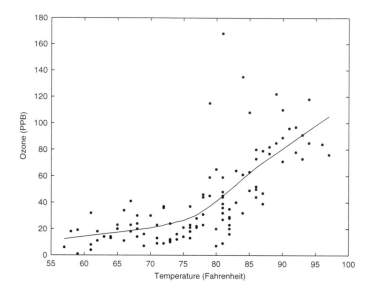

FIGURE 7.4
This shows the scatterplot of **ozone** and **temperature** along with the accompanying loess smooth.

distance function in the space of independent variables, which in most cases is taken to be Euclidean distance. As discussed in Chapter 1, it might make sense to divide each of the independent variables by an estimate of its scale before calculating the distance.

Now that we have the distance, we define the weight function for a p-dimensional point x_0 as

$$w_i\left(\mathbf{x}_0\right) = W\left(\frac{d\left(\mathbf{x}_0, \mathbf{x}_i\right)}{\Delta_k\left(\mathbf{x}_0\right)}\right),$$

where $d(\bullet)$ represents our distance function and $\Delta_k(\mathbf{x}_0)$ is the distance between the k-th nearest neighbor to $\mathbf{x}_0$ using the same definition of distance. W is the tri-cube function, as before. Once we have the weights, we construct either a multivariate linear or multivariate quadratic fit in the neighborhood of $\mathbf{x}_0$. Linear fits are less of a computational burden, but quadratic fits perform better in applications where the regression surface has a lot of curvature. The Visualizing Data Toolbox (may be downloaded for free) described in Appendix B contains a function for producing loess smooths for bivariate predictors. We illustrate its use in the next example.

Example 7.2

For this example, we use the **galaxy** data set that was analyzed in Cleveland and Devlin [1988]. Buta [1987] measured the velocity of the NGC 7531 spiral galaxy in the Southern Hemisphere at a set of points in the celestial sphere that covered approximately 200 arc sec in the north/south direction and around 135 arc sec in the east/west direction. As usual, we first load the data and set some parameters.

```
% First load the data and set some parameters.
load galaxy
% The following is for the contour plot.
contvals = 1420:30:1780;
% These are the parameters needed for loess.
alpha = 0.25;
lambda = 2;
```

Next we get some (x, y) points in the domain. We will get the estimated surface using loess at these points. Then we call the **loess2** function, which was downloaded as part of the Data Visualization Toolbox.

```
% Now we get the points in the domain.
% The loess surface will be evaluated
% at these points.
XI = -25:2:25;
YI = -45:2:45;
[newx,newy] = meshgrid(XI,YI);
newz = loess2(EastWest,NorthSouth,...
     Velocity,newx,newy,alpha,lambda,1);
```

To plot this in a contour plot, we use the following. The plot is shown in Figure 7.5.

```
% Now do the contour plot and add labels.
[cs,h] = contour(newx,newy,newz,contvals);
clabel(cs,h)
xlabel('East-West Coordinate (arcsec)')
ylabel('North-South Coordinate (arcsec)')
```

❏

We now discuss some of the issues for choosing the parameters for loess. These include the order λ of the polynomial that is estimated at each point of the smooth, the weight function W, and the smoothing parameter α. We have already touched on some of the issues with the degree of the polynomial. Any degree polynomial $\lambda = 0, 1, 2, \ldots$ can be used. $\lambda = 0$ provides a constant fit, but this seems too restrictive and the resulting curves/surfaces could be too rough. $\lambda = 1$ is adequate in most cases and is better computationally, but $\lambda = 2$ should be used in situations where there is a lot of curvature or many local maxima and minima.

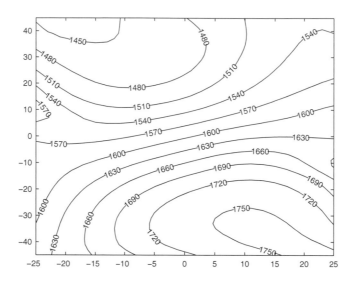

FIGURE 7.5
This is the contour plot showing the loess surface for the **galaxy** data.

As for the function W, we return to the four conditions specified previously. The first condition states that the weights must be positive, because negative weights do not make sense. The second requirement says that the weight function must be symmetric and that points to any side of x_0 should be treated the same way. The third property provides greater weight for those that are closer to x_0. The last property is not required, although it makes things simpler computationally. We save on computations if the weights outside the neighborhood are zero, because those observations do not have to be included in the least squares (see the indices on the summations in Equations 7.3 and 7.5—they only go up to k, not n). We could use a weight function that violates condition four, such as the normal probability density function, but then we would have to include all n observations in the least squares fit at every point of the smooth.

Perhaps the hardest parameter to come up with is the smoothing parameter α. If α is small, then the loess smooth tends to be wiggly and to overfit the data (i.e., the bias is less). On the other hand, if α is large, then the curve is smoother (i.e., the variance is less). In EDA, where the main purpose of the smooth is to enhance the scatterplot and to look for patterns, the choice of α is not so critical. In these situations, Cleveland [1979] suggests that one choose α in the range 0.2 to 0.8. Different values for α (and λ) could be used to obtain various loess curves. Then the scatterplot with superimposed loess curve and residuals plots (discussed in Section 7.4) can be examined to determine whether or not the model adequately describes the relationship. We will also return to this topic in Section 7.6 where we present a more analytical approach to choosing the smoothing parameter that is based on cross-validation.

7.3 Robust Loess

The loess procedure we described in the previous section is not robust, because it relies on the method of least squares for the local fits. A method is called *robust* if it performs well when the associated underlying assumptions (e.g., normality) are not satisfied [Kotz and Johnson, Vol. 8, 1986]. There are many ways in which assumptions can be violated. A common one is the presence of *outliers* or extreme values in the response data. These are points in the sample that deviate from the pattern of the other observations. Least squares regression is vulnerable to outliers, and it takes only one extreme value to unduly influence the result. The nonrobustness of least squares will be illustrated and explored in the exercises.

Cleveland [1993, 1979] and Cleveland and McGill [1984] present a method for smoothing a scatterplot using a robust version of loess. This technique uses the bisquare method [Hoaglin, Mosteller, and Tukey, 1983; Mosteller and Tukey, 1977; Huber, 1973; Andrews, 1974] to add robustness to the weighted least squares step in loess. The idea behind the bisquare is to re-weight points based on their residuals. If the residual for a given point in the neighborhood is large (i.e., it has a large deviation from the model), then the weight for that point should be decreased, since large residuals tend to indicate outlying observations. Alternatively, if the point has a small residual, then it should be weighted more heavily.

Before showing how the bisquare method can be incorporated into loess, we first describe the general bisquare least squares procedure. First a linear regression is used to fit the data, and the residuals $\hat{\varepsilon}_i$ are calculated from

$$\hat{\varepsilon}_i = y_i - \hat{y}_i.$$ (7.7)

The residuals are used to determine the weights from the bisquare function given by

$$B(u) = \begin{cases} \left(1-u^2\right)^2; & |u| < 1 \\ 0; & \text{otherwise.} \end{cases}$$ (7.8)

The robustness weights are obtained from

$$r_i = B\left(\frac{\hat{\varepsilon}_i}{6\hat{q}_{0.5}}\right),\tag{7.9}$$

where $\hat{q}_{0.5}$ is the median of $|\hat{\varepsilon}_i|$. A weighted least squares regression is performed using weights adjusted by r_i.

To add bisquare to loess, we first fit the loess smooth, using the same procedure as before. We then calculate the residuals using Equation 7.7 and determine the robust weights from Equation 7.9. The loess procedure is repeated using weighted least squares, but the weights are now $r_i w_i(x_0)$. Note that the points used in the fit are the ones in the neighborhood of x_0, as before. This is an iterative process and is repeated until the loess curve converges or stops changing. Cleveland and McGill [1984] suggest that two or three iterations are sufficient to get a reasonable model.

Procedure – Robust Loess

1. Fit the data using the loess procedure with weights w_i.
2. Calculate the residuals, $\hat{\varepsilon}_i = y_i - \hat{y}_i$, for each observation.
3. Determine the median of the absolute value of the residuals, $\hat{q}_{0.5}$.
4. Find the robustness weights from

$$r_i = B\left(\frac{\hat{\varepsilon}_i}{6\hat{q}_{0.5}}\right),$$

   using the bisquare function in Equation 7.8.
5. Repeat the loess procedure using weights of $r_i w_i$.
6. Repeat steps 2 through 5 until the loess curve converges.

In essence, the robust loess iteratively adjusts the weights based on the residuals. We illustrate the robust loess procedure in the next example, noting that while our example for robust loess involves only one predictor variable, we can easily apply it to the multivariate case.

Example 7.3
We now illustrate the robust loess using an example from Simonoff [1996]. The data represent the size of the annual spawning stock (x values) and the corresponding production of new fish of catchable size, called recruits (y values). The observations (in thousands of fish) were taken for the Skeena River sockeye salmon from 1940 to 1967. We provide a function called

loessr that implements the procedure outlined above, and its use is shown below.

```
% Load the data and set the values
% for x and y.
load salmon
x = salmon(:,1);
y = salmon(:,2);
% Obtain a domain over which to get
% the loess curve.
xo = linspace(min(x),max(x));
% Get both the regular loess.
yhat = loess(x,y,xo,0.6,2);
% Get the robust loess curve.
yhatr = loessr(x,y,xo,0.6,2);
% Plot both curves.
plot(xo,yhat,'-',xo,yhatr,':',x,y,'o')
legend({'Loess';'Robust Loess'})
xlabel('Spawners')
ylabel('Recruits')
```

The curves are shown in Figure 7.6, where we see a potential outlier in the upper right area of the plot. The robust loess curve is different in the area influenced by this observation.
❑

7.4 Residuals and Diagnostics with Loess

In this section, we address several ways to assess the output from the loess scatterplot smooth (and other types of smoothing), which can also be used to guide the analyst in choosing the values for α and λ. Since we are concerned with EDA methods in this text, we will cover only the graphical methods, such as residual plots, spread smooths, and upper/lower loess smooths. Cleveland and Devlin [1988] describe several statistics that are defined analogously with those used in fitting parametric functions by least squares; so some of the familiar techniques for making inferences in that setting can also be used in loess. Among other things, Cleveland and Devlin describe distributions of residuals, fitted values, and residual sum of squares.

7.4.1 Residual Plots

It is well known in regression analysis that checking the assumptions made about the residuals is important [Draper and Smith, 1981]. This is equally

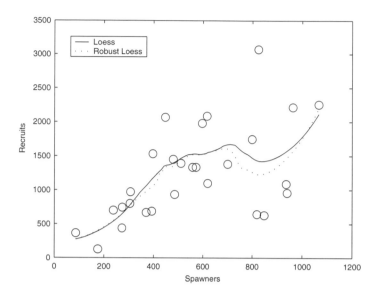

FIGURE 7.6
This shows the regular and robust loess curves for the **salmon** data. We fit a locally quadratic polynomial with $\alpha = 0.6$. Note the potential outlier in the upper right corner. The robust loess fit downweights the effect of this observation.

true when applying any smoothing technique, and we can use similar diagnostic plots. Recall that the true error ε_i is assumed to be normally distributed with mean zero and equal variance, and that the estimated residuals (or errors) are given by

$$\hat{\varepsilon}_i = y_i - \hat{y}_i.$$

We can construct a *normal probability plot* to determine whether the normality assumption is reasonable. We describe normal probability plots in more detail in Chapter 9 and just mention them here in the context of loess. The normal probability plot can be used when we have just one predictor or many predictors. One could also construct a histogram of the residuals to visually assess the distribution.

To check the assumption of constant variance, we can plot the absolute value of the residuals against the fitted values or $\hat{y}_i$. Here we expect to see a horizontal band of points with no patterns or trends.

Finally, to see if there is bias in our estimated curve, we can graph the residuals against the independent variables, where we would also expect to see a horizontal band of points. This is called a *residual dependence plot* [Cleveland, 1993]. If we have multiple predictors, then we can construct one of these plots for each variable. As the next example shows, we can enhance the diagnostic power of these scatterplots by superimposing a loess smooth.

Example 7.4

We turn to the software inspection data that was described in Chapter 1 to illustrate the various residual plots. Here we are interested in determining the relationship between the number of defects found as a function of the time spent inspecting the code or document. We have data that consists of 491 observations, with the x value representing the preparation or inspection time per page, and the response y is the number of defects found per page. After loading the data, we transform it because both variables are skewed. The initial scatterplot of the transformed data is shown in Figure 7.7, where we see that the relationship between the variables seems to be roughly linear.

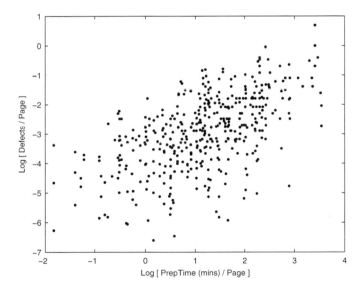

FIGURE 7.7
This is the scatterplot of observations showing the number of defects found per page versus the time spent inspecting each page. We see that the relationship is approximately linear.

```
load software
% Transform the data.
X = log(prepage);
Y = log(defpage);
% Get an initial plot.
plot(X,Y,'.')
xlabel('Log [ PrepTime (mins) / Page ]')
ylabel('Log [ Defects / Page ]')
```

Next we set up the parameters ($\alpha = 0.5$, $\lambda = 2$) for a loess smooth and show the smoothed scatterplot in Figure 7.8.

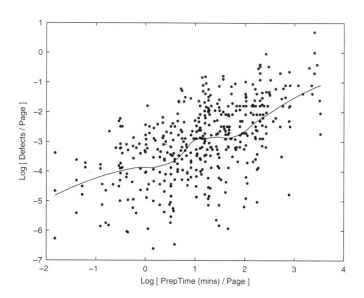

FIGURE 7.8

After we add the loess curve ($\alpha = 0.5$, $\lambda = 2$), we see that the relationship is not completely linear.

```
% Set up the parameters.
alpha = 0.5;
lambda = 2;
% Do the loess on this.
x0 = linspace(min(X),max(X));
y0 = loess(X,Y,x0,alpha,lambda);
% Plot the curve and scatterplot.
plot(X,Y,'.',x0,y0)
xlabel('Log[PrepTime (mins)/Page]')
ylabel('Log[Defects/Page]')
```

We can assess our results by looking at the residual plots. First we find the residuals and plot them in Figure 7.9 (top), where we see that they are roughly symmetric about zero. Then we plot the absolute value of the residuals against the fitted values (Figure 7.9 (bottom)). A loess smooth of these observations shows that the variance does not seem to be dependent on the fitted values.

```
% Get the residuals.
% First find the loess values at the observed X values.
yhat = loess(X,Y,X,alpha,lambda);
```

```
resid = Y - yhat;
% Now plot the residuals.
plot(1:length(resid),resid,'.')
ax = axis;
axis([ax(1), ax(2), -4 4])
xlabel('Index')
ylabel('Residuals')
% Plot the absolute value of the residuals
% against the fitted values.
r0 = linspace(min(yhat),max(yhat),30);
rhat = loess(yhat,abs(resid),r0,0.5,1);
plot(yhat,abs(resid),'.',r0,rhat)
xlabel('Fitted Values')
ylabel('| Residuals |')
```

The following code constructs a residual dependence plot for this loess smooth. We include a loess smooth for this scatterplot to better understand the results. This is shown in Figure 7.10; we do not see any indication of bias.

```
% Now plot the residuals on the vertical
% and the independent values on the
% horizontal. This is the residual
% dependence plot. Include a loess curve.
rhat = loess(X,resid,x0,.5,1)
plot(X,resid,'.',x0,rhat)
xlabel('Log[PrepTime (mins)/Page]')
ylabel('Residuals')
```

We continue our analysis of these results in the next example.
❏

7.4.2 Spread Smooth

It might be important in certain applications to understand the spread of y given x. We can try to ascertain this just by looking at the scatterplot of the variables, but as we have seen, it is sometimes hard to judge these relationships just from a scatterplot. Cleveland and McGill [1984] describe *spread smoothing* as a way of graphically addressing this issue.

Procedure – Spread Smooths

1. Compute the fitted values $\hat{y}_i$, using loess or some other appropriate estimation procedure.
2. Calculate the residuals $\hat{\varepsilon}_i$ using Equation 7.7.
3. Plot $|\hat{\varepsilon}_i|$ against x_i in a scatterplot.

FIGURE 7.9
The upper plot shows the residuals based on the loess curve from Figure 7.8, and we see a nice horizontal band of points. In the lower panel, we have the absolute value of the residuals versus the fitted values. While not perfect, these indicate that the variance is approximately constant.

FIGURE 7.10
This shows the residual dependence plot with a superimposed loess curve. We do not see any indication of bias in the estimated curve.

4. Smooth the scatterplot using loess and add the curve to the plot.

The smoothed values found in step 4 comprise the spread smoothing. We illustrate its use in Example 7.5.

Example 7.5
We show the spread smooth using the same data and residuals as in the previous example. Note that this is similar to the plot we have in Figure 7.9 (bottom), but this time we fit the absolute value of the residuals to the *observed predictor* values. The scatterplot with loess curve given in Figure 7.11 shows that the variance is fairly constant for the observed values of x.

```
% The y values in this plot will be
% the absolute value of the residuals.
% Superimpose a loess curve to better
% assess results.
r0 = linspace(min(X),max(X),30);
rhat = loess(X,abs(resid),r0,0.5,1);
plot(X,abs(resid),'.',r0,rhat)
xlabel('Log [ PrepTime (mins) / Page ]')
ylabel('| Residuals |')
```

❏

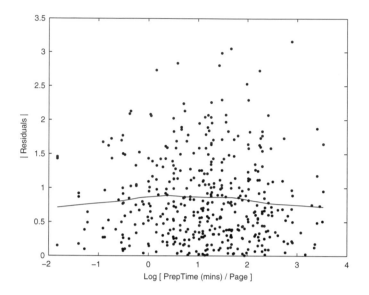

FIGURE 7.11
Here we have the spread smooth plot for the residuals found in Example 7.4. We see that
the variance is fairly constant.

7.4.3 Loess Envelopes — Upper and Lower Smooths

The loess smoothing method provides a model of the middle of the
distribution of y given x. This can be extended to give us upper and lower
smooths [Cleveland and McGill, 1984], where the distance between the upper
and lower smooths indicates the spread. This shows similar information to
the spread smooth, but in a way that is more in keeping with the typical error
bar plots. The procedure for obtaining the upper and lower smooths follows.

Procedure – Upper and Lower Smooths (Loess)

1. Compute the fitted values $\hat{y}_i$ using loess or robust loess.

2. Calculate the residuals $\hat{\varepsilon}_i = y_i - \hat{y}_i$.

3. Find the positive residuals $\hat{\varepsilon}_i^+$ and the corresponding x_i and $\hat{y}_i$
   values. Denote these pairs as $(x_i^+, \hat{y}_i^+)$.

4. Find the negative residuals $\hat{\varepsilon}_i^-$ and the corresponding x_i and $\hat{y}_i$
   values. Denote these pairs as $(x_i^-, \hat{y}_i^-)$.

5. Smooth the $(x_i^+, \hat{\varepsilon}_i^+)$ and add the fitted values from that smooth to
   $\hat{y}_i^+$. This is the upper smoothing.

6. Smooth the $(x_i^-, \hat{\varepsilon}_i^-)$ and add the fitted values from this smooth to $\hat{y}_i^-$. This is the lower smoothing.

Example 7.6

We do not show all of the MATLAB code to implement the upper and lower envelopes. Instead, we include a function that will provide them and just show how to use it in this example. We return to the same software inspection data used in the previous examples. The following code invokes the **loessenv** function and plots the curves.

```
% Get the envelopes and plot.
[yhat,ylo,xlo,yup,xup] = loessenv(X,Y,x0,0.5,2,1);
plot(X,Y,'.',x0,y0,xlo,ylo,xup,yup)
xlabel('Log [ PrepTime (mins) / Page ]')
ylabel('Log [ Defects / Page ]')
```

The loess curve with envelopes is given in Figure 7.12. The lower, middle, and upper smooths indicate that the distribution of y given x is symmetric at most values of x and that the variance is fairly constant.
☐

It is important to note that the methods described in this section for assessing the output of loess can also be used when other smoothing techniques are used. This includes the smoothing spline, which is the topic of the next section.

7.5 Smoothing Splines

We now turn our attention to another method called *smoothing splines*. The word spline comes from the engineering community, where draftsmen used long thin strips of wood called splines. They used these splines to draw a smooth curve between points; different global curves are produced when the positions of the points are changed. As we will see, a smoothing spline is a solution to a constrained optimization problem, where we optimally trade off fidelity of the fit to the data with smoothness of the estimate. To set the stage, we will first briefly describe parametric *spline regression* models, where a piecewise polynomial model is used to find the fit between the predictor variable and the response. This will be followed by a discussion of how these ideas can be extended to provide scatterplot smooths based on splines.

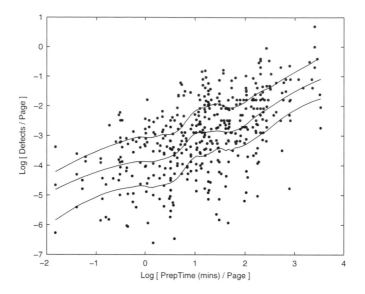

FIGURE 7.12
The lower, middle, and upper smooths indicate that the variance is constant and that the distribution of y given x is symmetric.

7.5.1 Regression with Splines

Regression splines use a piecewise polynomial fit to model the relationship between a predictor (or independent) variable and a response (or dependent) variable. These spline models can provide a way to adequately model the relationship between the variables, while avoiding negative consequences such as collinearity and high complexity. Some good references for the following discussion on splines are Marsh and Cormier [2002], Wahba [1990], Hastie and Tibshirani [1990], de Boor [2001], and Martinez and Martinez [2015].

Since we use pieces of different polynomials to construct the fit, we have to define the regions for each of the polynomials. These regions are determined by a sequence of points called **knots**. The set of **interior knots** will be denoted by t_i, where $t_1 < \ldots < t_K$, and the **boundary knots** will be represented by t_0 and t_{K+1}. Depending on the model, different conditions are satisfied at the knot points. For example, in the piecewise linear case, the function is continuous at the knot points, but the first (and higher) derivative is discontinuous.

In what follows, we assume that the number of knots and their locations are known. If the number and the location of knots have to be estimated, then one has to use more advanced methods, such as nonlinear least squares or stepwise regression. We refer the interested reader to Marsh and Cormier [2002] or Lee [2002] for details on these approaches and others.

The spline model is given by

$$f(X) = \beta_0 + \sum_{j=1}^{D} \beta_j X^j + \sum_{j=1}^{K} \beta_{j+D}\delta(t_j)(X-t_j)^D , \tag{7.10}$$

where

$$\delta(t_j) = \begin{cases} 0; & X \le t_j \\ 1; & X > t_j. \end{cases}$$

The quantity $\delta(t_j)$ is sometimes known as a **dummy variable** often used in regression analysis to distinguish different groups. They allow us to turn off the parameters in our model, yielding the desired piecewise polynomial. It should be clear from equation 7.10 that our function $f(X)$ is a linear combination of monomials.

Another way to represent this model is to use a truncated power basis of degree D:

$$f(X) = \beta_0 + \beta_1 X + \ldots + \beta_D X^D + \sum_{j=1}^{K} \beta_{j+D}(X-t_j)_+^D , \tag{7.11}$$

where $(a)_+$ is the positive part of a, which has the same effect as our dummy variable in equation 7.10 [Lee, 2002]. Other basis functions such as B-splines or radial basis functions can also be used [Green and Silverman, 1994].

Martinez and Martinez [2015] provide an example of using a regression spline model to estimate the fit from simulated data that uses the following true piecewise linear function [Marsh and Cormier, 2002]:

$$f(X) = \begin{cases} 55-1.4x; & 0 \le x \le 12 \\ 15+1.8x; & 13 \le x \le 24 \\ 101-1.7x; & 25 \le x \le 36 \\ -24+1.7x; & 37 \le x \le 48. \end{cases} \tag{7.12}$$

We can see from these equations that we would have three internal knots and two boundary knots. The scatterplot and regression fit for the simulated data are shown in Figure 7.13.

The polynomials in Equation 7.12 were linear ($D = 1$). We can see from Figure 7.13 that the function is continuous at the knot points, but the first derivative is not continuous. If we use splines with higher degree (e.g.,

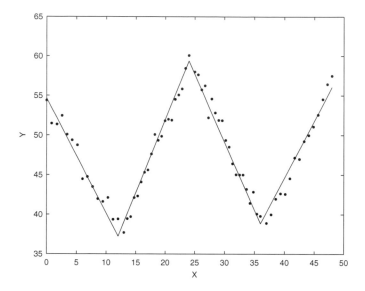

FIGURE 7.13
This is the spline regression fit and scatterplot for the simulated data based on the true relationship given by Equation 7.12 [Martinez and Martinez, 2015].

quadratic or cubic), then we will have smoother functions because additional constraints are placed on the higher derivatives at the knot points. If we use piecewise cubic polynomials, then we will have a continuous function with continuous first and second derivatives. In general, spline regression of degree $D \geq 1$ will provide continuous functions that have continuous derivatives up to order $D - 1$ at the knots.

7.5.2 Smoothing Splines

We now explain the idea of modeling relationships using piecewise polynomials to form the basis of a method called *smoothing splines*. We use the treatment found in Green and Silverman [1994] and Martinez and Martinez [2015] for the description of smoothing splines given below. We are going to switch the notation slightly in what follows to match these references. So, our estimate will now be denoted by $\hat{f}$ instead of $\hat{y}$.

Smoothing splines take a roughness penalty approach. This means that we find our estimate $\hat{f}$ by optimizing an objective function that includes a penalty for the roughness of $\hat{f}$. Given any twice-differentiable function f defined on the interval $[a, b]$, we define the penalized sum of squares by

$$S(f) = \sum_{i=1}^{n} \left[Y_i - f(x_i) \right]^2 + \alpha \int_a^b \left[f''(t) \right]^2 dt \ , \qquad (7.13)$$

where $\alpha > 0$ is the smoothing parameter. The estimate $\hat{f}$ is defined to be the minimizer of $S(f)$ over the class of all twice-differentiable functions f. The only constraint we have on the interval $[a, b]$ is that it must contain all of the observed x_i [Green and Silverman, 1994].

The term on the left in equation 7.13 is the familiar residual sum of squares that determines the goodness-of-fit to the data. The term on right is a measure of variability in the fit, and it represents the roughness penalty term. The smoothing parameter α governs the trade-off between smoothness and goodness-of-fit. For large values of α, we will have a very smooth curve. When α approaches infinity, the curve $\hat{f}$ approaches the fit we get from linear regression. As α approaches zero, we get closer to a fit that interpolates the data.

A function f defined on some interval $[a, b]$ is a ***cubic spline*** with knots given by $a < t_1 < \ldots < t_n < b$ if both of the following are true:

1. The function f is a cubic polynomial on each of the intervals (a, t_1), $(t_1, t_2), \ldots, (t_n, b)$.
2. The polynomials connect at the knots such that f and its first two derivatives are continuous at each t_i.

The second condition implies that the entire function f is continuous over the interval $[a, b]$ [Green and Silverman, 1994].

A ***natural cubic spline*** is a cubic spline with additional conditions at the boundary that must be satisfied. A cubic spline over the interval $[a, b]$ is a *natural* cubic spline if its second and third derivatives are zero at the end points a and b. This means that the function f is linear on the two boundary intervals $[a, t_1]$ and $[t_n, b]$.

In what follows, we will assume that $n \geq 3$, and we also assume that the observed predictor values x_i are *ordered*. Green and Silverman [1994] show that a natural cubic spline can be represented by the value of f and the second derivative at each of the knots t_i. If f is a natural cubic spline with knots $t_1 < \ldots < t_n$, then we define the value at the knot as

$$f_i = f(t_i) \, ,$$

and the second derivative at the knot as

$$\gamma_i = f''(t_i) \, ,$$

for $i = 1, \ldots, n$. We know from the definition of a natural cubic spline that the second derivatives of the end points are zero, so we have $\gamma_1 = \gamma_n = 0$.

We put the values of the function and the second derivative into vectors $\mathbf{f}$ and γ, where $\mathbf{f} = (f_1, ..., f_n)^T$ and $\gamma = (\gamma_2, ..., \gamma_{n-1})^T$. It is important to note that the vector γ is defined and indexed in a nonstandard way.

We now define two band matrices $\mathbf{Q}$ and $\mathbf{R}$ that will be used in the algorithm for smoothing splines. A *band matrix* is one whose nonzero entries are contained in a band along the main diagonal. The matrix entries outside this band are zero. First, we let $h_i = t_{i+1} - t_i$, for $i = 1, ..., n-1$. Then $\mathbf{Q}$ is the matrix with nonzero entries q_{ij}, given by

$$q_{j-1,j} = \frac{1}{h_{j-1}} \qquad q_{jj} = -\frac{1}{h_{j-1}} - \frac{1}{h_j} \qquad q_{j+1,j} = \frac{1}{h_j}, \qquad (7.14)$$

for $j = 2, ..., n-1$. Since this is a band matrix, the elements q_{ij} are zero for index values where $|i - j| \geq 2$. From this definition, we see that the columns of the matrix $\mathbf{Q}$ are indexed in a nonstandard way with the top left element given by q_{12} and that the size of $\mathbf{Q}$ is $n \times (n-2)$.

The matrix $\mathbf{R}$ is an $(n-2) \times (n-2)$ symmetric matrix with nonzero elements given by

$$r_{ii} = \frac{h_{i-1} + h_i}{3}; \qquad i = 2, ..., n-1, \qquad (7.15)$$

and

$$r_{i,i+1} = r_{i+1,i} = \frac{h_i}{6}; \qquad i = 2, ..., n-2. \qquad (7.16)$$

The elements r_{ij} are zero for $|i - j| \geq 2$.

It turns out that not all vectors $\mathbf{f}$ and γ (as defined above) will represent natural cubic splines. Green and Silverman [1994] show that the vectors $\mathbf{f}$ and γ specify a natural cubic spline if and only if

$$\mathbf{Q}^T\mathbf{g} = \mathbf{R}\gamma. \qquad (7.17)$$

We will use these matrices and relationships to help us find the estimated curve $\hat{f}$ that minimizes the penalized sum of squares in equation 7.13.

Reinsch [1967] proved a remarkable result that the function $\hat{f}$ that uniquely minimizes the penalized sum of squares is a natural cubic spline with knots given by the observed values x_i [Hastie and Tibshirani, 1990; Green and Silverman, 1994]. So, we will use the (ordered) observations x_i in place of our knot notation in the following discussion.

Reinsch also developed an algorithm to construct the smoothing spline. This is done by forming a system of linear equations for the γ_i at the knot locations. We can then obtain the values of the smoothing spline in terms of the γ_i and the observed y_i. See Green and Silverman [1994] for details on the derivation of the algorithm.

Reinsch procedure for smoothing splines

1. Given the observations $\{x_i, y_i\}$, where $a < x_1 < ... < x_n < b$ and $n \geq 3$, we collect them into vectors

$$\mathbf{x} = (x_1, ..., x_n)^T \text{ and } \mathbf{y}_i = (y_1, ..., y_n)^T.$$

2. Find the matrices $\mathbf{Q}$ and $\mathbf{R}$, using equations 7.14 through 7.16.
3. Form the vector $\mathbf{Q}^T\mathbf{y}$ and the matrix $\mathbf{R} + \alpha\mathbf{Q}^T\mathbf{Q}$.
4. Solve the following equation for γ:

$$\mathbf{R} + \alpha\mathbf{Q}^T\mathbf{Q}\gamma = \mathbf{Q}^T\mathbf{y}. \tag{7.18}$$

5. Find our estimate $\hat{\mathbf{f}}$ from

$$\hat{\mathbf{f}} = \mathbf{y} - \alpha\mathbf{Q}\gamma. \tag{7.19}$$

Example 7.7

We will use the **ozone** data set to illustrate the Reinsch method. The data set represents the daily maximum ozone concentrations (ppb) during the summer of 1974. There are two vectors in the file—one for Yonkers, New York and one for Stamford, Connecticut. We will use the Yonkers data in this example. First we load the data and set the smoothing parameter. Note that in many cases, we will have to sort the x_i values, so they can be knots. In this example, our x_i are an index and are already sorted.

```
load ozone
% Assign an index value for X
X = 1:length(Yonkers);
Y = Yonkers;
% In other cases, you might need to sort X
% so they can be knots.
% Make them column vectors-just in case.
x = x(:); y = y(:);
n = length(x);
alpha = 30;
```

The next step is to find the **Q** and **R** matrices using equations. Note that we keep the matrices at a full size of n by n to keep the indices correct. We then remove the columns or rows that are not needed.

```
% Next we get the Q and R matrices.
h = diff(x);
% Find 1/h_i;
hinv = 1./h;
% Keep the Q matrix as n by n originally,
% so the subscripts match the book.
% Then remove the first and last column.
qDs = -hinv(1:n-2) - hinv(2:n-1);
I = [1:n-2, 2:n-1, 3:n];
J = [2:n-1,2:n-1,2:n-1];
S = [hinv(1:n-2), qDs, hinv(2:n-1)];
% Create a sparse matrix.
Q = sparse(I,J,S,n,n);
% Delete the first and last columns.
Q(:,n) = []; Q(:,1) = [];
% Now find the R matrix.
I = 2:n-2;
J = I + 1;
tmp = sparse(I,J,h(I),n,n);
t = (h(1:n-2) + h(2:n-1))/3;
R = tmp'+tmp+sparse(2:n-1,2:n-1,t,n,n);
% Get rid of the rows/cols that are not needed.
R(n,:) = []; R(1,:) = [];
R(:,n) = []; R(:,1) = [];
```

The final step is to find the smoothing spline using steps 3 through 5 of the Reinsch method.

```
% Get the smoothing spline.
S1 = Q'*y;
S2 = R + alpha*Q'*Q;
% Solve for gamma;
gam = S2\S1;
% Find fhat.
fhat = y - alpha*Q*gam;
```

The smoothing spline is shown in Figure 7.14. We can see some interesting fluctuations in the smooth that are not apparent in the scatterplot alone. There is a function (**splinesmth**) included in the EDA Toolbox that constructs a fit based on the smoothing spline.
❑

    The previous example showed how to get a smoothing spline at the knot locations. However, we might also want to get values of the spline at

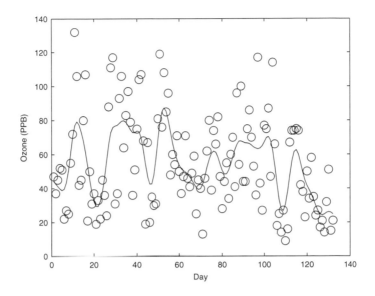

FIGURE 7.14
This shows the scatterplot with the smoothing spline for the ozone levels in Yonkers, New York from May to September 1974. It is interesting to note that the smooth shows some interesting structure and fluctuations that are not apparent in the scatterplot alone.

arbitrary target values so we can make predictions or plot the curve. Green and Silverman [1994] show how one can plot a full cubic spline using the vectors **f** and γ and knots $t_1 < \ldots < t_n$.

First, we give the expression for finding the value of the cubic spline for any value of x in intervals with endpoints given by the knots:

$$f(x) = \frac{(x - t_i) f_{i+1} + (t_{i+1} - x) f_i}{h_i} -$$

$$\frac{1}{6}(x - t_i)(t_{i+1} - x)\left[\left(1 + \frac{x - t_i}{h_i}\right)\gamma_{i+1} + \left(1 + \frac{t_{i+1} - x}{h_i}\right)\gamma_{ii}\right]$$

(7.20)

for $t_i \le x \le t_{i+1}$, $i = 1, \ldots, n - 1$. So, we see that equation 7.20 is valid over the intervals between each of the knots.

To find the value of the cubic spline outside of the interval $[t_1, t_n]$, we first need the first derivative at the outside knots:

$$f'(t_1) = \frac{f_2 - f_1}{t_2 - t_1} - \frac{1}{6}(t_2 - t_1)\gamma_2$$

$$f'(t_n) = \frac{f_n - f_{n-1}}{t_n - t_{n-1}} + \frac{1}{6}(t_n - t_{n-1})\gamma_{n-1}.$$

(7.21)

Since f is linear outside $[t_1, t_n]$, we have

$$f(x) = f_1 - (t_1 - x)f'(t_1); \qquad \text{for } a \le x \le t_1$$

$$f(x) = f_n + (x - t_n)f'(t_n); \qquad \text{for } t_n \le x \le b.$$

(7.22)

We used the more general notation in the derivation given above, but these can also be used to plot the estimated function $\hat{\mathbf{f}}$ in equation 7.19.

7.5.3 Smoothing Splines for Uniformly Spaced Data

Several approaches exist for smoothing when one has data that are evenly sampled over the domain, (e.g., time series). Some examples of these approaches are the moving average and exponential smoothing. Garcia [2010] recently developed a fast smoothing spline methodology for these type of data. His method is based on the discrete cosine transform and is suitable for equally spaced data in one dimension and higher. His methodology also includes steps to produce a robust estimate, to automatically choose the smoothing parameter, and to handle missing values.

We first present the case for one-dimensional data and then discuss how it can be extended to multi-dimensions. Assume one has an observed noisy one-dimensional signal $y = \hat{y} + \varepsilon$, where ε is Gaussian noise with mean zero and unknown variance. We also assume that $\hat{y}$ is smooth. Garcia takes the penalized least squares regression approach to smoothing which minimizes an objective function that balances fitting the data well and a term that penalizes the roughness of the fit. Thus, we seek to minimize

$$F(\hat{y}) = \|\hat{y} - y\|^2 + \alpha P(\hat{y}),$$

(7.23)

where $\|\cdot\|$ denotes the Euclidean norm, α is our smoothing parameter, and P is the penalty term for roughness. We can see the connection with smoothing splines when we compare Equation 7.23 with Equation 7.13.

Another way to model the roughness is to use a second-order divided difference, so

$$P(\hat{y}) = \|\mathbf{D}\hat{y}\|^2,$$

(7.24)

where **D** is a tridiagonal square matrix that has a simplified form when the data are observed on a uniform grid. Using Equations 7.23 and 7.24 and minimizing $F(\hat{y})$ yields

$$(\mathbf{I}_n + \alpha \mathbf{D}^T \mathbf{D})\hat{y} = y, \tag{7.25}$$

where $\mathbf{I}_n$ is the $n \times n$ identity matrix.

Since the data are evenly spaced, then the matrix **D** can be written as

$$\begin{bmatrix} -1 & 1 & & & & \\ 1 & -2 & 1 & & & \\ & \cdots & \cdots & \cdots & & \\ & & 1 & -2 & 1 \\ & & & 1 & -1 \end{bmatrix}.$$

We can perform an eigen decomposition of **D** to obtain

$$\mathbf{D} = \mathbf{U} \Lambda \mathbf{U}^{-1},$$

where Λ is a diagonal matrix. The diagonal elements contain the eigenvalues of **D** that are in the following form [Yuch, 2005]

$$\lambda_{ii} = -2 + 2\cos((i-1)\pi/n). \tag{7.26}$$

The matrix **U** is a unitary matrix, which means that

$$\mathbf{U}^{-1} = \mathbf{U}^T$$
$$\mathbf{U}\mathbf{U}^T = \mathbf{I}_n.$$

This allows us to write Equation 7.25 as

$$\hat{y} = \mathbf{U}(\mathbf{I}_n + \alpha \Lambda^2)^{-1} \mathbf{U}^T y \equiv \mathbf{U}\Gamma \mathbf{U}^T y, \tag{7.27}$$

where, given Equation 7.26, the elements of the diagonal matrix Γ are given by

$$\Gamma_{ii} = [1 + \alpha(2 - 2\cos((i-1)\pi/n))^2]^{-1}.$$

Garcia notes that **U** is an $n \times n$ inverse discrete cosine transform (IDCT) matrix, and $\mathbf{U}^T$ is a discrete cosine transform matrix (DCT).

$$\hat{y} = \mathbf{U}\boldsymbol{\Gamma}\mathbf{U}^T y = \text{IDCT}(\boldsymbol{\Gamma}\text{DCT}(y)).\qquad\qquad(7.28)$$

This idea of using the discrete cosine transform in smoothing has also been introduced by Buckley [1994].

In his paper, Garcia [2010] describes a procedure for automatically selecting a good value of the smoothing parameter α using generalized cross-validation (GCV). The GCV was developed by Craven and Wahba [1979] in the context of smoothing splines. A full discussion of this approach is not provided here, but some references are provided at the end of this chapter.

Missing data can occur in many applications, or we might want to weight observations based on their quality (e.g., outliers). Garcia addresses these issues in his methodology by incorporating weights into the smoothing process. If there are missing values, then the weight is given by zero, and an arbitrary finite value is assigned to the corresponding value of y. In this case, Garcia's algorithm will interpolate and smooth, and the values assigned to the missing data are estimated using the whole data set.

Garcia also uses this weighted smoothing to deal with outliers, resulting in a robust smooth. These are similar to what we discussed with loess, where the residuals are calculated, followed by the calculation of the bisquare weights. The smooth is recalculated using these weighted residual values. The following example uses an illustration and function provided by Garcia [2009].

Example 7.8

We downloaded the function called **smoothn** from Garcia [2009], as well as the two functions for performing the DCT and IDCT (**dctn** and **idctn**). These have been included with the EDA Toolbox for convenience. Garcia provides MATLAB code and figures for several examples, which can also be seen by typing help **smoothn** at the command line. We will use a data set from Simonoff [1996] to illustrate this approach. The data represents the grape yields in number of lugs for harvests from 1989 through 1991. The independent variable is the row number, which is uniformly spaced; so we can use this smoothing approach. The following code loads the data and performs the smooth.

```
% Load the data. There are two variables.
% The x is the row number, and y is the number of lugs.
load vineyard
yhat = smoothn(totlugcount);
plot(row, totlugcount,'o',row,yhat)
```

A scatterplot of the data, along with the superimposed smooth, is provided in Figure 7.15, where we see that the procedure did a good job of smoothing the data.
❑

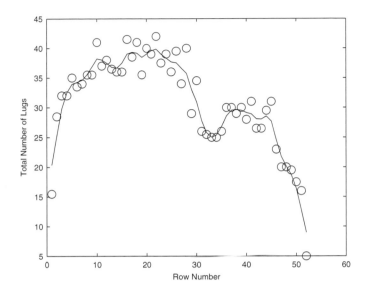

FIGURE 7.15
This shows a scatterplot of the **vineyard** data. A smooth based on Garcia's approach for uniformly spaced data is superimposed on the plot. The function **smoothn** was used to obtain the smooth, and it automatically chooses an appropriate smoothing parameter.

Garcia's approach generalizes easily to higher-dimensional evenly sampled data. This comes about because a multidimensional discrete cosine transform is a composition of one-dimensional DCTs along each of the dimensions [Strang, 1999; Garcia, 2010]. The **smoothn** function that we used in the previous example will provide smooths for multidimensional data.

7.6 Choosing the Smoothing Parameter

Most smoothing methods have some parameter that determines the smoothness of the estimate, e.g., $\hat{f}$. For example, this might be the number of regions used in the bin smoother method, the size of the neighborhood in the running line smoother,[1] or the smoothing parameter α in loess and smoothing splines [Martinez and Martinez, 2015]. In what follows, we will use the term *smoothing parameter* denoted by s to represent all of these quantities that drive the smoothness of our estimated curve.

In general, the smoothing parameter s can take on a range of values that depends on the method that is used. When s takes on values at one end of the range, then the curve will more closely follow the data, yielding a rougher or

---

[1] See the exercises for a description of the bin smoother and the running line smoother.

wiggly curve. As the value of the parameter moves to the other end of the range of parameter values, the curve becomes smoother. The choice of the smoothing parameter is an important one, because we might get estimates that show different types of structures (e.g., modes, trends, etc.) as we change the value of the smoothing parameter.

We could try an exploratory approach to choose s, where we construct estimates $\hat{f}$ for different values of the smoothing parameter and plot the curve on the scatterplot. This yields a set of curves and plots that are then examined by the analyst. It is possible that these estimates $\hat{f}$ with different degrees of smoothing will yield different structures or features. It is also possible that if one sees the *same* structure in many of the curves (i.e., the structure has persistence), then it is likely that the structure is *not* an artifact of the smoothing parameter that is used and might be meaningful.

We could also use a way that is more data-driven. This second approach uses cross-validation based on the prediction error, which is defined as the average squared error between the predicted response $\hat{f}$ and the observed response [Martinez and Martinez, 2015].

Cross-validation is a general technique that can be used to estimate the accuracy of a model. The basic idea is to partition the data into K subsets of approximately equal size. One partition is reserved for testing, and the rest of the data are used to get the estimated fit $\hat{f}_k$. Each of the observations in the test set is then used to calculate the squared error of the estimate $\hat{f}_k$. This is repeated K times, yielding n errors.

Perhaps the most accurate form of cross-validation is when $K = n$. In this case, we leave only one data point out at a time and use the $n - 1$ remaining observations in our smoothing. We will denote the estimated function at x_i with smoothing parameter s and leaving out the i-th observation as $\hat{f}_s^{(-i)}(x_i)$. Then, the **cross-validation sum of squares** (sometimes called the **cross-validation score function**) as a function of the smoothing parameter can be written as

$$CV(s) = \frac{1}{n}\sum_{i=1}^{n}\left(y_i - \hat{f}_s^{(-i)}(x_i)\right)^2 . \tag{7.29}$$

We then minimize the cross-validation function to estimate the smoothing parameter. The procedure is to first select a range of suitable values for s. Then, we find $CV(s)$ for each value of s and select the value of s that minimizes $CV(s)$.

Keep in mind that this procedure produces an *estimate* of the smoothing parameter s that optimizes the prediction error. Thus, it is subject to sampling variability, just like any estimate. Fox [2000b] points out that using the $CV(s)$ to choose a value for s can produce estimates that are too small when the sample size n is small. Also, it is important to keep in mind that the value for s from cross-validation might be a starting point for an exploratory procedure as we described earlier.

We can always use the cross-validation procedure and the expression in Equation 7.29, but a faster way to find the value of $CV(s)$ is available [Hastie and Tibshirani, 1990; Green and Silverman, 1994]. This alternative expression for the cross-validation function is valid when the smoothers are linear. A *linear smoother* is one that can be written in terms of a smoother matrix $\mathbf{S}$, as follows

$$\hat{\mathbf{f}} = \mathbf{S}(s)\mathbf{y},$$

where we have the fit $\hat{\mathbf{f}}$ at the observations x_i. $\mathbf{S}$ is called the *smoother matrix* with n rows and n columns, where the notation indicates its dependence on the smoothing parameter s.

The following smoothers are linear: running-mean, running-line, cubic smoothing spline, kernel methods, and loess. The robust version of loess is nonlinear. Another example of a nonlinear smoother would be the running-median, where the median is used in each neighborhood instead of the mean. In the case of smoothing splines, the smoother matrix is

$$\mathbf{S}(\alpha) = (\mathbf{I} + \alpha \mathbf{Q}\mathbf{R}^{-1}\mathbf{Q}^{T})^{-1},$$

where we have $s = \alpha$.

The cross-validation score for smoothing parameter s can be written as

$$CV(s) = \frac{1}{n}\sum_{i=1}^{n}\left(\frac{y_i - \hat{f}(x_i)}{1 - S_{ii}(s)}\right)^2, \tag{7.30}$$

where $S_{ii}(s)$ is the diagonal elements of the smoother matrix, and $\hat{f}(x_i)$ is the value of the fit with smoothing parameter s at x_i using the *full data set*. This saves a lot of computations, because we do not have to find n regression fits for each value of the smoothing parameter s. We will illustrate the cross-validation function for smoothing splines in the next example.

Example 7.9
In this example, we show how to calculate the cross-validation function from Equation 7.30 for smoothing splines. The steps given below use the function **splinesmth** that is included with the EDA Toolbox. This function returns the smoother matrix in addition to the values for $\hat{f}$. We return to the **ozone** data of the previous example, so we can see whether or not our choice for α was a good one.

```
load ozone
% Assign an index value for X
x = 1:length(Yonkers);
```

```
y = Yonkers;
% Set some possible values for alpha.
alpha = 1:0.5:50;
CV = zeros(1,length(alpha));
for i = 1:length(CV)
    i
    [x0,fhat,S] = splinesmth(x,y,alpha(i));
    num = y - fhat;
    den = 1 - diag(S);
    CV(i) = mean((num./den).^2);
end
% Find the minimum value of CV(alpha).
[m,ind] = min(CV);
% Find the corresponding alpha.
cvalpha = alpha(ind);
```

The value of the smoothing parameter that corresponds to the minimum of the cross-validation function is $\alpha = 8$. Our choice of $\alpha = 30$ in the previous example was not a very good one, according to this approach. The reader is asked in the exercises to construct the smoothing spline for $\alpha = 8$ and to compare the smooths. A plot of the cross-validation function is shown in Figure 7.16.
❏

FIGURE 7.16
This shows the cross-validation function for a smoothing spline fit to the **ozone** data of Example 7.9. According to this, the optimal value for the smoothing parameter is 8.

7.7 Bivariate Distribution Smooths

We now discuss some smoothings that can be used to graphically explore and summarize the distribution of two variables. Here we are not only looking to see how y depends on x, but also how x depends on y. Plotting pairs of loess smoothings on a scatterplot is one way of understanding this relationship.[2] Polar smoothing can also be used to understand the bivariate distribution between x and y by smoothing the edges of the point cloud.

7.7.1 Pairs of Middle Smoothings

In our previous examples of loess, we were in a situation where y was our response variable and x was the predictor variable, and we wanted to model or explore how y depends on the predictor. However, there are many situations where none of the variables is a factor or a response, and the goal is simply to understand the bivariate distribution of x and y.

We can use pairs of loess curves to address this situation. The idea is to smooth y given x, as before, and also smooth x given y [Cleveland and McGill, 1984; Tukey, 1977]. Both of these smooths are plotted simultaneously on the scatterplot. We illustrate this in Example 7.10, where it is applied to the **software** data.

Example 7.10

We now look at the bivariate relationships between three variables: number of defects per SLOC, preparation time per SLOC, and meeting time per SLOC (single line of code). First we get some of the parameters needed for obtaining the loess smooths and for constructing the scatterplot matrix. Notice also that we are transforming the data first using the natural logarithm, because the data are skewed.

```
% Get some things needed for plotting.
vars = ['log Prep/SLOC';...
   ' log Mtg/SLOC';' log Def/SLOC'];
% Transform the data using logs.
X = log(prepsloc);
Y = log(mtgsloc);
Z = log(defsloc);
% Set up the parameters.
alpha = 0.5;
```

---

[2] There are other ways to convey and understand a bivariate distribution, such as probability density estimation. The finite mixture method is one way (Chapter 6), and histograms is another (Chapter 9).

```
lambda = 2;
n = length(X);
```

Next we obtain all pairs of smooths; there are six of them.

```
% Get the pairs of middle smoothings.
% There should be 6 unique cases of these.
% First get domains.
x0 = linspace(min(X),max(X),50);
y0 = linspace(min(Y),max(Y),50);
z0 = linspace(min(Z),max(Z),50);
% Now get the curves.
xhatvy = loess(Y,X,y0,alpha,lambda);
yhatvx = loess(X,Y,x0,alpha,lambda);
xhatvz = loess(Z,X,z0,alpha,lambda);
zhatvx = loess(X,Z,x0,alpha,lambda);
yhatvz = loess(Z,Y,z0,alpha,lambda);
zhatvy = loess(Y,Z,y0,alpha,lambda);
```

Finally, we construct the scatterplot matrix. We use MATLAB's Handle Graphics to add the lines to each plot.

```
% Now do the plotmatrix.
data = [X(:),Y(:),Z(:)];
% gplotmatrix is in the Statistics Toolbox.
[H,AX,BigAx] = gplotmatrix(data,[],[],'k','.',...
      0.75,[],'none',vars,vars);
% Use Handle Graphics to construct the lines.
axes(AX(1,2));
line(y0,xhatvy);line(yhatvx,x0,'LineStyle','--');
axes(AX(1,3));
line(z0,xhatvz);line(zhatvx,x0,'LineStyle','--');
axes(AX(2,1));
line(x0,yhatvx);line(xhatvy,y0,'LineStyle','--');
axes(AX(2,3));
line(z0,yhatvz);line(zhatvy,y0,'LineStyle','--');
axes(AX(3,1));
line(x0,zhatvx);line(xhatvz,z0,'LineStyle','--');
axes(AX(3,2));
line(y0,zhatvy);line(yhatvz,z0,'LineStyle','--');
```

The results are shown in Figure 7.17. Note that we have the smooths of y given x shown as solid lines, and the smooths of x given y plotted using dashed lines. In the lower left plot, we see an interesting relationship between the preparation time and number of defects found, where we see a local maximum, possibly indicating a higher rate of defects found.
❑

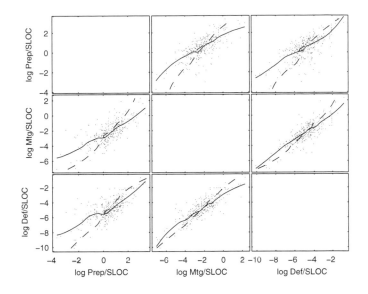

FIGURE 7.17
This shows the scatterplot matrix with superimposed loess curves for the **software** data, where we look at inspection information per SLOC. The solid lines indicate the smooths for y given x, and the dashed lines are the smooths for x given y.

7.7.2 Polar Smoothing

The goal of *polar smoothing* is to convey where the central portion of the bivariate point cloud lies. This summarizes the position, shape, and orientation of the cloud of points. Another way to achieve this is to find the *convex hull* of the points, which is defined as the smallest convex polygon that completely encompasses the data. However, the convex hull is sensitive to outliers because it must include them, so it can overstate the area covered by the data cloud. Polar smoothing by loess does not suffer from this problem.

We now describe the procedure for polar smoothing. The first three steps implement one of the many ways to center and scale x and y. The smooth is done using polar coordinates based on a form of these pseudovariates. At the end, we transform the results back to the original scale for plotting.

Procedure – Polar Smoothing

1. Normalize x_i and y_i using

$$x_i^* = (x_i - \text{median}(x)) \div \text{MAD}(x)$$

$$y_i^* = ((y_i - \text{median}(y)) \div \text{MAD}(y)),$$

where MAD is the median absolute deviation.

2. Calculate the following values:

$$s_i = y_i^* + x_i^*$$

$$d_i = y_i^* - x_i^*.$$

3. Normalize the s_i and d_i as follows:

$$s_i^* = s_i \div \text{MAD}(s)$$

$$d_i^* = (d_i \div \text{MAD}(d)).$$

4. Convert the (s_i^*, d_i^*) to polar coordinates (θ_i, m_i).
5. Transform the m_i as follows

$$z_i = m_i^{2/3}.$$

6. Smooth z_i as a function of θ_i. This produces the fitted values $\hat{z}_i$.
7. Find the fitted values for m_i by transforming the $\hat{z}_i$ using

$$\hat{m}_i = \hat{z}_i^{3/2}.$$

8. Convert the coordinates $(\theta_i, \hat{m}_i)$ to Cartesian coordinates $(\hat{s}_i^*, \hat{d}_i^*)$.
9. Transform these coordinates back to the original x and y scales by the following:

$$\hat{s}_i = \hat{s}_i^* \times \text{MAD}(s)$$

$$\hat{d}_i = \hat{d}_i^* \times \text{MAD}(d)$$

$$\hat{x}_i = [(\hat{s}_i - \hat{d}_i) \div 2] \times \text{MAD}(x) + \text{median}(x)$$

$$\hat{y}_i = [(\hat{s}_i + \hat{d}_i) \div 2] \times \text{MAD}(y) + \text{median}(y).$$

10. Plot the $(\hat{x}_i, \hat{y}_i)$ coordinates on the scatterplot by connecting each point by a straight line and then closing the polygon by connecting the first point to the n-th point.

More information on the smooth called for in Step 6 is in order. Cleveland and McGill [1984] suggest the following way to use the regular loess procedure to get a circular smooth. Recall that θ_i is an angle for $i = 1, ..., n$. We order the θ_i in ascending order, and we let $j = n/2$ be rounded *up* to an integer. We then give the following points to loess

$$(-2\pi + \theta_{n-j+1}, z_{n-j+1}), ..., (-2\pi + \theta_n, z_n),$$
$$(\theta_1, z_1), ..., (\theta_n, z_n),$$
$$(2\pi + \theta_1, z_1), ..., (2\pi + \theta_j, z_j).$$

The actual smoothed values that we need are those in the second row.

Example 7.11

We use the BPM data to illustrate polar smoothing. Results from Martinez [2002] show that there is some overlap between topics 8 (death of North Korean leader) and topic 11 (helicopter crash in North Korea). We apply the polar smoothing to these two topics to explore this issue. We use ISOMAP nonlinear dimensionality reduction with the IRad dissimilarity matrix to produce bivariate data. The scatterplot for the data after dimensionality reduction is shown in Figure 7.18.

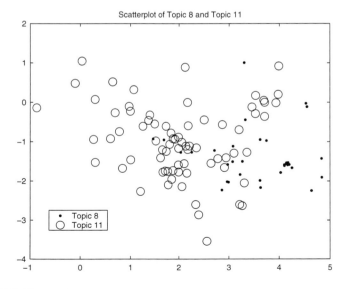

FIGURE 7.18
This is the scatterplot of topics 8 and 11 after we use ISOMAP to reduce the data to 2–D. We see some overlap between the two topics.

```
load L1bpm
% Pick the topics we will look at.
t1 = 8;
t2 = 11;
% Reduce the dimensionality using Isomap.
options.dims = 1:10;      % These are for ISOMAP.
options.display = 0;
[Yiso, Riso, Eiso] = isomap(L1bpm, 'k', 7, options);
% Get the data out.
X = Yiso.coords{2}';
% Get the data for each one and plot
ind1 = find(classlab == t1);
ind2 = find(classlab == t2);
plot(X(ind1,1),X(ind1,2),'.',X(ind2,1),X(ind2,2),'o')
title('Scatterplot of Topic 8 and Topic 11')
```

Now we do the polar smoothing for the first topic.

```
% First let's do the polar smoothing just for
% the first topic. Get the x and y values.
x = X(ind1,1);
y = X(ind1,2);
% Step 1.
% Normalize using the median absolute deviation.
% We will use the Matlab 'inline' functionality.
md = inline('median(abs(x - median(x)))');
xstar = (x - median(x))/md(x);
ystar = (y - median(y))/md(y);
% Step 2.
s = ystar + xstar;
d = ystar - xstar;
% Step 3. Normalize these values.
sstar = s/md(s);
dstar = d/md(d);
% Step 4. Convert to polar coordinates.
[th,m] = cart2pol(sstar,dstar);
% Step 5. Transform radius m.
z = m.^(2/3);
% Step 6. Smooth z given theta.
n = length(x);
J = ceil(n/2);
% Get the temporary data for loess.
tx = -2*pi + th((n-J+1):n);
% So we can get the values back, find this.
ntx = length(tx);
tx = [tx; th];
tx = [tx; th(1:J)];
```

```
ty = z((n-J+1):n);
ty = [ty; z];
ty = [ty; z(1:J)];
tyhat = loess(tx,ty,tx,0.5,1);
% Step 7. Transform the values back.
% Note that we only need the middle values.
tyhat(1:ntx) = [];
mhat = tyhat(1:n).^(3/2);
% Step 8. Convert back to Cartesian.
[shatstar,dhatstar] = pol2cart(th,mhat);
% Step 9. Transform to original scales.
shat = shatstar*md(s);
dhat = dhatstar*md(d);
xhat = ((shat-dhat)/2)*md(x) + median(x);
yhat = ((shat+dhat)/2)*md(y) + median(y);
% Step 10. Plot the smooth.
% We use the convex hull to make it easier
% for plotting.
K = convhull(xhat,yhat);
plot(X(ind1,1),X(ind1,2),'.',X(ind2,1),X(ind2,2),'o')
hold on
plot(xhat(K),yhat(K))
```

We wrote a function called **polarloess** that includes these steps. We use it to find the polar smooth for the second topic and add that smooth to the plot.

```
% Now use the polarloess function to get the
% other one.
[xhat2,yhat2] = ...
    polarloess(X(ind2,1),X(ind2,2),0.5,1);
plot(xhat2,yhat2)
hold off
```

The scatterplot with polar smooths is shown in Figure 7.19, where we get a better idea of the overlap between the groups.
❑

7.8 Curve Fitting Toolbox

We now briefly describe the Curve Fitting Toolbox[3] in this section. The reader is not required to have this toolbox for MATLAB functionality described outside this section. The Curve Fitting Toolbox is a collection of

---

[3] This toolbox is available from The MathWorks, Inc.

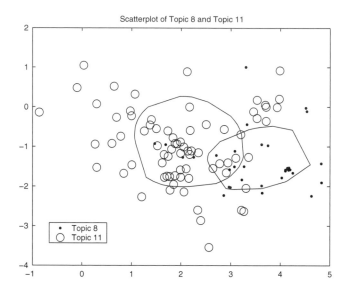

FIGURE 7.19
This is the scatterplot of the two topics with the polar smooths superimposed.

graphical user interfaces (GUIs) and M–file functions that are written for the MATLAB environment, just like other toolboxes.

The toolbox has the following major features for curve fitting:

- One can perform data preprocessing, such as sectioning, smoothing, and removing outliers.
- The analyst can fit curves to points using parametric and nonparametric methods. The parametric approaches include polynomials, exponential, rationals, sums of Gaussians, and custom equations. Nonparametric fits include spline smoothing and other interpolation methods.
- It does standard least squares, weighted least squares, and robust fitting procedures.
- Statistics indicating the goodness-of-fit are also available.

The toolbox can only handle fits between y and x; multivariate predictors are not supported.

The Curve Fitting Toolbox also has a stand-alone function for getting various smooths called **smooth**. The general syntax for this function is

```
x = smooth(x,y,span,method);
```

where **span** governs the size of the neighborhood. There are six methods available, as outlined below:

```
'moving'     - Moving average (default)
'lowess'     - Lowess (linear fit)
'loess'      - Loess (quadratic fit)
'sgolay'     - Savitzky-Golay
'rlowess'    - Robust Lowess (linear fit)
'rloess'     - Robust Loess (quadratic fit)
```

One possible inconvenience with the **smooth** function (as well as the GUI) is that it provides values of the smooth only at the observed x values.

7.9 Summary and Further Reading

Several excellent books on smoothing and nonparametric regression (a related approach) are available. Perhaps the most comprehensive one on loess is called *Visualizing Data* by Cleveland [1993]. This book also includes extensive information on visualization tools such as contour plots, wire frames for surfaces, coplots, multiway dot plots, and many others. For smoothing methods in general, we recommend Simonoff [1996]. It surveys the use of smoothing methods in statistics. The book has an applied focus, and it includes univariate and multivariate density estimation, nonparametric regression, and categorical data smoothing. A compendium of contributions in the area of smoothing and regression can be found in Schimek [2000]. A monograph on nonparametric smoothing in statistics was written by Green and Silverman [1994]; it emphasizes methods rather than the theory.

Loader [1999] provides an overview of local regression and likelihood, including theory, methods, and applications. It is easy to read, and it uses S–Plus code to illustrate the concepts. Efromovich [1999] provides a comprehensive account of smoothing and nonparametric regression, and he also includes time series analysis. Companion software in S–Plus for the text is available over the internet, but he does not include any code in the book itself. Another smoothing book with S–Plus is Bowman and Azzalini [1997]. For the kernel smoothing approach, see Wand and Jones [1995].

Generalized additive models is a way to handle multivariate predictors in a nonparametric fashion. In classical (or parametric) regression, one assumes a linear or some other parametric form for the predictors. In generalized additive models, these are replaced by smooth functions that are estimated by a scatterplot smoother, such as loess. The smooths are found in an iterative procedure. A monograph by Hastie and Tibshirani [1990] that describes this approach is available, and it includes several chapters on scatterplot smoothing, such as loess, splines, and others. There is also a survey paper on this same topic for those who want a more concise discussion [Hastie and

Tibshirani, 1986]. Another recent book that introduces generalized additive models using R is by Wood [2006].

For an early review of smoothing methods, see Stone [1977]. The paper by Cleveland [1979] describes the robust form of loess, and includes some information on the sampling distributions associated with locally weighted regression. Next we have Cleveland and McGill [1984], which is a wonderful paper that includes descriptions of many tools for scatterplot enhancements. Cleveland, Devlin, and Grosse [1988] discuss methods and computational algorithms for local regression. Titterington [1985] provides an overview of various smoothing techniques used in statistical practice and provides a common unifying structure. We also have Cleveland and Devlin [1988] that shows how loess can be used for exploratory data analysis, diagnostic checking of parametric models, and multivariate nonparametric regression. For an excellent summary and history of smoothing methods, see Cleveland and Loader [1996]. As for other smoothing methods, including ones based on kernel functions, the reader can refer to Scott [2015] and Hastie and Loader [1993].

There is a lot of literature on spline methods. For a nice survey article that synthesizes much of the work on splines in statistics, we refer the reader to Wegman and Wright [1983]. See Green and Silverman [1994] and Martinez and Martinez [2015] for a procedure to construct weighted smoothing splines and the generalized cross-validation for choosing the smoothing parameter. These are useful in situations where assumptions are violated, such as ties in predictor values or dependent errors. Hastie and Tibshirani [1990] have a good discussion of this topic, also.

Exercises

7.1 First consult the **help** files on the MATLAB functions **polyfit** and **polyval**, if you are unfamiliar with them. Next, for some given domain of points **x**, find the **y** values using **polyval** and degree 3. Add some normally distributed random noise (use **normrnd** with 0 mean and some small σ) to the **y** values. Fit the data using polynomials of degrees 1, 2, and 3 and plot these curves along with the data. Construct and plot a loess curve. Discuss the results.

7.2 Generate some data using **polyval** and degree 1. Add some small random noise to the points using **randn**. Use **polyfit** to fit a polynomial of degree 1. Add one outlying point at either end of the range of the **x** values. Fit these data to a straight line. Plot both of these lines, along with a scatterplot of the data, and comment on the differences.

7.3 Do a scatterplot of the data generated in Problem 7.1. Activate the **Tools** menu in the **Figure** window and click on the **Basic Fit-**

ting option. This brings up a GUI that has some options for fitting data. Explore the capabilities of this GUI.

7.4 Load the **abrasion** data set. Construct loess curves for abrasion loss as a function of tensile strength and abrasion loss as a function of hardness. Comment on the results. Repeat the process of Example 7.10 using this data set and assess the results.

7.5 Repeat the process outlined in Example 7.10 using the **environmental** data. Comment on your results.

7.6 Construct a sequence of loess curves for the **votfraud** data set. Each curve should have a different value of $\alpha = 0.2, 0.5, 0.8$. First do this for $\lambda = 1$ and then repeat for $\lambda = 2$. Discuss the differences in the curves. Just by looking at the curves, what α and λ would you choose?

7.7 Repeat Problem 7.6, but this time use the residual plots to help you choose the values for α and λ.

7.8 Repeat Problems 7.6 and 7.7 for the **calibrat** data.

7.9 Verify that the tri-cube weight satisfies the four properties of a weight function.

7.10 Using the data in Example 7.11, plot the convex hulls (use the function **convhull**) of each of the topics. Compare these with the polar smooth.

7.11 Choose some other topics from the BPM and repeat the polar smooths of Example 7.11.

7.12 Using one of the gene expression data sets, select an experiment (tumor, patient, etc.), and apply some of the smoothing techniques from this chapter to see how the genes are related to that experiment. Choose one of the genes and smooth as a function of the experiments. Construct the loess upper and lower smooths to assess the variation. Comment on the results.

7.13 Construct a normal probability plot of the residuals in the **software** data analysis. See Chapter 9 for information or the MATLAB Statistics Toolbox function **normplot**. Construct a histogram (use the **hist** function). Comment on the distributional assumptions for the errors.

7.14 Repeat Problem 7.13 for the analyses in Problems 7.6 and 7.8.

7.15 Choose two of the dimensions of the **oronsay** data. Apply polar smoothing to summarize the point cloud for each class. Do this for both classifications of the **oronsay** data. Comment on your results.

7.16 Do a 3–D scatterplot (see **scatter3**) and a scatterplot matrix (see **plotmatrix**) of the **galaxy** data of Example 7.2. Does the contour plot in Figure 7.5 match the scatterplots?

7.17 Describe what type of function you would get when fitting a spline model with $D = 0$.

7.18 The **bin smoother** [Hastie and Tibshirani, 1990; Martinez and Martinez, 2015] is a simple approach to smoothing. We divide the observed predictor variables x_i into disjoint and exhaustive regions represented by cut points $-\infty = c_0 < \ldots < c_K = \infty$. We define the indices of the data points belonging to the regions R_k as

$$R_k = \{i; \ (c_k \le x_i < c_{k+1})\},$$

for $k = 0, \ldots, K-1$. Then, we have the value of the estimated smooth for x_0 in R_k given by the average of the responses in that region:

$$\hat{f}(x_0) = \text{average}_{i \ in \ R_k}(y_i).$$

Write a MATLAB function that implements the bin smoother and apply it to the **salmon** data. Compare your results to Figure 7.6.

7.19   The *running mean smooth* [Martinez and Martinez, 2015] expands on the idea of the bin smoother by taking overlapping regions or bins $N^S(x_i)$, making the estimate smoother. Note that the following definition of a running mean smoother is for target values that are equal to one of our observed x_i:

$$\hat{f}(x_i) = \text{average}_{j \ in \ N^S(x_i)}(y_j).$$

This smooth is also known as the *moving average* and is used quite often in time series analysis where the observed predictor values are evenly spaced. Write a MATLAB function that implements the running mean smoother and apply it to the **salmon** data. Compare your results to previous methods.

7.20   The *running line smoother* [Martinez and Martinez, 2015] is a generalization of the running mean, where the estimate of f is found by fitting a local line using the observations in the neighborhood. The definition is given by the following

$$\hat{f}(x_i) = \hat{\beta}_0^i + \hat{\beta}_1^i x_i,$$

where $\hat{\beta}_0^i$ and $\hat{\beta}_1^i$ are estimated via ordinary least squares using the data in the neighborhood of x_i. Write a MATALB function that will compute the running line smoother and apply it to the **salmon** data. Compare the resulting smooth to the previous methods.

7.21   Repeat Example 7.7 using different values of the smoothing parameter. What happens as α approaches zero? What happens as α gets larger?

7.22   Construct residual plots for each of the smoothing spline fits obtained in the previous problem. Explain what you observe when analyzing the residual plots. (See Example 7.4.)

7.23   Apply the spread smooth techniques to some of the smoots of problem 7.21.

7.24   Repeat Example 7.7 using $\alpha = 8$. Compare the smooths.

7.25 The following example and code was taken from Garcia [2009]. This provides a demonstration of the discrete cosine smooth for uniformly spaced data. In the case outlined below demonstrate how the methodology works in the robust case. Run the code and produce plots of **z** (regular smoothing) and **zr** (robust smoothing).

```
x = linspace(0,100,2^8);
y = cos(x/10)+(x/50).^2 + randn(size(x))/5;
y([70 75 80]) = [5.5 5 6];
[z,s] = smoothn(y); % Regular smoothing
zr = smoothn(y,'robust'); % Robust smoothing
```

Call the function **smoothn** with your own smoothing parameter values, where you under-smooth and over-smooth (see the **help** for the function).

7.26 The following example and code was taken from Garcia [2009]. This provides a demonstration of how the discrete cosine transform smoothing methodology will work on 2–D data with missing values. This will generate noisy 2–D data based on the MATLAB peaks function. This is done by adding Gaussian noise, randomly removing half of the data, and making one area of it empty. Run this code and plot the original data (**y0**), the corrupted data (**y**), and the smooth (**z**). Use the function **imagesc** to plot them.

```
n = 256;
y0 = peaks(n);
y = y0 + rand(size(y0))*2;
I = randperm(n^2);
y(I(1:n^2*0.5)) = NaN; % Lose half of the data
y(40:90,140:190) = NaN; % Create a hole
z = smoothn(y,'MaxIter',250); % Smooth the data
```

Try moving the hole around to different places and compare the results.

Part III

Graphical Methods for EDA

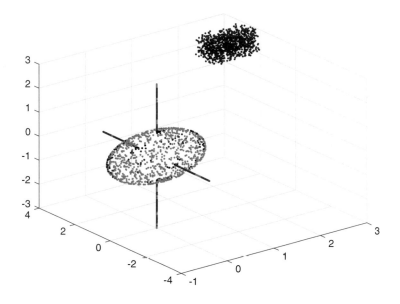

FIGURE 2.12
This is a scatterplot of the data in Figure 2.11 where we colored the points based on their estimated local dimension. We see that some points on the sphere and line are assigned an incorrect intrinsic dimension.

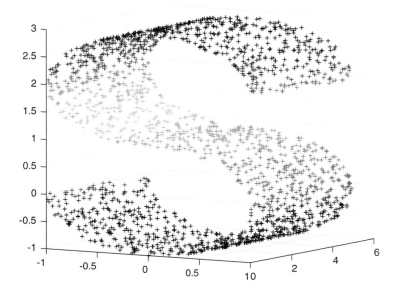

FIGURE 3.7
These points represent a random sample from the S-curve manifold. The colors are mapped to the height of the surface and are an indication of the neighborhood of a point.

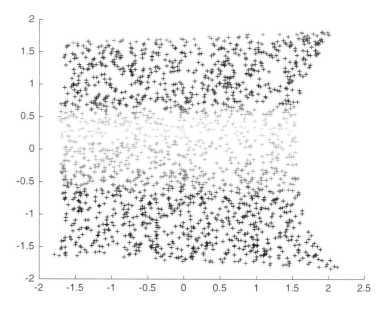

FIGURE 3.8
This is the 2-D embedding recovered using LLE. Note by the color that the neighborhood structure is preserved.

FIGURE 3.10
This is a U-matrix visualization of the SOM for the **oronsay** data. The distance of a map unit to each of its neighbors is shown using the color scale.

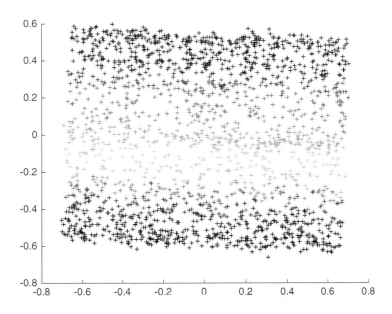

FIGURE 3.16
This shows the 2-D embedding of the S-curve manifold data using an autoencoder. Compare this with Figure 3.8. Some of the neighborhood structure is lost, but the rectangular shape of the 2–D manifold is captured better.

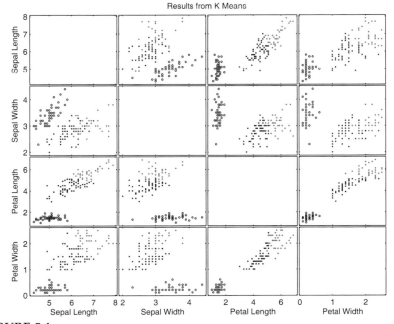

FIGURE 5.4
This is a scatterplot matrix of groups found in the **iris** data using k-means ($k = 3$).

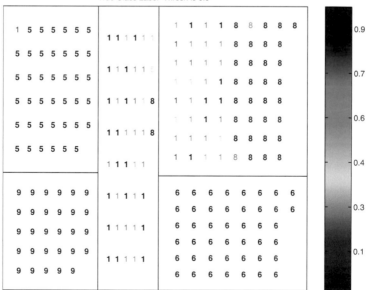

FIGURE 8.7
The ReClus plot at the top shows the cluster configuration based on the best model chosen from model-based clustering. Here we plot the true class label with the color indicating the probability that the observation belongs to that cluster. The next ReClus plot is for the same data and model-based clustering output, but this time we request that probabilities above 0.9 be shown in bold black font.

True Class Label

FIGURE 8.8

Here is the ReClus plot when we split the text data into two groups using hierarchical clustering. In this case, the scale corresponds to the silhouette values.

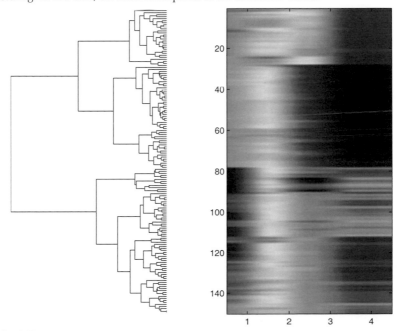

FIGURE 8.9

The data image for the **iris** data is shown on the right. The rows have been re-ordered based on the leaves of the dendrogram at the left.

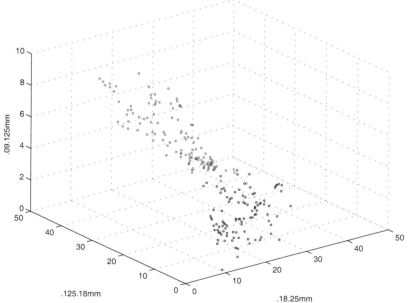

FIGURE 10.3

The top figure shows the 2-D scatterplot for two variables (i.e., columns 8 and 9) of the **oronsay** data set. The color of the plot symbols indicates the midden class membership for both plots. The lower plot shows the 3-D scatterplot for columns 8 through 10.

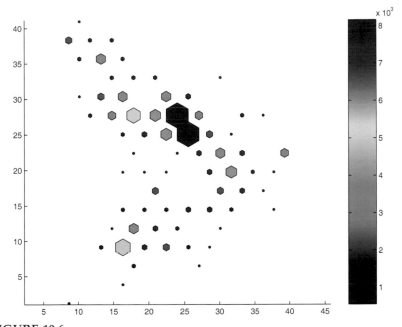

FIGURE 10.6
This shows a plot of the **oronsay** data using hexagonal binning. The color of the hexagons represents the value of the probability density at that bin.

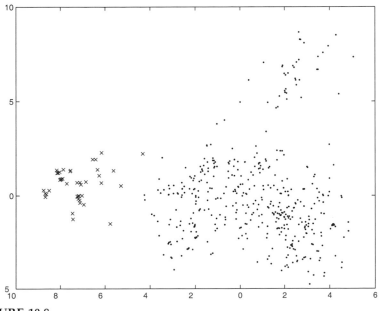

FIGURE 10.8
The red points in this scatterplot were highlighted using the **scattergui** function.

FIGURE 10.10

This is a scatterplot matrix showing brushing and linking in the transient mode.

FIGURE 10.19

This is the Andrews' image for the **iris** data after the rows have been re-ordered based on the dendrogram. The colorbar indicates the true class membership of the observations.

Chapter 8

Visualizing Clusters

In Chapters 5 and 6, we presented various methods for clustering, including agglomerative clustering, k-means clustering, and model-based clustering. In the process of doing that, we showed some visualization techniques such as dendrograms for visualizing hierarchical structures and scatterplots. We now turn our attention to other methods that can be used for visualizing the results of clustering. These include a space-filling version of dendrograms called treemaps, an extension of treemaps called rectangle plots, a novel rectangle-based method for visualizing nonhierarchical clustering called ReClus, and data images that can be used for viewing clusters, as well as outlier detection. We begin by providing more information on dendrograms.

8.1 Dendrogram

The *dendrogram* (also called the *tree diagram*) is a mathematical, as well as a visual representation of a hierarchical procedure that can be divisive or agglomerative. Thus, we often refer to the results of the hierarchical clustering as the dendrogram itself.

We start off by providing some terminology for a dendrogram; the reader can refer to Figure 8.1 for an illustration. The tree starts at the *root*, which can either be at the top for a vertical tree or on the left side for a horizontal tree. The *nodes* of the dendrogram represent clusters, and they can be *internal* or *terminal*. The internal nodes contain or represent all observations that are grouped together based on the type of linkage and distance used. In most dendrograms, terminal nodes contain a single observation. We will see shortly that this is not always the case in MATLAB's implementation of the dendrogram. Additionally, terminal nodes usually have labels attached to them. These can be names, letters, numbers, etc. The MATLAB **dendrogram** function labels the terminal nodes with numbers.

The *stem* or *edge* shows children of internal nodes and the connection with the clusters below it. The length of the edge represents the distances at which clusters are joined. The dendrograms for hierarchical clustering are **binary** *trees*; so they have two edges emanating from each internal node. The *topology* of the tree refers to the arrangement of stems and nodes.

The dendrogram illustrates the process of constructing the hierarchy, and the internal nodes describe particular partitions, once the dendrogram has been cut at a given level. Data analysts should be aware that the same data and clustering procedure can yield 2^{n-1} dendrograms, each with a different appearance depending on the order used to display the nodes. Software packages choose the algorithm for drawing this automatically, and they usually do not specify how they do this. Some algorithms have been developed for optimizing the appearance of dendrograms based on various objective functions [Everitt, Landau, and Leese, 2001]. We will see in the last section where we discuss the data image that this can be an important consideration.

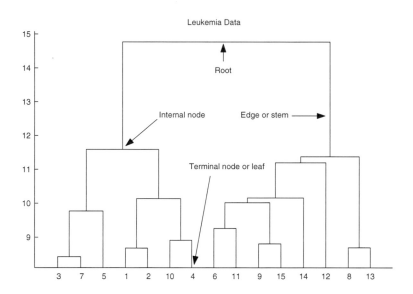

FIGURE 8.1
We applied agglomerative clustering to the **leukemia** data using Euclidean distance and complete linkage. The dendrogram with 15 leaf nodes is shown here. If we cut this dendrogram at level 10.5 (on the vertical axis), then we would obtain 5 clusters or groups.

Example 8.1

We first show how to construct the dendrogram in Figure 8.1 using the **leukemia** data set. The default for the **dendrogram** function is to display a maximum of 30 nodes. This is to prevent the displayed leaf nodes from being too cluttered. The user can specify the number to display, as we do below, or one can display all nodes by using **dendrogram(Z,0)**.

```
% First load the data.
```

```
load leukemia
[n,p] = size(leukemia);
x = zeros(n,p);
% Standardize each row (gene) to be mean
% zero and standard deviation 1.
for i = 1:n
    sig = std(leukemia(i,:));
    mu = mean(leukemia(i,:));
    x(i,:) = (leukemia(i,:) - mu)/sig;
end
% Do hierarchical clustering.
Y = pdist(x);
Z = linkage(Y,'complete');
% Display with only 15 nodes.
% Output arguments are optional.
[H,T] = dendrogram(Z,15);
title('Leukemia Data')
```

Note that the leaf nodes in this example of a MATLAB dendrogram do not necessarily represent one of the original observations; they likely contain several observations. We can (optionally) request some output variables from the **dendrogram** function to help us determine what observations are in the individual nodes. The output vector **T** contains the leaf node number for each object in the data set and can be used to find out what is in node 6 as follows:

```
ind = find(T==6)
ind =

    26
    28
    29
    30
    46
```

Thus, we see that terminal node 6 on the dendrogram in Figure 8.1 contains the original observations 26, 28, 29, 30, and 46.
❏

8.2 Treemaps

Dendrograms are very familiar to data analysts working in hierarchical clustering applications, and they are easy to understand because they match our concept of how trees are laid out in a physical sense with branches and leaves (except that the tree root is sometimes in the wrong place!). Johnson and Shneiderman [1991] point out that the dendrogram does not efficiently use the existing display space, since most of the display consists of white space with very little ink. They proposed a space-filling (i.e., the entire display space is used) display of hierarchical information called *treemaps*, where each node is a rectangle whose area is proportional to some characteristic of interest [Shneiderman, 1992].

The original application and motivation for treemaps was to show the directory and file structure on hard drives. It was also applied to the visualization of organizations such as the departmental structure at a university. Thus, the treemap visualization can be used for an arbitrary number of splits or branches at internal nodes, including the binary tree structure that one gets from hierarchical clustering.

Johnson and Shneiderman [1991] note that hierarchical structures contain two types of information. First, they contain structural or organizational information that is associated with the hierarchy. Second, they have content information associated with each node. Dendrograms can present the structural information, but do not convey much about the leaf nodes other than a text label. Treemaps can depict both the structure and content of the hierarchy.

The treemap displays hierarchical information and relationships by a series of nested rectangles. The parent rectangle (or root of the tree) is given by the entire display area. The treemap is obtained by recursively subdividing this parent rectangle, where the size of each sub-rectangle is proportional to the size of the node. The size could be representative of the size in bytes of the file or the number of employees in an organizational unit. In the case of clustering, the size would correspond to the number of observations in the cluster. We continue to subdivide the rectangles horizontally, vertically, horizontally, etc., until a given leaf configuration (e.g., number of groups in the case of clustering) is obtained.

We show an example of the treemap in Figure 8.2, along with its associated tree diagram. Note that the tree diagram is not the dendrogram that we talked about previously, because it has an arbitrary number of splits at each node. The root node has four children: a leaf node of size 12, a leaf node of size 8, an interior node with four children, and an interior node with three children. These are represented by the first vertical splits of the parent rectangle. We now split horizontally at the second level. The interior node with four children is split into four sub-rectangles proportional to their size. Each one of these is a terminal node, so no further subdivisions are needed. The next interior node has two children that are leaf nodes and one child that is an interior node. Note that this interior node is further subdivided into two leaf nodes of size 6 and 11 using a vertical split.

When we apply the treemap visualization technique to hierarchical clustering output, we must specify the number of clusters. Note also that there is no measure of distance or other objective function associated with the clusters, as there is with the dendrogram. Another issue with the treemap (as

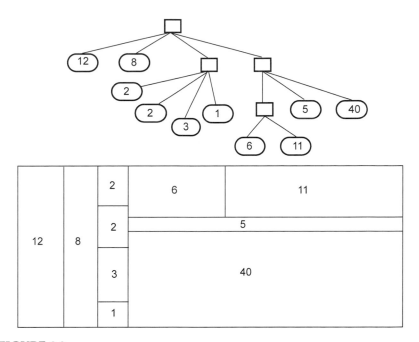

FIGURE 8.2
At the top of this figure, we show a tree diagram with nodes and links. Each leaf node has
a number that represents the size of the nodes. An alternative labeling might be just the
node number or some text label. The corresponding treemap diagram is shown below. Note
that the divisions from the root node are shown as vertical splits of the parent rectangle,
where the size of each sub-rectangle is proportional to the total size of all children. The next
split is horizontal, and we continue alternating the splits until all leaf nodes are displayed
[Shneiderman, 1992].

well as the dendrogram) is the lack of information about the original data,
because the rectangles are just given labels. We implemented the treemap
technique in MATLAB; its use is illustrated in the next example.

Example 8.2
In this example, we show how to use the **treemap** provided with the EDA
Toolbox. We return to the hierarchical clustering of the **leukemia** data
displayed as a dendrogram in Figure 8.1. The function we provide
implements the treemap display for binary hierarchical information only and
requires the output from the MATLAB **linkage** function. The inputs to the
function **treemap** include the **Z** matrix (output from **linkage**) and the
desired number of clusters. The default display is to show the leaf nodes with
the same labels as in the dendrogram. There is an optional third argument
that causes the treemap to display without any labels. The following syntax
constructs the treemap that corresponds to the dendrogram in Figure 8.1.

```
% The matrix Z was calculated in
```

```
% Example 8.1.
treemap(Z,15);
```

The treemap is given in Figure 8.3. Notice that we get an idea of the size of the nodes with the treemap, whereas this is not evident in the dendrogram (Figure 8.1). However, as noted before, we lose the concept of distance with treemaps; thus it is perhaps easier to see the number of groups one should have in a dendrogram instead of a treemap.
❑

FIGURE 8.3
Here we have the treemap that corresponds to the dendrogram in Figure 8.1.

8.3 Rectangle Plots

Recall that in the dendrogram, the user can specify a value along the axis, and different clusters or partitions are obtained depending on what value is specified. Of course, we do not visualize this change with the dendrogram; i.e., the display of the dendrogram is not dependent on the chosen number of clusters or cutoff distance. However, it is dependent on the number of leaf nodes chosen (in the MATLAB implementation). If one specifies a different number of leaf nodes, then the dendrogram must be completely redrawn, and node labels change. This can significantly change the layout of the dendrogram, as well as the understanding that is gained from it.

To display the hierarchical information as a treemap, the user must specify the number of clusters (or one can think of this as number of leaf nodes) rather than the cutoff point for the distance. If the user wants to explore other cluster configurations by specifying a different number of clusters, then the display is redrawn, as it is with the dendrogram. As stated previously, there is no measure of distance associated with the treemap display, and there is a lack of information about the original data. It would be useful to know what cases are clustered where. The next cluster visualization method attempts to address these issues.

Recall from Chapters 5 and 6 that one of the benefits of hierarchical clustering is that the output can provide a partition of the data for any given number of clusters or, alternatively, for any given level of dissimilarity. This property is an advantage in EDA, because we can run the algorithm once on a large set of data and then explore and visualize the results in a reasonably fast manner. To address some of the issues with treemaps and to take advantage of the strengths of hierarchical clustering, Wills [1998] developed the *rectangle visualization method*. This method is similar to the treemap, but displays the points as glyphs and determines the splits in a different way.

To construct a rectangle plot, we split the rectangles along the longest side, rather than alternating vertical and horizontal splits as in the treemap method. The alternating splits in treemaps are good at showing the depth of the tree, but it has a tendency to create long skinny rectangles, if the trees are unbalanced [Wills, 1998]. The splits in the rectangle plot provide rectangles that are more square.

We keep splitting rectangles until we reach a leaf node or until the cutoff distance is reached. If a rectangle does not have to be split because it reaches this cutoff point, but there is more than one observation in the rectangle, the algorithm continues to split rectangles until it reaches a leaf node. However, it does not draw the rectangles. It uses this leaf-node information to determine the layout of the points or glyphs, where each point is now in its own rectangle. The advantage to this method is that other configurations (i.e., number of clusters or a given distance) can be shown without redisplaying the glyphs. Only the rectangle boundaries are redrawn.

We illustrate the rectangle plot for a simulated data set in Figures 8.4 and 8.5. The data set contains randomly generated bivariate data ($n = 30$) comprising two clusters, one centered at $(-2, 2)^T$ and the other at $(2, 2)^T$. In the top part of Figure 8.4, we have the dendrogram with all 30 nodes displayed. We see that the node labels are difficult to distinguish. The corresponding rectangle plot for 30 clusters is shown in the bottom of Figure 8.4, where we see each observation in its own rectangle or cluster.

We show another dendrogram and rectangle plot in Figure 8.5 using the same data set and hierarchical clustering information. We plot a dendrogram requesting 10 leaf nodes and show the results in the top of the figure. The leaf nodes are now easier to read, but they no longer represent the observation numbers. The rectangle plot for 10 clusters is given in the lower half of the

FIGURE 8.4
The top portion of this figure shows the dendrogram (Euclidean distance and complete linkage) for a randomly generated bivariate data set containing two clusters. All $n = 30$ leaf nodes are shown, so we see over plotting of the text labels. The rectangle plot that corresponds to this is shown in the bottom half, where we plot each observation number in its own rectangle or cluster. The original implementation in Wills [1998] plots the observations as dots or circles.

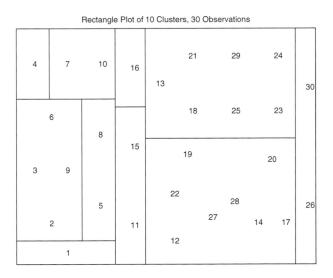

FIGURE 8.5

Using the same information that produced Figure 8.4, we now show the dendrogram when only 10 leaf nodes are requested. We see that the dendrogram has been completely redrawn when we compare it with the one in Figure 8.4. The rectangle plot for 10 clusters is given below the dendrogram. When the rectangle plots in Figures 8.4 and 8.5 are compared, we see that the positions of the glyphs have not changed; only the bounding boxes for the rectangles are different. We also see information about the original data via the observation numbers.

figure. When we compare Figures 8.4 and 8.5, we see that the dendrogram changed a lot, whereas only the bounding boxes change in the rectangle plot.

Example 8.3

We continue with the same **leukemia** data set for this example. First we show how to use the **rectplot** function to get a rectangle plot showing observation numbers. For comparison with previous results, we plot 15 clusters.

```
% Use same leukemia data set and matrix Z
% from Example 8.1. The second argument is the
% number of clusters. The third argument is a string
% specifying what information the second
% argument provides.
rectplot(Z,15,'nclus')
```

The plot is shown in the top of Figure 8.6. Next we show how to use the optional input argument to use the true class labels as the glyphs. First we have to convert the class labels to numbers, since the input vector must be numeric.

```
% We now show how to use the optional
% class labels, using the cancer type.
% The argument must be numeric, so we
% convert strings to numbers.
% First set all indices to 0 - this will
% be class ALL.
labs = zeros(length(cancertype),1);
% Now find all of the AML cancers.
% Set them equal to 1.
inds = strmatch('AML',cancertype);
labs(inds) = 1;
% Now do the rectangle plot.
rectplot(Z,15,'nclus',labs)
```

This plot is shown in the lower half of Figure 8.6. The observations labeled 0 are the ALL cancers, and those plotted with a 1 correspond to the AML cancer type.
❑

In our MATLAB implementation of this technique, we plot each point with its observation number or its true class label. This can cause some over plotting with large data sets. A future implementation will include other plot symbols, thus saving on display space. The user can also specify the number of clusters by providing a cutoff dissimilarity based on the dendrogram for the second input to the function. In this case, the third argument to **rectplot** is **'dis'**.

FIGURE 8.6
A rectangle plot for the `leukemia` data is shown in the top of this figure. Here we plot the observation numbers for a specified number of clusters or partitions. The rectangle plot shown in the bottom of the figure plots the observations using the true cancer labels. Class 0 corresponds to ALL and Class 1 is AML.

Wills' original motivation for the rectangle plot was to include the notion of distance in a treemap-like display. He did this by providing a supplemental line graph showing the number of clusters on the horizontal axis, and the dissimilarity needed to do the next split on the vertical axis. The user can interact with the display by dragging the mouse over the line graph and seeing the corresponding change in the number of clusters shown in the rectangle plot.

The rectangle method is also suitable for linking and brushing applications (see Chapter 10), where one can highlight an observation in one plot (e.g., a scatterplot) and see the same observation highlighted in another (e.g., a rectangle plot). A disadvantage with the rectangle plot is that some of the nesting structure seen in treemaps might be lost in the rectangle display. Another problem with the plots discussed so far is their applicability to the display of hierarchical information only. The plot we show next can be applied to nonhierarchical clustering procedures.

8.4 ReClus Plots

The **ReClus method** was developed by Martinez [2002] as a way to view the output of nonhierarchical clustering methods, such as k-means, model-based clustering, etc., that is reminiscent of the treemap and rectangle displays. We note that ReClus (standing for **RectangleClus**ters) can also be used to convey the results of any hierarchical clustering method once we have a given partition.

As in the previous methods, ReClus uses the entire display area as the parent rectangle. This is then partitioned into sub-rectangles, where the area of each one is proportional to the number of observations that belong to that cluster. The observations are plotted using either the observation number or the true class label, if known. The glyphs are plotted in a systematic way, either by order of the observation number or the class label.

There are some additional options. If the output is from model-based clustering, then we can obtain the probability that an observation belongs to the cluster. This additional information is displayed via the font color. For faster and easier comprehension of the cluster results, we can set a threshold so that the higher probabilities are shown in bold black type. We provide a similar capability for other cluster methods, such as k-means, using the silhouette values. We now outline the procedure for constructing a ReClus plot.

Procedure – ReClus Plot

1. Set up the parent rectangle. We will subdivide the rectangle along the longer side of the parent rectangle according to the proportion of observations that are in each group.
2. Find all of the points in each cluster and the corresponding proportion.
3. Order the proportions in ascending order.
4. Partition the proportions into two groups. If there is an odd number of clusters, then put more of the clusters into the 'left/lower' group.
5. Based on the total proportion in each group, split the longer side of the parent rectangle. We now have two children. Note that we have to re-normalize the proportions based on the current parent rectangle.
6. Repeat steps 4 through 5 until all rectangles represent only one cluster.
7. Find the observations in each cluster and plot, either as the case label or the true class label.

We illustrate the use of ReClus in the next example.

Example 8.4

For this example, we use the **L1bpm** interpoint distance matrix derived from the BPMS of the 503 documents discussed in Chapter 1. We first use ISOMAP to reduce the dimensionality and get our derived observations. We are going to use only five of the sixteen topics, so we also set up the indices to extract them.

```
load L1bpm
% Get just those topics on 5 through 11
ind = find(classlab == 5);
ind = [ind; find(classlab == 6)];
ind = [ind; find(classlab == 8)];
ind = [ind; find(classlab == 9)];
ind = [ind; find(classlab == 11)];
% Change the class labels for class 11
% for better plotting.
clabs = classlab(ind);
clabs(find(clabs == 11)) = 1;
% First do dimensionality reduction - ISOMAP
[Y, R, E] = isomap(L1bpm,'k',10);
% Choose 3 dimensions, based on residual variance.
XX = Y.coords{3}';
% Only need those observations in classes of interest.
```

```
X = XX(ind,:);
```

Next we do model-based clustering specifying a maximum of 10 clusters.

```
% Now do model-based clustering.
[bics,bestm,allm,Z,clabsmbc] = mbclust(X,10);
```

We see that model-based clustering found the correct number of groups (five), and model seven is the best fit (according to the BIC). Now that we have the cluster information, we need to find the probability that an observation belongs to the cluster or the silhouette values if using some other clustering procedure. In the case of model-based clustering, we use the function **mixclass** to get the probability that an observation does *not* belong to the cluster, based on the model.

```
% Must get the probability that an observation does
% not belong to the cluster.
[clabsB,uncB] = mixclass(X,bestm.pies,...
    bestm.mus,bestm.vars);
% Plot with true class labels.
% The function requires the posterior probability
% that it belongs to the cluster.
reclus(clabsB,clabs,1 - uncB)
```

Note that **mixclass** returns the *uncertainty* in the classification, so we must subtract this from one in the argument to **reclus**. The resulting plot is shown in the top of Figure 8.7. Note that topics 8 and 1 (formerly topic 11) are both about North Korea, and we see some mistakes in how these topics are grouped. However, topics 5, 6, and 9 are clustered together nicely, except for one document from topic 1 that is grouped with topic 5. The color coding of the posterior probabilities makes it easier to see the level of uncertainty. In some applications, we might gain more insights regarding the clusters by looking at a more *binary* application of color. In other words, we want to see those observations that have a high posterior probability separated from those with a lower one. An optional argument to **reclus** specifies a threshold, where we plot posterior probabilities above this value in a bold black font. This makes it easier to get a quick overall view of the quality of the clusters.

```
% Look at the other option in reclus.
% Plot points with posterior probability above
% 0.9 in bold, black font.
reclus(clabsB,clabs,1 - uncB,.9)
```

This ReClus plot is given in the bottom of Figure 8.7, where we see that documents in topics 5, 6, and 9 have posterior probabilities above the threshold. The one exception is the topic 1 document that was grouped with topic 5. We see that this document has a lower posterior probability that was not readily apparent in Figure 8.7 (top). We also see that the documents in the two groups with topics 1 and 8 mixed together have a large number of members with posterior probabilities below the threshold. It would be interesting to explore these two clusters further to see if this grouping is an indication of sub-topics.

❏

The next example shows how to use ReClus with the information from hierarchical clustering, in addition to some of the other options that are available with the **reclus** function. One of the advantages of ReClus is the ability to rapidly visualize the strength of the clustering for the observations when the true class membership is known. It would be useful to be able to find out which observations correspond to interesting ones in the ReClus plot. For instance, in Figure 8.7 (top), we might want to locate those stories that were in the two mixed-up groups to see if the way they were grouped makes sense. Or, we might want to find the topic 1 document that was mis-grouped with topic 5.

Example 8.5

We use the same data from the previous example, but now we only look at two topics: 8 and 11. These are the two that concern North Korea, and there was some confusion in the clustering of these topics using model-based clustering. We will use the ReClus plot to help us assess how hierarchical clustering works on these two topics. First we extract the observations that we need and get the labels.

```
% Continuing with same data used in
% Example 8.4.
ind = find(classlab == 8);
ind = [ind; find(classlab == 11)];
clabs = classlab(ind);
% Change the class labels for class 11
% for better plotting.
clabs = classlab(ind);
clabs(find(clabs == 11)) = 1;
% Only need those observations in classes of interest.
X = XX(ind,:);
```

Next we perform hierarchical clustering using Euclidean distance and complete linkage. We use the **cluster** function to request two groups, and then get the silhouette values. Note that this syntax for the **silhouette** function suppresses the plot and only returns the silhouette values.

```
% Get the hierarchical clustering.
Y = pdist(X);
Z = linkage(Y,'complete');
dendrogram(Z);
cids = cluster(Z,'maxclust',2);
```

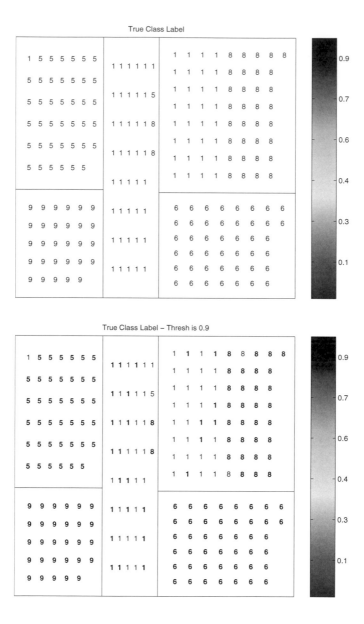

FIGURE 8.7
The ReClus plot at the top shows the cluster configuration based on the best model chosen from model-based clustering. Here we plot the true class label with the color indicating the probability that the observation belongs to that cluster. The next ReClus plot is for the same data and model-based clustering output, but this time we request that probabilities above 0.9 be shown in bold black font. (**SEE COLOR INSERT.**)

```
% Now get the silhouette values.
S = silhouette(X,cids);
```

Our initial ReClus plot uses the observation number as the plotting symbol. We have to include the true class labels as an argument, so they can be mapped into the same position in the second ReClus plot. The plots are shown in Figure 8.8, and the code to construct these plots is given below.

```
% The following plots the observation
% numbers, with 'pointers' to the
% true class labels plotted next.
reclus(cids, clabs)
% Now plot same thing, but with true class labels.
reclus(cids,clabs,S)
```

We see from the plots that this hierarchical clustering does not provide a good clustering, as measured by the silhouette values, since we have several negative values in the right-hand cluster.
❏

8.5 Data Image

The *data image* is a technique for visualizing high-dimensional data as if they comprised an image. We have seen an example of this already in Chapter 1, Figure 1.1, where we have the gene expression data displayed as a gray-scale image. The basic idea is to map the data into an image framework, using the gray-scale values or colors (if some other color map is desired) to indicate the magnitude of each variable for each observation. Thus, the data image for a data set of size n with p variables will be $n \times p$.

An early version of the data image can be found in Ling [1973], where the data are plotted as a matrix of characters with different amounts of ink indicating the gray scale value of the observation. However, Ling first plots the interpoint dissimilarity (or similarity) matrix in its raw form. He then reorders rows and columns of the dissimilarity matrix using the cluster labels after some clustering method has been applied. In other words, the original sequence of observations has been arranged such that the members of every cluster lie in consecutive rows and columns of the permuted dissimilarity matrix. Clearly defined dark (or light, depending on the gray scale) squares along the diagonal indicate compact clusters that are well separated from neighboring points. If the data do not contain significant clusters, then this is readily seen in the image.

Wegman [1990] also describes a version of the data image, but he calls it a *color histogram*. He suggests coloring pixels using a binned color gradient and presenting this as an image. Sorting the data based on one variable

Case Numbers

```
36  54  68  80  90  99   3        37  45  63  88   4  12  19  26  33

44  55  70  81  91 101  11        38  48  66  93   5  13  20  27  34

46  58  71  82  92 102             39  49  69 100   6  14  21  28  35

47  59  72  83  94 103             40  56  76 106   7  15  22  29

50  60  73  84  95 104             41  57  77 108   8  16  23  30

51  64  74  85  96 105             42  61  79   1   9  17  24  31

52  65  75  87  97 107             43  62  86   2  10  18  25  32

53  67  78  89  98 109
```

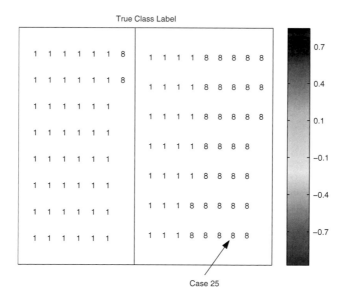

Case 25

FIGURE 8.8

In Figure 8.8 (top), we show the ReClus plot for topics 8 and 1 (topic 11) based on hierarchical clustering. The positions of these symbols correspond to the glyphs in Figure 8.8 (bottom). This makes it easy to see what case belongs to observations of interest. The scale in the lower ReClus plot corresponds to the silhouette values. (**SEE COLOR INSERT.**)

enables one to observe positive and negative associations. One could explore the data in a tour-like manner by performing this sort for each variable.

Minnotte and West [1998] use binning on a fine scale, so they coined the term *data image* to be more descriptive of the output. Rather than sorting on only one variable, they suggest finding an ordering such that points that are close in high-dimensional space are close to one another in the image. They suggest that this will help the analyst better visualize high-dimensional structure. We note that this is reminiscent of multidimensional scaling. In particular, they apply this method to understand the cluster structure in high-dimensional data, which should appear as vertical bands within the image.

Minnotte and West propose two methods for ordering the data. One approach searches for the shortest path through the cloud of high-dimensional points using traveling salesman type algorithms [Cook et al., 1998]. Another option is to use an ordering obtained from hierarchical clustering, which is what we do in the next example.

Example 8.6

We use the familiar **iris** data set for this example. We put the three classes of iris into one matrix and randomly reorder the rows. The data image for this arrangement is shown in Figure 8.9 (top). We see the four variables as columns or vertical bands in the image, but we do not see any horizontal bands indicating groups of observations.

```
load iris
% Put into one matrix.
data = [setosa;versicolor;virginica];
% Randomly reorder the data.
data = data(randperm(150),:);
% Construct the data image.
imagesc(-1*data)
colormap(gray(256))
```

We now cluster the data using agglomerative clustering with complete linkage. To get our ordering, we plot the dendrogram with *n* leaf nodes. The output argument **perm** from the function **dendrogram** provides the order of the observations left to right or bottom to top, depending on the orientation of the tree. We use this to rearrange the data points.

```
% Now get the ordering using hierarchical
% clustering and the dendrogram.
Y = pdist(data);
Z = linkage(Y,'complete');
% Plot both the dendrogram and the data
% image together.
figure
subplot(1,2,1)
```

```
[H, T, perm] = dendrogram(Z,0,'orientation','left');
axis off
subplot(1,2,2)
% Need to flip the matrix to show as an image.
imagesc(flipud(-1*data(perm,:)))
colormap(gray(256))
```

The data image for the reordered data and the associated dendrogram are given in Figure 8.9 (bottom). We can now see three horizontal bands that would indicate the presence of three groups. However, we know that each class of iris has 50 observations and that some are likely to be incorrectly clustered together.
❑

The data image displayed along with the dendrogram, as we saw in Example 8.6, is used extensively in the gene expression analysis literature. However, it is not generally known by that name. We now show how to apply the data image concept to locate clusters in a data set using Ling's method.

Example 8.7
We use the data (topics 6 and 9) from Example 8.4 to illustrate how the data image idea can be applied to the interpoint dissimilarity matrix. After extracting the topics we want to work with, we then find the distance matrix using **pdist** and **squareform**. The image of this is shown in Figure 8.10 (top). We randomly reordered the data; thus it is difficult to see any cluster structure or clusters. The code to do this follows.

```
% Continuing with same data used in Example 8.4.
% Use just Topics 6 and 9.
ind = find(classlab == 6);
ind = [ind; find(classlab == 9)];
n = length(ind);
clabs = classlab(ind);
data = XX(ind,:);
% Randomly reorder the data and view the
% interpoint distance matrix as an image.
data = data(randperm(n),:);
Y = pdist(data);
Ys = squareform(Y);
imagesc(Ys);
colormap(gray(256))
title('Interpoint Distance Matrix - Random Order')
axis off
```

We now have to obtain some clusters. We will use hierarchical clustering and request cluster labels based on a partition into two groups. Then we order the

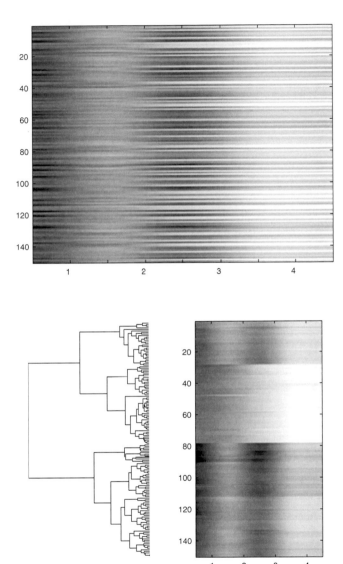

FIGURE 8.9
The data image at the top is for the **iris** data, where the observations have been randomly ordered. Clusters or horizontal bands are not obvious. We apply the data image concept by reordering the data according to the results of hierarchical clustering. The data image of the rearranged data and the corresponding dendrogram are shown at the bottom. Three horizontal bands or clusters are now readily visible. (**SEE COLOR INSERT.**)

Interpoint Distance Matrix – Random Order

Interpoint Distance Matrix – Cluster Order

FIGURE 8.10

The top image shows the interpoint distance matrix for topic 6 and topic 9 when the data are in random order. The presence of any groups or clusters is difficult to discern. Next we cluster the data according to our desired method and rearrange the points such that those observations in the same group are adjacent in the distance matrix. We show this re-ordered matrix in the bottom half of the figure, and the two groups can be clearly seen.

distances such that observations in the same group are adjacent and replot the distance matrix as an image.

```
% Now apply Ling's method. First need to
% get a partition or clustering.
Z = linkage(Y,'complete');
% Now get the ordering based on the cluster scheme.
T = cluster(Z,'maxclust',2);
% Sort these, so points in the same cluster
% are adjacent.
[Ts,inds] = sort(T);
% Sort the distance matrix using this and replot.
figure
imagesc(Ys(inds,inds))
title('Interpoint Distance Matrix - Cluster Order')
colormap(gray(256))
axis off
```

This image is shown in Figure 8.10 (bottom). We can now clearly see the two clusters along the diagonal.

❑

8.6 Summary and Further Reading

One of the main purposes of clustering is to organize the data; so graphical tools to display the results of these methods are important. These displays should enable the analyst to see whether or not the grouping of the data illustrates some latent structure. Equally important, if real structure is not present in the data, then the cluster display should convey this fact. The cluster visualization techniques presented in this chapter should enable the analyst to better explore, assess, and understand the results from hierarchical clustering, model-based clustering, *k*-means, etc.

Some enhancements to the dendrogram have been proposed. One is a generalization of dendrograms called *espaliers* [Hansen, Jaumard, and Simeone, 1996]. In the case of a vertical diagram, espaliers use the length of the horizontal lines to encode another characteristic of the cluster, such as the diameter. Other graphical aids for assessing and interpreting the clusters can be found in Cohen et al. [1977].

Johnson and Shneiderman [1991] provide examples of other types of treemaps. One is a *nested treemap*, where some space is provided around each sub-rectangle. The nested nature is easier to see, but some display space is wasted in visualizing the borders. They also show a *Venn diagram treemap*, where the groups are shown as nested ovals.

The treemap display is popular in the computer science community as a tool for visualizing directory structures on disk drives. Several extensions to the treemap have been developed. These include *cushion treemaps* [Wijk and Wetering, 1999] and *squarified treemaps* [Bruls, Huizing, and Wijk, 2000]. Cushion treemaps keep the same basic structure of the space-filling treemaps, but they add surface height and shading to provide additional insight into the hierarchical structure. The squarified treemap methodology attempts to construct squares instead of rectangles (as much as possible) to prevent long, skinny rectangles, but this sacrifices the visual understanding of the nested structure to some extent.

Marchette and Solka [2003] use the data image to find outliers in a data set. They apply this concept to the interpoint distance matrix by treating the columns (or alternatively the rows) of the matrix as observations. These observations are clustered using some hierarchical clustering scheme, and the rows and columns are permuted accordingly. Outliers show up as a dark *v* or cross (depending on the color scale).

Others in the literature have discussed the importance of ordering the data to facilitate exploration and understanding. A recent discussion of this can be found in Friendly and Kwan [2003]. They outline a general framework for ordering information in visual displays, including tables and graphs. They show how these *effect-ordered data displays* can be used to discover patterns, trends, and anomalies in the data.

Exercises

8.1 Do a **help** on **dendrogram** and read about the optional input argument **'colorthreshold'**. Repeat Example 8.1 using this option.

8.2 Compare and contrast the dendrogram, treemap, rectangle plots, and ReClus cluster visualization techniques. What are the advantages and disadvantages of each? Comment on the usefulness of these methods for large data sets.

8.3 Find the ReClus plot (without class labels and with class labels) for the **leukemia** partition with 15 clusters obtained in Example 8.1. Compare to the previous results using the hierarchical visualization techniques.

8.4 Repeat Example 8.3 using the distance input argument to specify the number of clusters displayed in the rectangle plot.

8.5 For the following data sets, use an appropriate hierarchical clustering method and visualize using the methods described in the chapter. Analyze the results.

   a. **geyser**

   b. **singer**

   c. **skulls**

   d. **sparrow**

   e. **oronsay**

   f. gene expression data sets

8.6   Repeat Example 8.7 using all of the data in the matrix **XX** (reduced from ISOMAP) from Examples 8.4, and 8.5. Use hierarchical clustering or *k*-means and ask for 16 groups. Is there any evidence of clusters?

8.7   Apply the methodology of Example 8.7 to the **iris** data.

8.8   Repeat Examples 8.4, 8.5 and 8.7 using other BPM data sets and report on your results.

8.9   Repeat Example 8.4 using the silhouette values for the model-based clustering classification.

8.10   Repeat Example 8.5 using other types of hierarchical clustering. Compare your results.

8.11   For the following data sets, use *k*-means or model-based clustering. Use the ReClus method for visualization. Analyze your results.

   a. **skulls**

   b. **sparrow**

   c. **oronsay** (both classifications)

   d. BPM data sets

   e. gene expression data sets

8.12   Looking at the data image in Figure 8.9, comment on what variables seem most useful for classifying the species of iris.

Chapter 9

Distribution Shapes

In this chapter, we show various methods for visualizing the shapes of distributions. The ability to visualize the distribution shape in exploratory data analysis is important for several reasons. First, we can use it to summarize a data set to better understand general characteristics such as shape, spread, or location. In turn, this information can be used to suggest transformations or probabilistic models for the data. Second, we can use these methods to check model assumptions, such as symmetry, normality, etc. We present several techniques for visualizing univariate and bivariate distributions. These include 1–D and 2–D histograms and kernel densities, boxplots and variants, violin plots, beanplots, quantile-based plots, bagplots, and rangefinder boxplots.

9.1 Histograms

A *histogram* is a way to graphically summarize or describe a data set by visually conveying its distribution using vertical bars. They are easy to create and are computationally feasible, so they can be applied to massive data sets. In this section, we describe several varieties of histograms. These include the frequency and relative frequency histogram, and what we are calling the density histogram.

9.1.1 Univariate Histograms

A *frequency histogram* is obtained by first creating a set of bins or intervals that cover the range of the data set. It is important that these bins do not overlap and that they have equal width. We then count the number of observations that fall into each bin. To visualize this information, we place a bar at each bin, where the height of the bar corresponds to the frequency. *Relative frequency histograms* are obtained by mapping the height of the bin to the relative frequency of the observations that fall into the bin.

The basic MATLAB package has a function for calculating and plotting a univariate frequency histogram called **hist**. This function is illustrated in the example given below, where we show how to construct both types of histograms.

Example 9.1

In this example, we look at the two univariate histograms showing relative frequency and frequency. We can obtain a simple histogram in MATLAB using these commands:

```
load galaxy
% The 'hist' function can return the
% bin centers and frequencies.
% Use the default number of bins - 10.
[n, x] = hist(EastWest);
% Plot and use the argument of width = 1
% to get bars that touch.
bar(x,n,1,'w');
title('Frequency Histogram - Galaxy Data')
xlabel('Velocity')
ylabel('Frequency')
```

Note that calling the **hist** function with no output arguments will find the pieces necessary to construct the histogram based on a given number of bins (default is 10 bins) and will also produce the plot. We chose to use the option of first extracting the bin locations and bin frequencies so we could get the relative frequency histogram using the following code:

```
% Now create a relative frequency histogram.
% Divide each box by the total number of points.
% We use bar to plot.
bar (x,n/140,1,'w')
title('Relative Frequency Histogram - Galaxy Data')
xlabel('Velocity')
ylabel('Relative Frequency')
```

These plots are shown in Figure 9.1. Notice that the shapes of the histograms are the same in both types of histograms, but the vertical axes are different. From the shape of the histograms, it seems reasonable to assume that the data are normally distributed (for this bin configuration).
❑

One problem with using a frequency or relative frequency histogram is that they do not represent meaningful probability densities, because the total area represented by the bars does not equal one. This can be seen by superimposing a corresponding normal distribution over the relative frequency histogram as shown in Figure 9.2. However, they are very useful for gaining a quick picture of the distribution of the data.

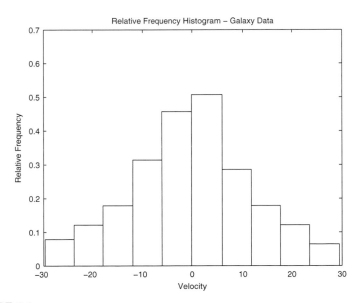

FIGURE 9.1
The bars in the top histogram show the number of observations that fall into each bin, while the bar heights in the bottom histogram correspond to the relative frequency. Note that the shape of the histogram is the same, even though the vertical axes represent different quantities.

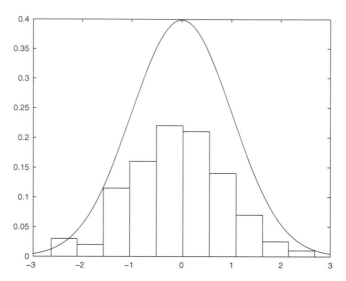

FIGURE 9.2
This shows a relative frequency histogram for some data generated from a standard normal distribution. Note that the curve is higher than the histogram, indicating that the histogram is not a valid probability density function.

A *density histogram* is a histogram that has been normalized so the area under the curve (where the curve is represented by the heights of the bars) is one. A density histogram is given by the following equation

$$\hat{f}(x) = \frac{v_k}{nh} \qquad x \text{ in } B_k, \qquad (9.1)$$

where B_k denotes the k-th bin, v_k represents the number of data points that fall into the k-th bin, and h represents the width of the bins.

Since our goal is to estimate a *bona fide* probability density, we want to have an estimate that is nonnegative and satisfies the constraint that

$$\int_{-\infty}^{\infty} \hat{f}(x)dx = 1.$$

It is left as an exercise to the reader to show that Equation 9.1 satisfies this condition.

The density histogram depends on two parameters: 1) an origin t_0 for the bins and 2) a bin width h. These two parameters define the mesh over which the histogram is constructed. The bin width h is sometimes referred to as the *smoothing parameter*, and it fulfills a similar purpose as that found in the chapter on smoothing scatterplots. The bin width determines the smoothness of the histogram. Small values of h produce histograms with a lot of variation in the heights of the bins, while larger bin widths yield smoother histograms. This phenomenon is illustrated in Figure 9.3, where we show histograms of the same data, but with different bin widths.

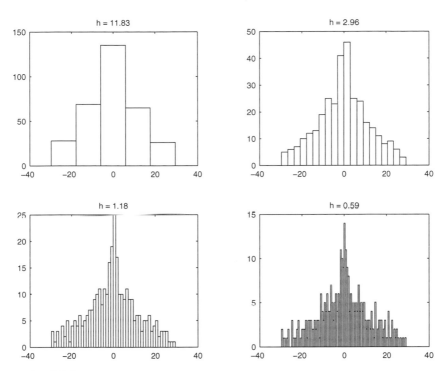

FIGURE 9.3
These are histograms for the **galaxy** data used in Example 9.1. Notice that for the larger bin widths, we have only one peak. As the smoothing parameter gets smaller, the histogram displays more variation and spurious peaks appear in the histogram estimate.

We now look at how we can choose the bin width h, in an attempt to minimize our estimation error. It can be shown that setting h small to reduce the bias increases the variance in our estimate. On the other hand, creating a smoother histogram reduces the variance, at the expense of worsening the bias. This is the familiar trade-off between variance and bias discussed previously. We now present some of the common methods for choosing the bin width, most of which are obtained by trying to minimize the squared error [Scott, 2015] between the true density and the estimate.

Histogram Bin Widths

1. <u>Sturges' Rule</u>

$$k = 1 + \log_2 n, \tag{9.2}$$

where k is the number of bins. The bin width h is obtained by taking the range of the sample data and dividing it into the requisite number of bins, k [Sturges, 1926].

2. <u>Normal Reference Rule - 1–D Histogram</u>

$$\hat{h}^* = \left\{\frac{24\sigma^3 \sqrt{\pi}}{n}\right\}^{\frac{1}{3}} \approx 3.5 \times \sigma \times n^{-\frac{1}{3}}. \tag{9.3}$$

Scott [1979, 2015] proposed the sample standard deviation as an estimate of σ in Equation 9.3 to get the following bin width rule.

3. <u>Scott's Rule</u>

$$\hat{h}^* = 3.5 \times s \times n^{-\frac{1}{3}}.$$

4. <u>Freedman–Diaconis Rule</u>

$$\hat{h}^* = 2 \times IQR \times n^{-\frac{1}{3}}.$$

This robust rule developed by Freedman and Diaconis [1981] uses the interquartile range (IQR) instead of the sample standard deviation.

It turns out that when the data are skewed or heavy-tailed, the bin widths are too large using the Normal Reference Rule. Scott [1979, 2015] derived the following correction factor for skewed data:

$$\text{skewness factor} = \frac{2^{1/3}\sigma}{e^{5\sigma^2/4}(\sigma^2 + 2)^{1/3}(e^{\sigma^2} - 1)^{1/2}}. \tag{9.4}$$

If one suspects the data come from a skewed distribution, then the Normal Reference Rule bin widths should be multiplied by the factor given in Equation 9.4.

So far, we have discussed histograms from a visualization standpoint only. We might also need an estimate of the density at a given point x, as we will see in the next section on boxplots. We can find a value for our density estimate for a given x, using Equation 9.1. We obtain the value $\hat{f}(x)$ by taking the number of observations in the data set that fall into the same bin as x and multiplying by $1/(nh)$.

Example 9.2

In this example, we provide MATLAB code that calculates the estimated value $\hat{f}(x)$ for a given x. We use the same data from the previous example and Sturges' Rule for estimating the number of bins.

```
load galaxy
n = length(EastWest);
% Use Sturges' Rule to get the number of bins.
k = round(1 + log2(n));
% Bin the data.
[nuk,xk] = hist(EastWest,k);
% Get the width of the bins.
h = xk(2) - xk(1);
% Plot as a density histogram.
bar(xk, nuk/(n*h), 1, 'w')
title('Density Histogram - Galaxy Data')
xlabel('Velocity')
```

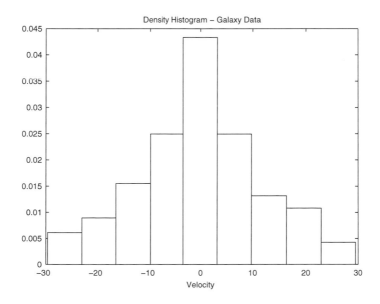

FIGURE 9.4
This shows the density histogram for the **galaxy** data.

The histogram produced by this code is shown in Figure 9.4. Note that we had to adjust the output from **hist** to ensure that our estimate is a *bona fide* density. Let's get the estimate of our function at a point $x_0 = 0$.

```
% Now return an estimate at a point xo.
xo = 0;
% Find all of the bin centers less than xo.
ind = find(xk < xo);
% xo should be between these two bin centers:
b1 = xk(ind(end));
b2 = xk(ind(end)+1);
% Put it in the closer bin.
if (xo-b1) < (b2-xo)    % then put it in the 1st bin
    fhat = nuk(ind(end))/(n*h);
else
    fhat = nuk(ind(end)+1)/(n*h);
end
```

Our result is **fhat = 0.0433**. Looking at Figure 9.4, we see that this is the correct estimate. The main MATLAB software now has a enhanced function for creating histograms called **histogram**. This is in addition to the **hist** function. The **histogram** function has options for creating different types of histograms. The **'Normalization'** argument can be set to **'count'** to get a frequency histogram, **'probability'** to create the relative frequency histogram, and **'pdf'** to get the density version.
❑

9.1.2 Bivariate Histograms

We can easily extend the univariate density histogram to multivariate data, but we restrict our attention in this text to the bivariate case. The bivariate histogram is defined as

$$\hat{f}(\mathbf{x}) = \frac{v_k}{n h_1 h_2}; \qquad \mathbf{x} \text{ in } B_k, \tag{9.5}$$

where v_k denotes the number of observations falling into the bivariate bin B_k, and h_i is the width of the bin along the i-th coordinate axis. Thus, the estimate of the probability density would be given by the number of observations falling into that same bin divided by the sample size and the bin widths.

As before, we must determine what bin widths to use. Scott [2015] provides the following multivariate Normal Reference Rule.

<u>Normal Reference Rule - Multivariate Histograms</u>

$$h_i^* \approx 3.5 \times \sigma_i \times n^{\frac{-1}{2+p}}; \qquad i = 1, ..., p. \qquad (9.6)$$

Notice that this reduces to the same univariate Normal Reference Rule (Equation 9.3) when $p = 1$. As before, we can use a suitable estimate for σ_i, based on our data.

Example 9.3

We return to the data used in Example 7.11 to illustrate the bivariate histogram. Recall that we first reduced the BPM data to 2–D using ISOMAP and the L_1 proximity measure. The code to do this is repeated below.

```
load L1bpm
% Reduce the dimensionality using Isomap.
options.dims = 1:10;    % These are for ISOMAP.
options.display = 0;
[Yiso, Riso, Eiso] = isomap(L1bpm, 'k', 7, options);
% Get the data out.
XX = Yiso.coords{2}';
inds = find(classlab==8 | classlab==11);
x = [XX(inds,:)];
[n,p] = size(x);
```

We use the normal reference rule to find the density histogram of the data.

```
% Need bin origins.
bin0 = floor(min(x));
% The bin width h, for p = 2:
h = 3.5*std(x)*n^(-0.25);
% Find the number of bins
nb1 = ceil((max(x(:,1))-bin0(1))/h(1));
nb2 = ceil((max(x(:,2))-bin0(2))/h(2));
% Find the bin edges.
t1 = bin0(1):h(1):(nb1*h(1)+bin0(1));
t2 = bin0(2):h(2):(nb2*h(2)+bin0(2));
[X,Y] = meshgrid(t1,t2);
% Find bin frequencies.
[nr,nc] = size(X);
vu = zeros(nr-1,nc-1);
for i = 1:(nr-1)
    for j = 1:(nc-1)
        xv = [X(i,j) X(i,j+1) X(i+1,j+1) X(i+1,j)];
        yv = [Y(i,j) Y(i,j+1) Y(i+1,j+1) Y(i+1,j)];
        in = inpolygon(x(:,1),x(:,2),xv,yv);
```

```
        vu(i,j) = sum(in(:));
    end
end
% Find the proper height of the bins.
Z = vu/(n*h(1)*h(2));
% Plot using bars.
bar3(Z,1,'w')
```

The plot for histogram is shown in Figure 9.5. We used some additional MATLAB code to get axes labels that make sense. Please refer to the M-file for this example to see the code to do that. The Statistics Toolbox has a function called **hist3** for constructing histograms of bivariate data.
❑

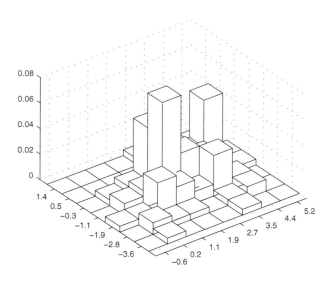

FIGURE 9.5
This is the histogram for the data representing topics 8 and 11. The normal reference rule was used for the bin widths. Please see Figure 7.18 for the corresponding scatterplot.

9.2 Kernel Density

We mentioned kernel density estimation in Chapter 2, where we used it to visualize the density of univariate data (see Example 2.6). We now provide some details on the approach and explore how it can be used to understand how our data are distributed. Our treatment of the topic follows that found in Silverman [1986], Scott [2015], Martinez and Martinez [2015], and Martinez and Cho [2014].

9.2.1 Univariate Kernel Density Estimation

The univariate kernel estimator is given by

$$\hat{f}_{Ker}(x) = \frac{1}{nh} \sum_{i=1}^{n} K\left(\frac{x - X_i}{h}\right), \tag{9.7}$$

where the function $K(t)$ is called a **kernel**. This must satisfy the condition that $\int K(t)dt = 1$ to ensure that our estimate in Equation 9.7 is a *bona fide* density estimate. Usually, the kernel is a symmetric probability density function, and often a standard normal density is used. Other common choices include the triangle and the Epanechnikov [Scott, 2015]. All of these kernels and more are available in the MATLAB function **ksdensity**, which we illustrate in the next example.

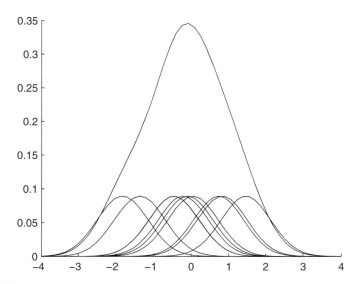

FIGURE 9.6
We obtain the above kernel density estimate for $n = 10$ random variables. A weighted kernel is centered at each data point, and the curves are averaged together to obtain the estimate. [Martinez and Martinez, 2015]

From Equation 9.7, the estimated probability density function is obtained by placing a weighted kernel function, centered at each data point and then taking the average. See Figure 9.6 for an illustration of this procedure. Notice that the places where there are more kernels yield higher density regions in the final estimate.

As in the histogram, the parameter h determines the amount of smoothing we have in the estimate $\hat{f}_{Ker}(x)$. In kernel density estimation, the h is usually called the **window width** or the **bandwidth**. A small value of h yields a rough curve, while a large value of h produces a smoother curve. In practice, it is recommended that the analyst examine kernel density estimates for different window widths to search for structures such as modes or bumps.

Scott [2015] has an extensive discussion of ways to choose an appropriate value for h. One approach is to choose a value that minimizes the asymptotic mean integrated standard error (AMISE). Scott shows that, under certain conditions, the AMISE for a nonnegative univariate kernel density estimator is

$$\text{AMISE}_{Ker}(h) = \frac{R(K)}{nh} + \frac{1}{4}\sigma_k^4 h^4 R(f''), \tag{9.8}$$

where the kernel K is a continuous probability density function with $\mu_K = 0$ and $0 < \sigma_K^2 < \infty$. The bandwidth h that minimizes this is given by

$$h_{Ker}^* = \left(\frac{R(K)}{n\sigma_k^4 R(f'')} \right)^{1/5}. \tag{9.9}$$

Parzen [1962] and Scott [2015] provide the conditions under which this holds.

For a kernel that is equal to the normal density $R(f'') = 3/(8\sqrt{\pi}\sigma^5)$, we have the following **normal reference rule** for the window width h.

$$h_{Ker}^* = \left(\frac{4}{3} \right)^{1/5} \sigma n^{-1/5} \approx 1.06\sigma n^{-1/5}.$$

We can use some suitable estimate for σ, such as the standard deviation, or $\hat{\sigma} = IQR/1.348$. The latter yields a bandwidth of

$$\hat{h}_{Ker}^* = 0.786 \times IQR \times n^{-1/5}.$$

Silverman [1986] recommends that one use whichever is smaller, the sample standard deviation or $IQR/1.348$ as an estimate for σ.

It is known [Scott, 2015] that the choice of smoothing parameter h is more important than choosing the kernel because any effects from the type of kernel are reduced by the averaging process. What determines the choice of kernel are computational considerations or the amount of differentiability required in the estimate. The reader is asked to explore the effect of the kernel choice and bandwidth in the exercises.

Example 9.4

The MATLAB Statistics Toolbox has a function called **ksdensiy** that will create kernel density estimates for both univariate and bivariate data. This function will return the estimate or just create the plot. We are going to use the Buffalo **snowfall** data. These data [Scott, 2015] contain the annual snowfall (inches) in Buffalo, New York from 1910 to 1972. First, we load the data.

```
% Load the snowfall data.
load snowfall
```

Next, we specify the bandwidth. We could use the default bandwidth given by the normal reference rule, but this turns out to be too large for these data. It over smooths and obscures additional bumps. So, we will use a bandwidth that is approximately one-half the one given by the normal reference rule.

```
% Get the bandwidth.
n = length(snowfall);
% Approximately one-half the bandwidth for
% a normal distribution.
b = 0.5*std(snowfall)*n^(-1/5);
```

We are now ready to estimate the probability density and plot it using the following code.

```
% Estimate and plot the density.
ksdensity(snowfall,'Bandwidth',b);
```

We can add a 1–D scatterplot or *rug plot* with these steps. The rug plot displays the data along the horizontal axis as tick marks.

```
% Add a rug plot.
ax = axis;
xr = snowfall;
yr = ax(4)/25;
hold on
h = stem(xr,yr*ones(n,1));
set(h,'markersize',0.01,'MarkerEdgeColor','white')
set(gca,'TickDir','out')
hold off
```

We provide a function in the EDA Toolbox called **rug** that will add this type of 1–D scatterplot. The resulting plot is shown in Figure 9.7.
❑

9.2.2 Multivariate Kernel Density Estimation

In the multivariate case, each observation x_i is a p-dimensional vector. The simplest case for the multivariate kernel estimator is the product kernel.

FIGURE 9.7
Here is a plot of the estimated probability density function for the Buffalo snowfall data.
We used a smaller bandwidth than the one given by the normal reference rule in order to
explore other structure or intervals of higher density. We added a rug plot or 1–D scatterplot
to provide additional information about the data.

Descriptions of the general kernel density estimate can be found in Scott
[2015] and in Silverman [1986]. The ***product kernel*** is

$$\hat{f}_{Ker}(\mathbf{x}) = \frac{1}{nh_1 \ldots h_p} \sum_{i=1}^{n} \left\{ \prod_{j=1}^{p} K\left(\frac{x_j - X_{ij}}{h_j}\right) \right\}, \tag{9.10}$$

where X_{ij} is the j-th component of the i-th observation. Note that this is the
product of the same univariate kernel, with possibly a different window
width in each dimension.

Scott [2015] gives expressions for the asymptotic integrated squared bias
and asymptotic integrated variance for the multivariate product kernel. If the
normal kernel is used, then minimizing these yields a normal reference rule
for the multivariate case, as follows

$$h_{j_{Ker}}^* = \left(\frac{4}{n(p+2)}\right)^{\frac{1}{p+4}} \sigma_j; \qquad j = 1, \ldots, p, \tag{9.11}$$

A suitable estimate for σ_j can be used in Equation 9.11. If there is any skewness or kurtosis evident in the data, then the window widths should be narrower.

Example 9.5

We will use the MATLAB **ksdensity** function to estimate the probability density function of the **faithful** data. This data set is one version of the Old Faithful geyser data [Härdle, 1991]. Each data point includes the eruption and waiting time in minutes.

```
% We use the Old Faithful data.
load faithful

% Use the default bandwidth and kernel.
ksdensity(faithful)
axis tight
xlabel('Eruption Time (mins)')
ylabel('Waiting Time (mins)')
```

A surface plot of the estimated probability density function is given in Figure 9.8.

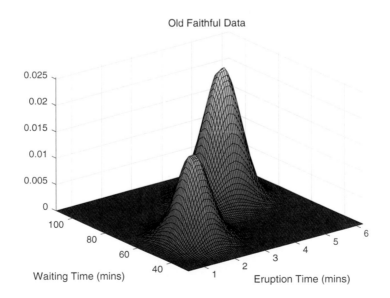

FIGURE 9.8
This is the estimated bivariate probability density function for the Old Faithful data. We see two high density regions or bumps.

9.3 Boxplots

Boxplots (sometimes called *box-and-whisker diagrams*) have been in use for many years [Tukey, 1977]. They are an excellent way to visualize summary statistics such as the median, to study the distribution of the data, and to supplement multivariate displays with univariate information. Benjamini [1988] outlines the following characteristics of the boxplot that make them useful:

1. Statistics describing the data are visualized in a way that readily conveys information about the location, spread, skewness, and longtailedness of the sample.
2. The boxplot displays information about the observations in the tails, such as potential outliers.
3. Boxplots can be displayed side-by-side to compare the distribution of several data sets.
4. The boxplot is easy to construct.
5. The boxplot is easily explained to and understood by users of statistics.

In this section, we first describe the basic boxplot. This is followed by several enhancements and variations of the boxplot. These include variable-width boxplots, the histplot, box-percentile plots, violin plots, beeswarm plots, and the beanplot.

9.3.1 The Basic Boxplot

Before we describe the boxplot, we need to define some terms. In essence, three statistics from a data set are needed to construct all of the pieces of the boxplot. These are the *sample quartiles*: $q(0.25)$, $q(0.5)$, $q(0.75)$. The sample quartiles are based on *sample quantiles*, which are defined next [Kotz and Johnson, 1986].

Given a data set $x_1, \ldots, x_n$, we order the data from smallest to largest. These are called the *order statistics*, and we denote them as

$$x_{(1)}, \ldots, x_{(n)}.$$

The u $(0 < u < 1)$ quantile $q(u)$ of a random sample is a value belonging to the range of the data such that a fraction u (approximately) of the data are less than or equal to u.

The quantile denoted by $q(0.25)$ is also called the *lower quartile*, where approximately 25% of the data are less than or equal to this number. The quantile $q(0.5)$ is the *median*, and $q(0.75)$ is the *upper quartile*. We need to define a form for the u in quantile $q(u)$. For a random sample of size n, we let

$$u_i = \frac{i - 0.5}{n}. \qquad (9.12)$$

The form for u_i given in Equation 9.12 is somewhat arbitrary [Cleveland, 1993] and is defined for u_i, $i = 1, ..., n$. This definition can be extended for all values of u, $0 < u < 1$ by interpolation or extrapolation, given the values of u_i and $q(u_i)$. We will study the quantiles in more detail in the next section.

Definitions of quartiles can vary from one software package to another. Frigge, Hoaglin, and Iglewicz [1989] describe a study on how quartiles are implemented in some popular statistics programs such as Minitab, S, SAS, SPSS, and others. They provide eight definitions for quartiles and show how this can affect the appearance of the boxplots. We will use the one defined in Tukey [1977] called standard *fourths* or *hinges*.

Procedure – Finding Quartiles

1. Order the data from smallest to largest.
2. Find the median $q(0.5)$, which is in position $(n+1)/2$ in the list of ordered data:
   a. If n is odd, then the median is the middle data point.
   b. If n is even, then the median is the average of the two middle points.
3. Find the lower quartile, which is the median of the data that lie at or below the median:
   a. If n is odd, then $q(0.25)$ is the median of the ordered data in positions 1 through $(n+1)/2$.
   b. If n is even, then $q(0.25)$ is the median of the ordered data in positions 1 through $n/2$.
4. Find the upper quartile, which is the median of the data that lie at or above the median:
   a. If n is odd, then $q(0.75)$ is the median of the ordered data in positions $(n+1)/2$ through n.
   b. If n is even, then $q(0.75)$ is the median of the ordered data in positions $n/2 + 1$ through n.

Thus, we see that the lower quartile is the median of the lower half of the data set, and the upper quartile is the median of the upper half of the sample. We show how to find these using MATLAB in the next example.

Example 9.6

We will use the **geyser** data to illustrate the MATLAB code for finding the quartiles. These data represent the time (in minutes) between eruptions of the Old Faithful geyser at Yellowstone National Park.

```
load geyser
% First sort the data.
geyser = sort(geyser);
% Get the median.
q2 = median(geyser);
% First find out if n is even or odd.
n = length(geyser);
if rem(n,2) == 1
    odd = 1;
else
    odd = 0;
end
if odd
    q1 = median(geyser(1:(n+1)/2));
    q3 = median(geyser((n+1)/2:end));
else
    q1 = median(geyser(1:n/2));
    q3 = median(geyser(n/2:end));
end
```

The sample quartiles are: 59, 76, 83. The reader is asked in the exercises to verify that these make sense by looking at a histogram and other plots. We provide a function called **quartiles** that implements this code for general use.
❑

Recall from introductory statistics that the sample *interquartile range* (IQR) is the difference between the first and the third sample quartiles. This gives the range of the middle 50% of the data. It is found from the following:

$$IQR = q(0.75) - q(0.25).$$

We need to define two more quantities to determine what observations qualify as potential outliers. These limits are the *lower limit* (LL) and the *upper limit* (UL). They are calculated from the IQR as follows

$$LL = q(0.25) - 1.5 \times IQR$$
$$UL = q(0.75) + 1.5 \times IQR.$$

(9.13)

Observations outside these limits are *potential outliers*. In other words, observations smaller than the LL and larger than the UL are flagged as interesting points because they are outlying with respect to the bulk of the data. *Adjacent values* are the most extreme observations in the data set that are within the lower and the upper limits. If there are no potential outliers, then the adjacent values are simply the maximum and the minimum data points.

Just as the definition of quartiles varies with different software packages, so can the definition of outliers. In some cases multipliers other than 1.5 are used in Equation 9.13, again leading to different boxplots. Hoaglin, Iglewicz, and Tukey [1986] examine this problem and show how it affects the number of outliers displayed in a boxplot.

The original boxplot, as defined by Tukey, did not include the display of outliers, as it does today. He called the boxplot with outliers the *schematic plot*. It is the default now in most statistics packages (and text books) to construct the boxplot with outliers as we outline below.

To construct a boxplot, we place horizontal lines at each of the three quartiles and draw vertical lines at the edges to create a box. We then extend a line from the first quartile to the smallest adjacent value and do the same for the third quartile and largest adjacent value. These lines are sometimes called the whiskers. Finally, any possible outliers are shown as an asterisk or some other plotting symbol. An example of a boxplot with labels showing the various pieces is shown in Figure 9.9.

Boxplots for different univariate samples can be plotted together for visually comparing the corresponding distributions, and they can also be plotted horizontally rather than vertically.

Example 9.7

We now show how to do a boxplot by hand using the **defsloc** variable in the **software** data because it has some potential outliers. First we load the data and transform it using the logarithm.

```
load software
% Take the log of the data.
x = log(sort(defsloc));
n = length(x);
```

The next step is to find the quartiles, the interquartile range, and the upper and lower limits.

```
% First get the quartiles.
q = quartiles(x);
% Find the interquartile range.
```

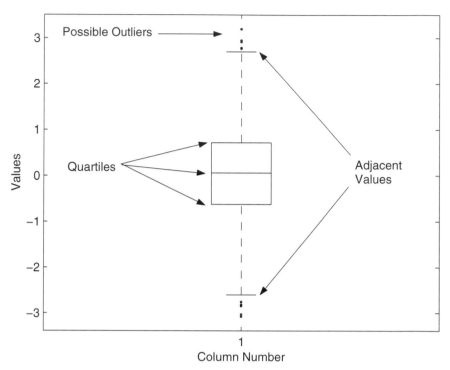

FIGURE 9.9
This is an example of a boxplot with possible outliers.

```
iq = q(3) - q(1);
% Find the outer limits.
UL = q(3) + 1.5*iq;
LL = q(1) - 1.5*iq;
```

We can find observations that are outside these limits using the following code:

```
% Find any outliers.
ind = [find(x > UL); find(x < LL)];
outs = x(ind);
% Get the adjacent values. Find the
% points that are NOT outliers.
inds = setdiff(1:n,ind);
% Get their min and max.
adv = [x(inds(1)) x(inds(end))];
```

Now we have all of the quantities necessary to draw the plot.

```
% Now draw the necessary pieces.
% Draw the quartiles.
```

```
plot([1 3],[q(1),q(1)])
hold on
plot([1 3],[q(2),q(2)])
plot([1 3],[q(3),q(3)])
% Draw the sides of the box
plot([1 1],[q(1),q(3)])
plot([3 3],[q(1),q(3)])
% Draw the whiskers.
plot([2 2],[q(1),adv(1)],[1.75 2.25],[adv(1) adv(1)])
plot([2 2],[q(3),adv(2)],[1.75 2.25],[adv(2) adv(2)])
% Now draw the outliers with symbols.
plot(2*ones(size(outs)), outs,'o')
hold off
axs = axis;
axis([-1 5 axs(3:4)])
set(gca,'XTickLabel',' ')
ylabel('Defects per SLOC (log)')
```

The boxplot is shown in Figure 9.10, where we see that the distribution is not exactly symmetric.

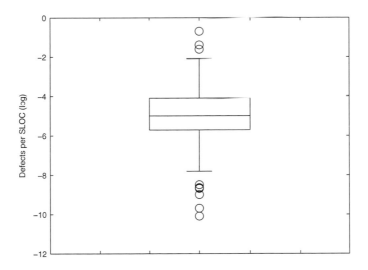

FIGURE 9.10
This shows the boxplot for the number of defects per SLOC (log) from the **software** data set.

We provide a function called **boxp** that will construct boxplots, including some of the variations discussed below. The MATLAB Statistics Toolbox also has a **boxplot** function that the reader is asked to explore in the exercises.

9.3.2 Variations of the Basic Boxplot

We now describe some enhancements and variations of the basic boxplot described above. When we want to understand the significance of the differences between the medians, we can display boxplots with notches [McGill, Tukey, and Larsen, 1978]. The notches in the sides of the boxplots represent the uncertainty in the locations of central tendency and provide a rough measure of the significance of the differences between the values. If the intervals represented by the notches do not overlap, then there is evidence that the medians are significantly different. The MATLAB Statistics Toolbox function **boxplot** will produce boxplots with notches, as explained in the exercises.

Vandervieren and Hubert [2004] present a robust version of the boxplot for skewed distributions. In this type of distribution, too many observations can be classified as outliers. Their generalization to the boxplot has a robust measure of skewness that is used to find the whiskers. They show that their adjusted boxplot provides a more accurate representation of the data distribution than the basic boxplot. See Appendix B for information on where to download functions for this and other robust analysis methods.

Another enhancement of the boxplot also comes from McGill, Tukey, and Larsen [1978]. This is called the *variable-width boxplot*, and it incorporates a measure of the sample size. Instead of having boxplots with equal widths, we could make the widths proportional to some function of n. McGill, Tukey, and Larsen recommend using widths proportional to the square root of n and offer it as the standard. They suggest others, such as making the widths directly proportional to sample size or using a logit scale.

Benjamini [1988] uses the width of the boxplot to convey information about the density of the data rather than the sample size. He offers two types of boxplots that incorporate this idea: the *histplot* and the *vaseplot*. In a histplot, the lines at the three quartiles are drawn with a width that is proportional to an estimate of the associated density at these positions. His implementation uses the density histogram, but any other density estimation method can also be used. He then extends this idea by drawing the width of the box at each point proportional to the estimated density at that point. This is called the vaseplot, because it produces a vase-like shape. One can no longer use the notches to show the confidence intervals for the medians in these plots, so Benjamini uses shaded bars instead. All of these extensions adjust the width of the boxes only; the whiskers stay the same.

The *box-percentile plot* [Esty and Banfield, 2003] also uses the sides of the boxplot to convey information about the distribution of the data over the range of data values. They no longer draw the whiskers or the outliers, so there is no ambiguity about how to define these characteristics of the boxplot.

To construct a box-percentile plot, we do the following. From the minimum value up to the 50th percentile, the width of the 'box' is proportional to the percentile of that height. Above the 50th percentile, the width is proportional to 100 minus the percentile. The box-percentile plots are wide in the middle, like boxplots, but narrowing as they get further away from the middle.

We now describe the procedure in more detail. Let w indicate the maximum width of the box-percentile plot, which will be the width at the median. We obtain the order statistics of our observed random sample, $x_{(1)}$, ..., $x_{(n)}$. Then the sides of the box-percentile plot are obtained as follows:

1. For $x_{(k)}$ less than or equal to the median, we plot the observation at height $x_{(k)}$ at a distance $kw/(n + 1)$ on either side of a vertical axis of symmetry.

2. For $x_{(k)}$ greater than the median, we plot the point at height $x_{(k)}$ at a distance $(n + 1 - k)w/(n + 1)$ on either side of the axis of symmetry.

We illustrate the box-percentile plot and the histplot in the next example. Constructing a variable-width boxplot is left as an exercise to the reader.

Example 9.8

We use some simulated data sets similar to those in Esty and Banfield [2003] to illustrate the histplot and the box-percentile plots. We did not have their exact distributional models, but we tried to reproduce them as much as possible. The first data set is a standard normal distribution. The second one is uniform in the range [−1.2, 1.2], with some outliers close to −3 and 3. The third random sample is trimodal. These three data sets have similar quartiles and ranges, as illustrated in the boxplots in Figure 9.11. The code used to generate the samples follows; we saved the data in a MAT-file for your use.

```
% Generate some standard normal data.
X(:,1) = randn(400,1);
% Generate some uniform data.
tmp = 2.4*rand(398,1) - 1.2;
% Add some outliers to it.
X(:,2) = [tmp; [-2.9 2.9]'];
tmp1 = randn(300,1)*.5;
tmp2 = randn(50,1)*.4-2;
tmp3 = randn(50,1)*.4+2;
X(:,3) = [tmp1; tmp2; tmp3];
save example96 X
```

We show the side-by-side boxplots in Figure 9.11, where we used the MATLAB Statistics Toolbox **boxplot** function with long whiskers so no outliers will be shown.

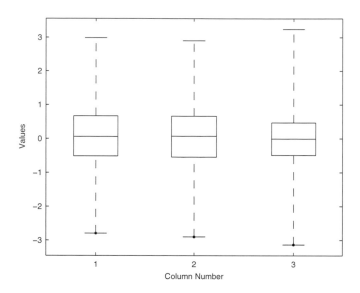

FIGURE 9.11
This shows the boxplots for the generated data in Example 9.6. The first one is for a standard normal distribution. The second is uniform with outliers at the extremes of –3 and 3. The third is from a trimodal distribution. Notice that these distributions do not seem very different based on these boxplots.

```
% This is from the Statistics Toolbox:
figure,boxplot(X,0,[],1,10)
```

We can gain more insights into the distributions by looking at the histplots. The same function called **boxp** mentioned earlier includes this capability; its other functionality will be explored in the exercises.

```
% We can get the histplot. This function is
% included with the text.
boxp(X,'hp')
```

This plot is shown in Figure 9.12, where differences are now apparent. The histplot functionality provided with this text uses a kernel probability density estimate [Scott, 2015; Martinez and Martinez, 2015] for the density at the quartiles. Now, we show the box-percentile plot, which we have coded in the function **boxprct**. This function will construct both constant width boxes, as well as ones with variable width. See the **help** on the function for more information on its capabilities. The plain box-percentile plots are created with this syntax:

```
% Let's see what these look like using the
% box-percentile plot.
boxprct(X)
```

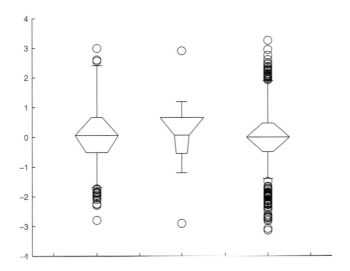

FIGURE 9.12
This is the histplot version of the data in Example 9.6. We now see some of the differences in the distributions.

This plot is shown in Figure 9.13. Note that the sides of the plots now give us more information and insights into the distributions, and that the differences between them are more apparent. In a box-percentile plot, outliers like we have in the second distribution will appear as long, skinny lines. To get a variable width box-percentile plot (the maximum width is proportional to the square root of *n*), use the syntax

```
boxprct(X,'vw')
```

❏

In our opinion, one of the problems with the histplot is its dependence on the estimated density. This density estimate is highly dependent on the bin width (or window width in the case of kernel estimation), so the histplots might be very different as these change. The nice thing about the box-percentile plot is that we gain a better understanding of the distribution, and we do not have to arbitrarily set parameters such as the limits to determine outliers or the bin widths to find the density.

9.3.3 Violin Plots

Violin plots were developed by Hintze and Nelson [1998] in an attempt to add information to the summary statistics displayed in a boxplot. It merges two graphical approaches—the boxplot and a plot of the estimated

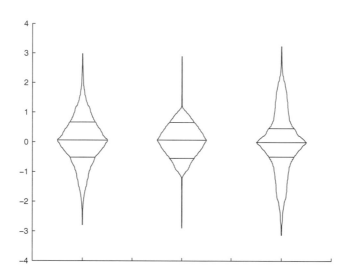

FIGURE 9.13

This is the box-percentile plot of the simulated data in Example 9.6. The plot for the normal distribution shows a single mode at the median, symmetry about the median, and concave sides. The uniform distribution has outliers at both ends, which are shown as long, thin lines. The center part of the plot is a diamond shape, since the percentile plot of a uniform distribution is linear. Although somewhat hard to see, we have several modes in the last one, indicated by valleys (with few observations) and peaks on the sides.

probability density function. He calls the later the ***density trace***. These are called violin plots because they often have the shape of a violin.

The first component of a violin plot is a compact boxplot without any potential outliers being displayed as points. This is at the center of the violin. The lengths of the boxplot whiskers extend from the quartiles to the adjacent values. An estimate of the probability density is displayed symmetrically on both sides of the boxplot. This marks the edges of the violin shape. The density traces surround the boxplot and convey a better idea of the magnitude of the density over the range of the data.

The probability density function can be estimated using any approach one would like to use. For instance, we could use the kernel density estimation method or histograms. Based on this, we see that there is some similarity between violin plots and the variations of the boxplot described previously.

Example 9.9

We return to the **snowfall** data from our previous example showing how to do kernel density estimation for univariate data. First, we load the data, and we set the bandwidth to one-half the normal reference rule.

```
% Load the data.
```

```
load snowfall

% Let's use the same bandwidth we had previously.
% Get the bandwidth.
n = length(snowfall);
% Approximately one-half the bandwidth for
% a normal distribution.
b = 0.5*std(snowfall)*n^(-1/5);
```

We found a function on MATLAB Central that produces part of a violin plot [Hoffmann, 2015]. It does not include the boxplot at the center, so we have to add it as shown below.

```
% This function was downloaded from MATLAB Central.
violin(snowfall','bw',b);

% Hold the plot to add the boxplot.
% Get the axis limits to reset.
ax = axis;
hold on
boxplot(snowfall,'PlotStyle','compact');
hold off
axis(ax)
ylabel('Snowfall (inches)')
xlabel('Buffalow Snowfall, 1910 - 1972')
set(gca,'Xticklabel',' ')
```

❏

The violin plot is shown in Figure 9.14. We can see additional details that are not visible in the boxplot, such as the two additional high density areas on the sides of the violin. Peaks and valleys in the probability density are not seen in boxplots, but they can be seen in violin plots [Hintze and Nelson, 1998]. Of course, the visibility of such structures also depends on the amount of smoothing one uses (e.g., histograms bin widths, kernel bandwidths, etc.), as discussed elsewhere.

9.3.4 Beeswarm Plot

We showed previously how to add a 1–D scatterplot or rug superimposed on an existing line plot, such as a probability density function. There is another type of 1–D scatterplot called a *beeswarm plot*. This is also called a *column scatterplot* or *stripchart*. It is similar to a jittered or textured dot plot that is available in GGobi, which is open-source software for dynamic data visualization [Cook and Swayne, 2007]. See Appendix B for details.

The beeswarm plot has a vertical orientation like most boxplots rather than the horizontal display of a rug plot. Usually, the beeswarm plot is displayed

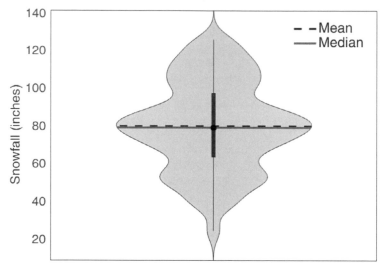

Buffalow Snowfall, 1910 - 1972

FIGURE 9.14

Here we show a violin plot of the Buffalo **snowfall** data. The bumps reflecting high and low snowfall amounts can be seen via the symmetric density traces marking the sides of the violin. Compare the density on both sides of the violin to Figure 9.7. The boxplot in the middle shows additional summary statistics.

as a stand-alone plot, but it can also be added to other distribution plots, as we discuss in the next section. Each observation is represented by a point in the plot. They are jittered where a small amount of noise is added to the data value to allow overlapping data points to be seen. A wide spread of the data indicates more observations are located in the region.

Example 9.10

We obtained a function called **plotSpread** from MATLAB Central that will create a beeswarm plot [Jonas, 2012]. We use the following code to construct a beeswarm plot for the **geyser** data. Recall that these data correspond to the waiting times between eruptions of the Old Faithful geyser.

```
% Load the geyser data.
load geyser

subplot(1,2,1)
% The beeswarm plot function requires the data
% in column format, so we convert it.
plotSpread(geyser(:))
ax1 = axis;
```

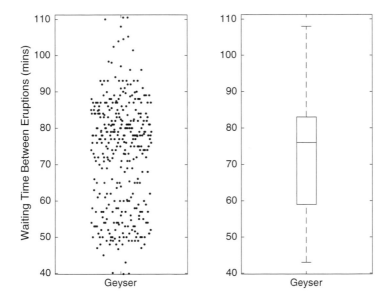

FIGURE 9.15
This is a beeswarm plot for the **geyser** data. These are waiting times in minutes between eruptions of the Old Faithful geyser. We included a boxplot for comparison purposes. The boxplot provides some descriptive statistics, while the beeswarm plot gives us a better idea of the data density.

```
box on
ylabel('Waiting Time Between Eruptions (mins)')
set(gca,'XTickLabel','Geyser')
```

We see that there is some evidence of two areas of higher density—for shorter and longer waiting times. We can also construct a boxplot for comparison. The plots are shown in Figure 9.15.

```
% Next, we display a boxplot for comparison.
subplot(1,2,2)
boxplot(geyser)
ax2 = axis;
axis([ax2(1:2),ax1(3:4)])
set(gca,'XTickLabel','Geyser')
```

❏

We can see that the beeswarm plot and the boxplot each provide useful information about our data. Thus, it is a good idea to visualize these together. Next, we show how a beeswarm plot can be combined with boxplots and its variations.

9.3.5 Beanplot

Beanplots developed by Kampstra [2008] are a combination of the rug plot with the violin plot. He calls them a beanplot because the density shape can look like the pod of a green bean, and the rug plot looks like the seeds inside the pod.

Kampstra [2008] describes several enhancements of the beanplot, but we describe only the basic version here. This plot has symmetric estimated probability density traces with a rug-like plot added to the center of the violin shape. The code used to create this plot is illustrated in the next example.

Example 9.11

We use the **geyser** data again for this example. First, we load it, and we then create a violin plot.

```
% First load the data.
load geyser
n = length(geyser);

% Do a violin plot first.
subplot(1,2,1)
violin(geyser(:));
hold on
```

Next, we create a 1–D scatterplot as lines and add that to the center of the violin plot.

```
% Add a rug-like plot where we show the points as
% ticks in the center of the density traces.
x = [0.95, 1.05];
y = [geyser(:),geyser(:)];
for i = 1:n
    plot(x,y(i,:))
end
ax = axis;
hold off
```

Instead of showing the observations as lines, we could use a beeswarm plot. the MATLAB code is given here.

```
% Alternatively, we could superimpose a
% beeswarm over the violin plot.
subplot(1,2,2)
violin(geyser(:));
hold on
plotSpread(geyser(:));
axis(ax)
hold off
```

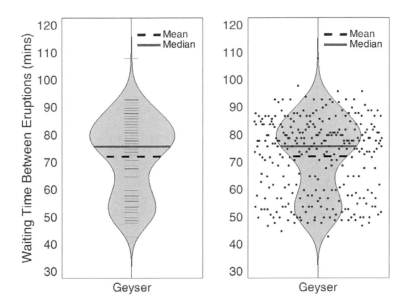

FIGURE 9.16
These are two beanplots for the **geyser** data. On the left is a version of the beanplot where observations are shown as ticks. A beanplot with jittered observations is on the right. The jittering is done via the **plotSpread** function.

These two versions of the beanplot are shown in Figure 9.16.
❏

There is a function called **distributionPlot** on the MATLAB Central file exchange [Jonas, 2009]. This allows comparison of multiple distributions and also creates variations of violin plots and beanplots. We illustrate some of its uses in the next example.

Example 9.12

The **distributionPlot** function on MATLAB Central has several options for comparing the distributions of multiple data sets using variations of the violin plot. These include using the histogram instead of the density for the trace, adding a 1–D scatterplot, showing density regions using the color map, and more. We demonstrate two of the options using the petal length variable of *Versicolor* and *Virgnica* species in the Fisher's iris data.

```
% Load the iris data.
load iris

% Extract the petal length from two species.
X1 = versicolor(:,3);
```

```
X2 = virginica(:,3);
```

The basic syntax **distributionPlot(X)** shows side-by-side violin plots of the columns in **X** with the mean shown as a cross and the median as a square. We can construct side-by-side beanplots with the following code.

```
% This displays multiple beanplots, where
% the observations are shown as points.
distributionPlot([X1,X2],'addSpread',...
     true,'showMM',false)
set(gca,'XTickLabel',{'Versicolor','Virginica'})
ylabel('Petal Length')
title('Fisher''s Iris Data')
```

The plot is shown in Figure 9.17. We can directly compare two distributions by plotting one density trace on the right and the other on the left. Here is the code.

```
% Plot Versicolor on the right.
distributionPlot(X1,'histOri','right',...
     'color','r','widthDiv',[2 2],...
     'showMM',0)
% Plot Virginica on the left.
distributionPlot(X2,'histOri','left',...
     'color','b','widthDiv',[2 1],...
     'showMM',0)
box on
title('Fisher''s Iris Data')
ylabel('Petal Length')
```

This is shown in Figure 9.18. We can see the multi-modal shapes and the overlap in the two distributions.
❑

9.4 Quantile Plots

As an alternative to the boxplots, we can use quantile-based plots to visually compare the distributions of two samples. These are also appropriate when we want to compare a known theoretical distribution and a sample. In making the comparisons, we might be interested in knowing how they are shifted relative to each other or to check model assumptions, such as normality.

In this section, we discuss several versions of quantile-based plots. These include *probability plots, quantile-quantile plots* (sometimes called *q-q plots*), and *quantile plots*. The probability plot has historically been used to

FIGURE 9.17
This shows side-by-side beanplots for the petal length of Fisher's iris *Versicolor* and *Virginica* species. The beanplot is a violin plot providing information about the density, along with a 1–D scatterplot showing the jittered data.

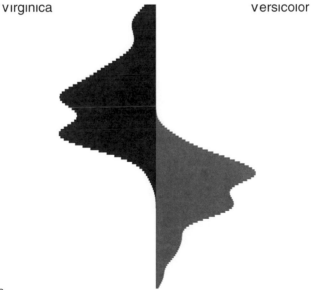

FIGURE 9.18
This is a comparison of petal length for Fisher's iris species *Virginica* and *Versicolor*. These are violin plots where the left trace corresponds to one group (*Virginica*), and the right trace corresponds to the another group (*Versicolor*). We see that each of the densities are multimodal, and they overlap.

compare sample quantiles with the quantiles from a known theoretical distribution, such as normal, exponential, etc. Typically, a *q-q* plot is used to determine whether two random samples were generated by the same distribution. The *q-q* plot can also be used to compare a random sample with a theoretical distribution by generating a sample from the theoretical distribution as the second sample. Finally, we have the quantile plot that conveys information about the sample quantiles.

9.4.1 Probability Plots

A probability plot is one where the theoretical quantiles are plotted against the ordered data, i.e., the sample quantiles. The main purpose is to visually determine whether or not the data could have been generated from the given theoretical distribution. If the sample distribution is similar to the theoretical one, then we would expect the relationship to follow an approximate straight line. Departures from a linear relationship are an indication that the distributions are different.

To get this display, we plot the $x_{(i)}$ on the vertical axis, and on the other axis we plot

$$F^{-1}\left(\frac{i - 0.5}{n}\right),\tag{9.14}$$

where $F^{-1}(\bullet)$ denotes the inverse of the cumulative distribution function for the hypothesized distribution. If the sample arises from the same distribution represented by Equation 9.14, then the theoretical quantiles and the sample quantiles should fall approximately on a straight line. As we discussed before, the 0.5 in the above argument can be different [Cleveland, 1993]. For example, we could use $i/(n + 1)$. See Kimball [1960] for other options. A well-known example of a probability plot is the ***normal probability plot***, where the theoretical quantiles from the normal distribution are used.

The MATLAB Statistics Toolbox has several functions for obtaining probability plots. One is called **normplot** to assess the assumption that a data set comes from a normal distribution. There is also a function for constructing a probability plot that compares a data set to the Weibull distribution. This is called **weibplot**. Probability plots for other theoretical distributions can be obtained using the MATLAB code given below, substituting the appropriate function to get the theoretical quantiles.

The Statistics Toolbox also has several functions that can be used to explore distribution shapes. One is the **probplot** function. The default for this function is to construct a normal probability plot with a reference line. However, this can be changed via an input argument for **'distname'** that specifies the desired distribution. Additionally, there is a GUI tool for fitting distributions. To start the tool, type **dfittool** at the command line. This tool allows you to load data from the workspace, fit distributions to the data, plot the distributions, and manage/evaluate different fits.

Example 9.13

This example illustrates how you can display a probability plot in MATLAB, where we return to the **galaxy** data. From previous plots, these data look approximately normally distributed, so we will check that assumption using the probability plot. First, we get the sorted sample, and then we obtain the corresponding theoretical quantiles for the normal distribution. The resulting quantile plot is shown in Figure 9.19.

```
load galaxy
% We will use the EastWest data again.
x = sort(EastWest);
n = length(x);
% Get the probabilities.
prob = ((1:n)-0.5)/n;
% Now get the theoretical quantiles for
% a normal distribution.
qp = norminv(prob,0,1);
% Now plot theoretical quantiles versus
% the sorted data.
plot(qp,x,'.')
ylabel('Sorted Data')
xlabel('Standard Normal Quantiles')
```

We see in the plot that there is slight curvature at the ends, indicating some departure from normality. However, these plots are exploratory only, and the results are subjective. Data analysts should use statistical inference methods (e.g., goodness-of-fit tests) to assess the significance of any departure from the theoretical distribution.
❑

9.4.2 Quantile-Quantile Plot

The q-q plot was originally proposed by Wilk and Gnanadesikan [1968] to visually compare two distributions by graphing the quantiles of one versus the quantiles of the other. Either or both of these distributions may be empirical or theoretical. Thus, the probability plot is a special case of the q-q plot.

Say we have two data sets consisting of univariate measurements. We denote the order statistics for the first data set by

$$x_{(1)}, x_{(2)}, \ldots, x_{(n)}.$$

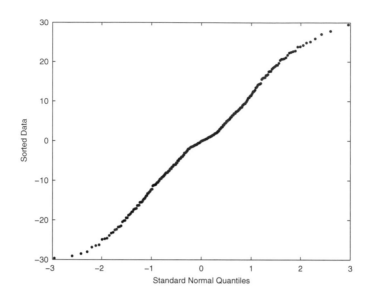

FIGURE 9.19
This is a probability plot for the **EastWest** variable in the **galaxy** data set. The curvature indicates that the data are not exactly normally distributed.

Let the order statistics for the second data set be

$$y_{(1)}, y_{(2)}, \ldots, y_{(m)},$$

where without loss of generality, $m \leq n$.

   In the next example, we show how to construct a *q-q* plot where the sizes of the data sets are equal, so $m = n$. In this case, we simply plot the sample quantiles of one data set versus the other data set as points.

Example 9.14

We will generate two sets of normal random variables and construct a *q-q* plot. Constructing a *q-q* plot for random samples from different distributions and different sample sizes will be covered in the next example. The first simulated data set is standard normal; the second one has a mean of 1 and a standard deviation of 0.75.

```
% Generate the samples - same size.
x = randn(1,300);
% Make the next one a different mean
% and standard deviation.
y = randn(1,300)*.75 + 1;
% Find the order statistics - sort them.
```

```
xs = sort(x);
ys = sort(y);
% Construct the q-q plot - do a scatterplot.
plot(xs, ys, '.')
xlabel('Standard Normal - Sorted Data')
ylabel('Normal - Sorted Data')
title('Q-Q Plot')
```

The *q-q* plot is shown in Figure 9.20. The data appear to be from the same family of distributions, since the relationship between them is approximately linear.
❏

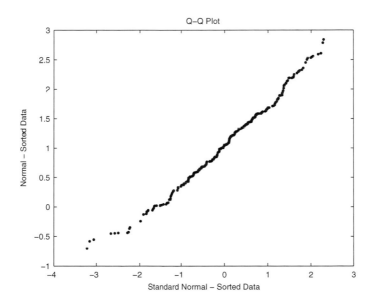

FIGURE 9.20
This is a *q-q* plot of two generated random samples, each one from a normal distribution. We see that the relationship between them is approximately linear, as expected.

We now look at the case where the sample sizes are not equal and $m < n$. To obtain the *q-q* plot, we graph the $y_{(i)}$, $i = 1, ..., m$ against the $(i-0.5)/m$ quantile of the other data set. The $(i-0.5)/m$ quantiles of the x data are usually obtained via interpolation.

Users should be aware that *q-q* plots provide only a rough idea of how similar the distribution is between two random samples. If the sample sizes are small, then a lot of variation is expected; so comparisons might be suspect. To help aid the visual comparison, some *q-q* plots include a reference line. These are lines that are estimated using the first and third quartiles of each data set and extending the line to cover the range of the data. The

MATLAB Statistics Toolbox provides a function called **qqplot** that displays this type of plot. We show below how to add the reference line.

Example 9.15

This example shows how to do a *q-q* plot when the samples do not have the same number of points. We use the function provided with this book called **quantileseda**[1] to get the required sample quantiles from the data set that has the larger sample size. We then plot these versus the order statistics of the other sample. Note that we add a reference line based on the first and third quartiles of each data set, using the function **polyfit**. We first generate the data sets, both from different distributions.

```
% We will generate some samples - one will be
% from the normal distribution, the other will
% be uniform.
n = 100;
m = 75;
x = randn(1,n);
y = rand(1,m);
```

Next we get the order statistics from the *y* values and the corresponding quantiles from the *x* data set.

```
% Sort y; these are the order statistics.
ys = sort(y);
% Now find the associated quantiles using the x.
% Probabilities for quantiles:
p = ((1:m) - 0.5)/m;
% The next function comes with this text.
xs = quantileseda(x,p);
```

Now we can construct the plot, adding a reference line based on the first and third quartiles to help assess the linearity of the relationship.

```
% Construct the plot.
plot(xs,ys,'.')
% Get the reference line. Use the 1st and 3rd
% quartiles of each set to get a line.
qy = quartiles(y);
qx = quartiles(x);
[pol, s] = polyfit(qx([1,3]),qy([1,3]),1);
% Add the line to the figure.
yhat = polyval(pol,xs);
hold on
plot(xs,yhat,'k')
```

---

[1] The Statistics Toolbox has similar functions called **quantile** and **prctile** to calculate the sample quantiles and percentiles.

```
xlabel('Sample Quantiles - X'),
ylabel('Sorted Y Values')
hold off
```

We see in Figure 9.21 that the data do not come from the same distribution, since the relationship is not linear.
❑

   A major benefit of the quantile-based plots discussed so far is that they do not require the two samples (or the sample and theoretical distribution) to have the same location and scale parameter. If the distributions are the same, but differ in location or scale, then we would still expect them to produce a straight line. This will be explored in the exercises.

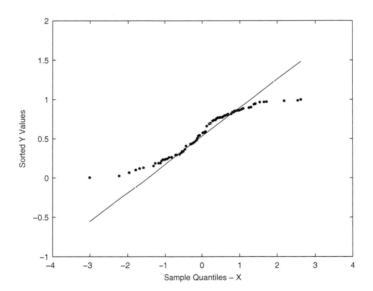

FIGURE 9.21
This shows the *q-q* plot for Example 9.9. The *x* values are normally distributed with $n = 100$. The *y* values are uniformly distributed with $m = 75$. The plot shows that these data do not come from the same distribution.

9.4.3 Quantile Plot

Cleveland [1993] describes another version of a quantile-based plot. He calls this the quantile plot, where we have the u_i values (Equation 9.12) along the horizontal axis and the ordered data $x_{(i)}$ along the vertical. These ordered pairs are plotted as points, joined by straight lines. Of course, this does *not* have the same interpretation or provide the same information as the probability plot or *q-q* plot described previously. This quantile plot provides an initial look at the distribution of the data, so we can search for unusual structure or behavior. We also obtain some information about the sample quantiles and their relationship to the rest of the data. This type of display is discussed in the next example.

Example 9.16

We use a data set from Cleveland [1993] that contains the heights (in inches) of singers in the New York Choral Society. The following MATLAB code constructs the quantile plot for the **Tenor_2** data.

```
load singer
n = length(Tenor_2);
% Sort the data for the y-axis.
ys = sort(Tenor_2);
% Get the associated u values.
u = ((1:n)-0.5)/n;
plot(u,ys,'-o')
xlabel('u_i value')
ylabel('Tenor 2 Height (inches)')
```

We see from the plot shown in Figure 9.22 that the values of the quartiles (and other quantiles of interest) are easy to see.
❑

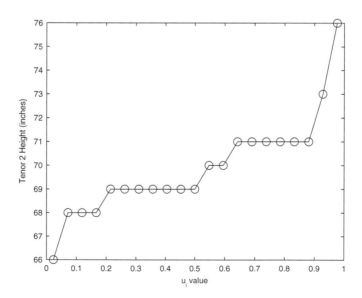

FIGURE 9.22
This is the quantile plot for the **Tenor_2** data, which is part of the **singer** data.

A word of caution is in order regarding the quantile-based plots discussed in this section. Some of the names given to them are not standard throughout statistics books. For example, the quantile plot is also called the *q-q* plot in Kotz and Johnson [Vol. 7, p. 233, 1986].

9.5 Bagplots

The *bagplot* is a bivariate box-and-whiskers plot developed by Rousseeuw, Ruts, and Tukey [1999]. This is a generalization of the univariate boxplot discussed previously. It uses the idea of the location depth of an observation with respect to a bivariate data set. This extends the idea of ranking or ordering univariate data to the bivariate case, so we can find quantities analogous to the univariate quartiles. The bagplot consists of the following components:

1. A *bag* that contains the inner 50% of the data, similar to the IQR;
2. A cross (or other symbol) indicating the depth median (described shortly);
3. A *fence* to determine potential outliers; and
4. A *loop* that indicates the points that are between the bag and the fence.

We need to define several concepts to construct the bagplot. First, we have the notion of the *halfspace location depth* introduced by Tukey [1975]. The halfspace location depth of a point θ relative to a bivariate data set is given by the smallest number of data points contained in a closed half-plane where the boundary line passes through θ. Next, we have the *depth region* D_k, which is the set of all θ with halfspace location depth greater than or equal to k. Donoho and Gasko [1992] define the *depth median* of the bivariate point cloud, which in most cases is the center of gravity of the deepest region. Rousseeuw and Ruts [1996] provide a time-efficient algorithm for finding the location depth and depth regions, as well as an algorithm for calculating the depth median [1998].

The bag is constructed in the following manner. We let $\# D_k$ be the number of data points in the depth region D_k. First, we find the value k for which

$$\# D_k \leq \left\lfloor \frac{n}{2} \right\rfloor < \# D_{k-1},$$

where the notation $\lfloor x \rfloor$ denotes the greatest integer less than or equal to x. Then, the bag is obtained by linearly interpolating between D_k and D_{k-1}, relative to the depth median. The fence is constructed by inflating the bag by some factor relative to the depth median. Any points that are outside the fence are flagged as potential outliers. Finally, the loop is the outer boundary of the bagplot; i.e., it is the convex hull of the bag and the nonoutlying points.

Example 9.17

Rousseuw, Ruts and Tukey provide Fortran and MATLAB code to construct the bagplot; see Appendix B for download information. However, a more user-friendly version for MATLAB is now available in the LIBRA Toolbox [Verboven and Hubert, 2005],[2] and we illustrate its use in this example. We will use the **environmental** data set, where the two variables of interest are temperature and ozone.

```
% First we load up some data to be used in the plot.
load environmental
% Now we put the data set together.
data = [Temperature,Ozone];
```

There are many options available in the **bagplot** function; use the **help** feature to get more information on them. We will use the default values in our function call.

```
% The bagplot function was taken from
% the LIBRA toolbox. It is used here
% with the permission of the authors.
bagplot(data)
title('Bagplot of Environmental Data')
xlabel('Temperature')
ylabel('Ozone')
```

The resulting plot is shown in Figure 9.23.
❑

9.6 Rangefinder Boxplot

Becketti and Gould [1987] introduced a bivariate extension of the boxplot called the *rangefinder boxplot*. These are used in conjunction with 2–D scatterplots, which are discussed in the next chapter. The rangefinder boxplot contains the same information for both variables x_1 and x_2 that one has with univariate boxplots, along with the additional information seen in a scatterplot.

A rangefinder boxplot is displayed using six lines that are superimposed on a scatterplot. Three of these lines are vertical, and three are horizontal. The

---

[2] The **bagplot** function is included with the EDA Toolbox (with permission) and is governed by the license found at **http://wis.kuleuven.be/stat/robust/LIBRA/LIBRA-home**

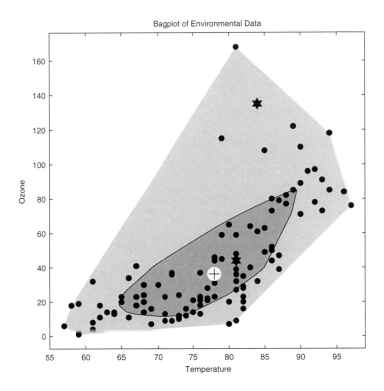

FIGURE 9.23

This is a bagplot showing two variables in the **environmental** data set. The depth median is displayed as a large circle with a cross inside.

positions and the lengths of the lines indicate the medians, the interquartile ranges, and the adjacent values for each of the variables. We provide the details of constructing a rangefinder boxplot below and show how to construct one in the next example.

Procedure – Rangefinder Boxplot

1. First, construct a scatterplot of the data.

2. Find the quartiles for each variable x_1 and x_2:

$$q_{x_1}(0.25), q_{x_1}(0.50), q_{x_1}(0.75)$$
$$q_{x_2}(0.25), q_{x_2}(0.50), q_{x_2}(0.75).$$

3. Find the interquartile ranges for each variable:

$$IQR_{x_1} = q_{x_1}(0.75) - q_{x_1}(0.25)$$
$$IQR_{x_2} = q_{x_2}(0.75) - q_{x_2}(0.25).$$

These values determine the lengths of the lines. All of the horizontal lines have length equal to IQR_{x_1}, and all vertical lines have length equal to IQR_{x_2}.

4. Find the adjacent values based on the upper limit and the lower limit for each of the variables (Equation 9.13). These values govern the placement of the outer lines.
5. Place one of the vertical lines and one of the horizontal lines at position $(q_{x_1}(0.50), q_{x_2}(0.50))$, forming a cross at the medians.
6. Place a vertical line at each of the adjacent values for variable x_1.
7. Place an horizontal line at each of the adjacent values for variable x_2.

Example 9.18

We use two variables of the **oronsay** data to illustrate the process of creating a rangefinder boxplot. First, we load the data and put the variables of interest into our data matrix.

```
% Use 2 variables of the oronsay data for this example.
load oronsay
X = oronsay(:,7:8);
```

Now, we construct a scatterplot of the variables, which is shown in Figure 9.16.

```
% Construct a scatterplot.
plot(X(:,1),X(:,2),'.')
xlabel(labcol{7})
ylabel(labcol{8})
title('Oronsay Data')
% Hold the plot, so we can add the lines.
hold on
```

Next, we find the quartiles and the interquartile range for each variable.

```
% Find the quartiles for each variable.
qx1 = quartiles(X(:,1));
qx2 = quartiles(X(:,2));
% Find the interquartile ranges.
iqr1 = qx1(3) - qx1(1);
iqr2 = qx2(3) - qx2(1);
```

We use this information to determine the upper and lower limits for both of the variables.

```
% Find the upper and lower limits.
LL1 = qx1(1) - 1.5*iqr1;
UL1 = qx1(3) + 1.5*iqr1;
LL2 = qx2(1) - 1.5*iqr2;
UL2 = qx2(3) + 1.5*iqr2;
```

Recall that the adjacent values are the most extreme (largest and smallest) observations that are not outliers. The following code will find the adjacent values.

```
% Now find the adjacent values.
Xs(:,1) = sort(X(:,1));
Xs(:,2) = sort(X(:,2));
ind1 = find(Xs(:,1) > LL1 & (Xs(:,1) < UL1));
adjv1 = [min(Xs(ind1,1)), max(Xs(ind1,1))];
ind2 = find(Xs(:,2) > LL2 & (Xs(:,2) < UL2));
adjv2 = [min(Xs(ind2,2)), max(Xs(ind2,2))];
```

We are ready to add the lines of the rangefinder boxplot to the scatterplot. First, we add the two lines that cross at the medians.

```
% Place the cross at the medians.
plot([qx1(1),qx1(3)],[qx2(2),qx2(2)])
plot([qx1(2),qx1(2)],[qx2(1),qx2(3)])
% Plot a circle at the medians to check.
plot(qx1(2),qx2(2),'o')
```

The two vertical lines are placed at the adjacent values for the variable along the horizontal axis.

```
% Plot the two vertical lines at the
% adjacent values for x_1.
plot([adjv1(1),adjv1(1)], [qx2(1),qx2(3)])
plot([adjv1(2),adjv1(2)], [qx2(1),qx2(3)])
```

Finally, the two horizontal lines are shown at the adjacent values for the variable displayed on the vertical axis.

```
% Plot the two horizontal lines at the
% adjacent values for x_2.
plot([qx1(1),qx1(3)], [adjv2(1),adjv2(1)])
plot([qx1(1),qx1(3)], [adjv2(2),adjv2(2)])
```

The resulting rangefinder boxplot is superimposed on the scatterplot shown in Figure 9.24. We also include a side-by-side boxplot of the two variables, so the reader can compare the information that is available in both plots. These boxplots are displayed in Figure 9.25.
❑

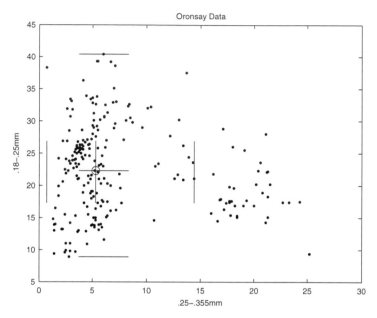

FIGURE 9.24
This plot shows the rangefinder boxplot for two variables of the **oronsay** data set. Note that this provides the same information one can get from the individual boxplots for each of the variables (e.g., interquartile range and potential outliers), but with the added benefit of seeing the actual data shown in a scatterplot.

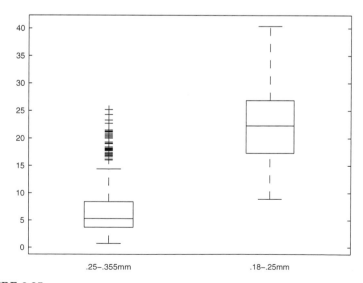

FIGURE 9.25
This displays the univariate boxplots for each variable in the rangefinder boxplot, so the reader can compare this with the information contained in Figure 9.16.

9.7 Summary and Further Reading

We presented several methods for visualizing the shapes of distributions for continuous random variables. The easiest and most intuitive methods to use are the histograms and boxplots. We discussed several variations of these plots. For histograms, we presented frequency, relative frequency, density, and 2–D histograms. Enhancements to the boxplot included histplots, box-percentile plots, variable-width boxplots, and others. We concluded the chapter with quantile-based plots and a generalization of the univariate boxplot to 2–D.

There is an extensive literature on probability density estimation, both univariate and multivariate. The best comprehensive book in this area is Scott [2015]. He discusses histograms, frequency polygons, kernel density estimation, and the average shifted histograms. He covers the theoretical foundation of these methods, how to choose the smoothing parameters (e.g., bin widths), and many practical examples. For a MATLAB perspective on these methods, please see Martinez and Martinez [2015]. The book called *Visualizing Data* by William Cleveland [1993] is an excellent resource on many aspects of data visualization, including quantile-based plots discussed in this chapter.

We only covered the quantile-based plots for continuous data in this chapter. Versions of these are also available for discrete data distributions, such as binomial or Poisson. MATLAB implementations of quantile-based plots for discrete distributions can be found in Martinez and Martinez [2015]. Another resource for the visualization of categorical data is Friendly [2000], where SAS software is used for implementation of the ideas. We recommend Hoaglin and Tukey [1985] for a nice summary of various methods for checking the shape of discrete distributions. The third edition of this book has a new chapter (Chapter 11) on visualizing categorical data, where the interested reader will discover ways to view and understand these types of variables.

Exercises

9.1   Repeat Example 9.1 using 5 and 50 bins. Compare these results with what we had in Figure 9.1.

9.2   Repeat Example 9.1 using the **forearm** data.

9.3   Apply the various bin width rules to the **forearm** and to the **galaxy** data. Discuss the results.

9.4   The histogram methods presented in the chapter call the **hist** function with the desired number of bins. Do a **help** on **hist** or **histc** to see how to set the bin centers instead. Use this option to construct histograms with a specified bin width and origin.

9.5   Using the code from Problem 9.4, show how changing the bin starting point affects histograms of the **galaxy** data.

9.6   Using the data in Example 9.2, construct a normal curve, using the sample mean and standard deviation. Superimpose this over the histogram and analyze your results.

9.7   Plot the histogram in Example 9.3 using the **surf** plot, using this code:

```
% Plot as a surface plot.
% Get some axes that make sense.
[XX,YY]=...
meshgrid(linspace(min(x(:,1)),max(x(:,1)),nb1),...
linspace(min(x(:,2)),max(x(:,2)),nb2));
% Z is the height of the bins in Example 9.3.
surf(XX,YY,Z)
```

9.8   Use a boxplot and a histogram to verify that the quartiles in Example 9.6 for the **geyser** data make sense.

9.9   Generate some standard normal data, with sample sizes $n = 30, n = 50$, and $n = 100$. Use the function **boxp** to first get a set of plain boxplots and then use it to get variable width boxplots. The following code might help:

```
% Generate some standard normal data with
% different sample sizes. Put into a cell array.
X{1} = randn(30,1);
X{2} = randn(50,1);
X{3} = randn(100,1);
% First construct the plain boxplot.
boxp(X)
% Next we get the boxplot with variable
% widths.
boxp(X,'vw')
```

9.10   Generate some random data. Investigate the **histfit** function in the Statistics Toolbox.

9.11   Show that the area represented by the bars in the density histogram sums to one.

9.12   Load the data used in Example 9.8 (**load example96**). Construct histograms (use 12 bins) of the columns of **X**. Compare with the boxplots and box-percentiles plots.

9.13   Type **help boxplot** at the MATLAB command line to learn more about this Statistics Toolbox Function. Do side-by-side boxplots of the **oronsay** data set. The length of the whisker is easily adjusted using optional input arguments to **boxplot**. Try various values for this option.

9.14 Explore the **'notches'** option of the **boxplot** function. Generate bivariate normal data, where the columns have the same means. Construct notched boxplots with this matrix. Now generate bivariate normal data where the columns have very different means. Construct notched boxplots with this matrix. Discuss your results.

9.15 Apply the **boxplot** with notches to the **oronsay** data and discuss the results.

9.16 Generate two data sets, each from a normal distribution with different location and scale parameters. Construct a *q-q* plot (see Example 9.15) and discuss your results.

9.17 Reconstruct Figure 9.2 using the density histogram. Superimpose the normal curve over this one. Discuss the difference between this and Figure 9.2.

9.18 Construct a bagplot using the BPM data from Example 7.11. Compare with the polar smoothing.

9.19 A *rootogram* is a histogram where the heights of the bins correspond to the square root of the frequency. Write a MATLAB function that will construct this type of plot. Use it on the **galaxy** data.

9.20 Generate data as shown in Example 9.8 and examine the distribution of the two sets of random variables using boxplots, variable width box-percentile plots, violin plots, and beanplots.

9.21 Use the MATLAB **boxplot** function to get side-by-side boxplots of the following data sets and discuss the results. Construct boxplots with and without notches.

a. **skulls**

b. **sparrow**

c. **pollen**

d. BPM data sets (after using ISOMAP)

e. gene expression data sets

f. **spam**

g. **iris**

h. **software**

9.22 Apply some of the other types of boxplots to the data in Problem 9.21.

9.23 Generate uniform random variables (use the **rand** function) and construct a normal probability plot (or a *q-q* plot with the other data set generated according to a standard normal). Do the same thing with random variables generated from an exponential distribution (see **exprnd** in the Statistics Toolbox). Discuss your results.

9.24 Construct a rangefinder boxplot using the BPM data from Example 7.11. Compare your results with the polar smoothing and the bagplot.

9.25 Create kernel density estimates of the **forearm, galaxy, snowfall**, and **geyser** data using the default bandwidth. Try smaller (e.g., half the default) and larger bandwidths (e.g., twice the default). Describe your results.

9.26   Create kernel density estimates of the **forearm**, **galaxy**, **snowfall**, and **geyser** data using different kernels and the bandwidths from the previous problem. Discuss the differences, if there are any.

9.27   Construct a 2–D scatterplot of the **faithful** data and compare it to the estimated density in Figure 9.8.

9.28   Explore some of the options in **distributionPlot** using the data in Example 9.12.

a. The default options create side-by-side violin plots.

```
distributionPlot([X1,X2])
```

b. This shows the quantiles as lines in the violin plot.

```
distributionPlot([X1,X2],'showMM',6)
```

c. This displays the distributions as a heatmap density.

```
distributionPlot([X1,X2],'colormap',copper,...
        'showMM',6,'variableWidth',false)
```

9.29   The **histogram** function has additional options for the **'Normal-ization'** argument. Consult the **help** file for more information, and apply them to the **galaxy** data. You can create a histogram with bin heights corresponding to the cumulative count and one where the heights are the count density.

Chapter 10

Multivariate Visualization

In this chapter, we present several methods for visualizing and exploring multivariate data. We have already seen some of these methods in previous chapters. For example, we had grand tours and projection pursuit in Chapter 4, where the dimensionality of the data is first reduced to 2–D and then visualized in scatterplots. The primary focus of this chapter is to look at ways to visualize and explore all of the dimensions in our data at once.

The first thing we cover is glyph plots. Then we present information about scatterplots, both 2–D and 3–D, as well as scatterplot matrices. Next, we talk about some dynamic graphics, such as linking and brushing. These techniques enable us to find connections between points in linked graphs, delete and label points, and highlight subsets of points. We then cover coplots, which convey information about conditional dependency between variables. This is followed by dot charts that can be used to visualize summary statistics and other data values. Next, we discuss how to view each of our observations as curves via Andrews' plots or as broken line segments in parallel coordinates. We then show how the concepts of the data image and Andrews' curves can be combined to reveal structure in high-dimensional data. Next, we describe how these methods can be combined with the plot matrix concept and the grand tour. Finally, we conclude the chapter with a discussion of biplots, which can be used to visualize the results of PCA, nonnegative matrix factorization and other similar dimensionality reduction methods.

10.1 Glyph Plots

We first briefly discuss some of the multivariate visualization methods that will *not* be covered in detail in this text. Most of these are suitable for small data sets only, so we do not think that they are in keeping with the trend towards analyzing massive, high-dimensional data sets, as seen in most applications of EDA and data mining.

The first method we present is due to Chernoff [1973]. His idea was to represent each observation (with dimensionality $p \le 18$) by a cartoon face. Each feature of the face, such as length of nose, mouth curvature, eyebrow shape, size of eyes, etc., would correspond to a value of the variable. This technique is useful for understanding the overall regularities and anomalies in the data, but it has several disadvantages. The main disadvantage is the lack of quantitative visualization of the variables; we just get a qualitative understanding of the values and trends. Another problem is the subjective assignment of features to the variables. In other words, assigning a different variable to the eyebrows and other facial features can have a significant effect on the final shape of the face. We show an example of *Chernoff faces* for the `cereal` data (described in Appendix C) in Figure 10.1.

Star diagrams [Fienberg, 1979] are a similar plot, in that we have one glyph or star for each observation, so they suffer from the same restrictions as the faces regarding sample size and dimensionality. Each observed data point in the sample is plotted as a star, with the value of each measurement shown as a radial line from a common center point. Thus, each measured value for an observation is plotted as a spoke that is proportional to the size of the measured variable with the ends of the spokes connected with line segments to form a star. We show the star plot for the same cereal data in Figure 10.2.

The Statistics Toolbox has a function called `glyphplot` that will construct either Chernoff faces or star diagrams for each observation. The use of this function will be explored in the exercises.

Other glyphs and similar plots have been described in the literature [Kleiner and Hartigan, 1981; du Toit, Steyn, and Stumpf, 1986], and most of them suffer from the same drawbacks. These include star-like diagrams, where the rays emanate from a circle, and the end points are not connected. We also have profile plots, where each observation is rendered as a bar chart, with the height of the bar indicating the value of the variable. Another possibility is to represent each observation by a box, where the height, length, and width correspond to the variables.

10.2 Scatterplots

We have already introduced the reader to *scatterplots* and *scatterplot matrices* in previous chapters, but we now examine these methods in more detail, especially how to construct them in MATLAB. We also present an enhanced scatterplot based on *hexagonal binning* that is suitable for massive data sets.

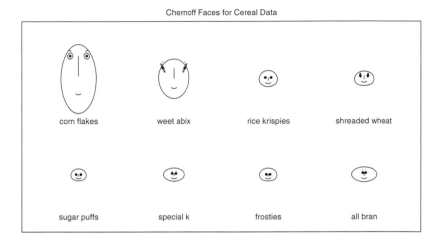

FIGURE 10.1

This shows the Chernoff faces for the `cereal` data, where we have 8 observations and 11 variables. The shape and size of various facial features (head, eyes, brows, mouth, etc.) correspond to the values of the variables. The variables represent the percent agreement to statements about the cereal. The statements are: comes back to, tastes nice, popular with all the family, very easy to digest, nourishing, natural flavor, reasonably priced, a lot of food value, stays crispy in milk, helps to keep you fit, fun for children to eat.

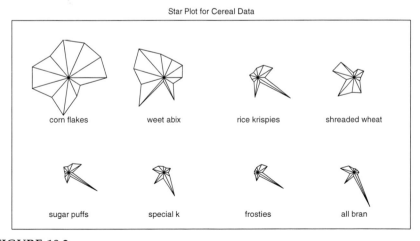

FIGURE 10.2

This shows the star plots for the same `cereal` data. There is one ray for each variable. The length of the ray indicates the value of the attributed.

10.2.1 2–D and 3–D Scatterplots

The scatterplot is a visualization technique that enjoys widespread use in data analysis and is a powerful way to convey information about the relationship between two variables. To construct one of these plots in 2–D, we simply plot the individual (x_i, y_i) pairs as points or some other symbol. For 3–D scatterplots, we add the third dimension and plot the (x_i, y_i, z_i) triplets as points.

The main MATLAB package has several ways we can construct 2–D and 3–D scatterplots, as shown in Example 10.1. The Statistics Toolbox also has a function to create 2–D scatterplots called **gscatter** that will construct a scatterplot, where different plotting symbols are used for each cluster or class. However, as we will see in Example 10.1 and the exercises, similar results can be obtained using the **scatter** and **plot** functions.

Example 10.1

In this example, we illustrate the 2–D and 3–D scatterplot functions called **scatter** and **scatter3**. Equivalent scatterplots can also be constructed using the basic **plot** and **plot3** functions, which will be explored in the exercises. Since a scatterplot is generally for 2–D or 3–D, we need to extract a subset of the variables in the **oronsay** data. So we've chosen variables 8, 9, and 10: 0.18–0.25mm, 0.125–0.18mm, and 0.09–0.125mm. We first use variables 8 and 9 to construct a basic 2–D scatterplot.

```
% First load up the data and get the
% variables of interest.
load oronsay
% Use the oronsay data set. Just plot two
% of the variables. Now for the plot:
scatter(oronsay(:,8),oronsay(:,9))
xlabel(labcol{8})
ylabel(labcol{9})
```

This plot is shown in Figure 10.3 (top). The basic syntax for the scatter function is

```
scatter(X,Y,S,C,M)
```

where **X** and **Y** are the data vectors to be plotted, and the other arguments are optional. **S** can be either a scalar or a vector indicating the area (in units of points-squared) of each marker. **M** is an alternative marker (default is the circle), and **C** is a vector of colors. Next we show how to use the color vector to plot the observations in different colors, according to their midden group membership. See color insert Figure 10.3 (top) for the resulting plot.

```
% If we want to use different colors for the groups,
% we can use the following syntax. Note that this
% is not the only way to do the colors.
```

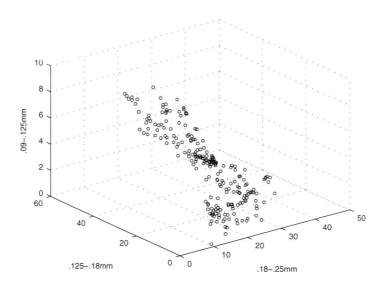

FIGURE 10.3
The top figure shows the 2–D scatterplot for two variables (i.e., columns 8 and 9) of the
oronsay data set. The color of the plot symbols indicates the midden class membership for
both plots. The lower plot shows the 3–D scatterplot for columns 8 through 10. (**SEE COLOR
INSERT.**)

```
ind0 = find(midden==0); % Red
ind1 = find(midden==1); % Green
ind2 = find(midden==2); % Blue
% This creates an RGB - 3 column colormap matrix.
C = zeros(length(midden),3);
C(ind0,1) = 1;
C(ind1,2) = 1;
C(ind2,3) = 1;
scatter(oronsay(:,8),oronsay(:,9),5,C)
xlabel(labcol{8})
ylabel(labcol{9})
zlabel(labcol{10})
```

3–D scatterplots can also be very useful, and they are easily created using the **scatter3** function, as shown below.

```
% Now show scatter3 function. Syntax is the same;
% just add third vector.
scatter3(oronsay(:,8),oronsay(:,9),oronsay(:,10),5,C)
xlabel(labcol{8})
ylabel(labcol{9})
zlabel(labcol{10})
```

The 3–D scatterplot is shown in the bottom panel of Figure 10.3 and in the corresponding color insert. MATLAB has another useful feature in the **Figure Window** toolbar buttons. This is the familiar *rotate* button ⟳ that, when selected, allows the user to click on the 3–D axis and rotate the plot. The user can see the current elevation and azimuth (in degrees) in the lower left corner of the figure window while the axes are being rotated.
❏

The Statistics Toolbox has a useful function called **scatterhist** that creates a 2–D scatterplot and adds univariate histograms to the horizontal and vertical axes. This provides additional useful information about the marginal distributions of the data. We illustrate the use of **scatterhist** in the next example.

Example 10.2

We return to the **oronsay** data that was used to produce the scatterplot in Figure 10.3 to create a 2–D scatterplot with marginal histograms.

```
% First load up the data and get the
% variables of interest.
load oronsay
% Now create the scatterplot with the
% marginal histograms.
% Note that we can provide an optional argument
% to the function that specifies the number of
```

```
% bins for each of the histograms.
scatterhist(oronsay(:,8),oronsay(:,9),[20,20])
xlabel(labcol{8})
ylabel(labcol{9})
```

The resulting plot is shown in Figure 10.4. The marginal histograms indicate some interesting structure that is not readily visible in the scatterplot. ❏

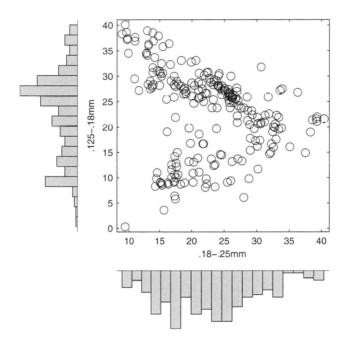

FIGURE 10.4
Here is the plot we get using **scatterhist** on two variables in the **oronsay** data set. The **scatterhist** function produces a 2–D scatterplot with univariate histograms along the horizontal and vertical axes. These histograms can show useful information about the marginal distributions.

10.2.2 Scatterplot Matrices

Scatterplot matrices are suitable for multivariate data, when $p > 2$. They show all possible 2–D scatterplots, where the axis of each plot is given by one of the variables. The scatterplots are then arranged in a matrix-like layout for easy viewing and comprehension. Some implementations of the scatterplot matrix show the plots in the lower triangular portion of the matrix layout only, since showing both is somewhat redundant. However, we feel that showing all of them makes it easier to understand the relationships between the variables. As we will see in Example 10.3, the MATLAB functions for scatterplot matrices show all plots.

One of the benefits of a scatterplot matrix is that one can look across a row or column and see the scatterplots of a given variable against all other variables. We can also view the scatterplot matrix as a way of partially linking points in different views, especially when observations of interest are shown with different marker styles (symbols and/or colors).

The main MATLAB package has a function called **plotmatrix** that will produce a scatterplot matrix for a data matrix **X**. The diagonal boxes of the scatterplot matrix contain histograms showing the distribution of each variable (i.e., the column of **X**). Please see the **help** on this function for its other uses. The Statistics Toolbox includes an enhanced version called **gplotmatrix**, where one can provide group labels; so observations belonging to different groups are shown with different symbols and colors.

Example 10.3
The **plotmatrix** function will construct a scatterplot matrix when the first argument is a matrix. An alternative syntax allows the user to plot the columns of one matrix against the columns of the other. We use the following commands to construct a scatterplot matrix of the three variables of the **oronsay** data used in the previous example.

```
% Use the same 3 variables from a previous example.
X = [oronsay(:,8),oronsay(:,9),oronsay(:,10)];
plotmatrix(X,'.');
% Let's make the symbols slightly smaller.
Hdots = findobj('type','line');
set(Hdots,'markersize',1)
```

We chose these from the scatterplot matrix of the full data set, because they seemed to show some interesting structure. It is also interesting to note the histograms of the variables that are shown along the diagonals, as they provide information about their distribution. The scatterplot matrix is shown in Figure 10.5.
❑

10.2.3 Scatterplots with Hexagonal Binning

Carr et al. [1987] introduced several scatterplot matrix methods for situations where the size of the data set *n* is large. In our view, *n* is large when the scatterplot can have a lot of overplotting, so that individual points are

FIGURE 10.5
This is the scatterplot matrix for columns 8 through 10 of the **oronsay** data. The first row of the plots shows us column 8 plotted against column 9 and then against column 10. We have similar plots for the other two rows.

difficult to see. When there is significant overplotting, then it is often more informative to convey an idea of the density of the points, rather than the observations alone. Thus, Carr et al. suggest displaying bivariate densities represented by gray scale or symbol area instead of individual observations.

To do this, we first need to estimate the density in our data. We have already introduced this issue in Chapter 9, where we looked at estimating the bivariate density using a histogram with bins that are rectangles or squares. Recall, also, that we used vertical bars to represent the value of the density, rather than a scatterplot. In this chapter, we are going to use bins that are hexagons instead of rectangles, and the graphic used to represent data density will be symbol size, as well as color.

Carr et al. recommended the hexagon as an alternative to using the square as a symbol, because the square tends to create bins that appear stretched out in the vertical or horizontal directions. The procedure we use for hexagonal binning is outlined below.

Procedure – Hexagonal Binning for Scatterplots

1. Find the length *r* of the side of the hexagon bins based on a given number of bins.
2. Obtain a set of hexagonal bins over the range of the data.
3. Bin the data.

4. Scale the hexagons with *nonzero* bin counts, such that the bin with maximum frequency has sides with length r and the smallest frequency (*nonzero*) has length $0.1r$.

5. Display a hexagon at the center of each bin, where the sides correspond to the lengths found in step 4.

We provide a MATLAB function called **hexplot** and show how it is used in the next example.

Example 10.4

The basic syntax for **hexplot** is:

```
hexplot(X,nbin,flag)
```

The first two arguments *must* be provided. **X** is a matrix with n rows and two columns, and **nbin** is the approximate number of bins for the dimension with the larger range. We use the **oronsay** data from the previous examples to illustrate the function.

```
X = [oronsay(:,8),oronsay(:,9)];
% Construct a hexagon scatterplot with
% 15 bins along the longer dimension.
hexplot(X,15);
```

This plot is shown in Figure 10.6 (top). Since the bins and resulting plot depend on the number of bins, it is useful to construct other scatterplots with different **nbin** values to see if other interesting density structure becomes apparent. The optional input argument **flag** (this can be any value) produces a scatterplot where the color of the hexagon symbols corresponds to the probability density at that bin. The probability density is found in a similar manner to a bivariate histogram with rectangular bins, except that we now normalize using the area of the hexagonal bin. An example of this is shown in Figure 10.6 (bottom), and it was produced with the following MATLAB statements:

```
hexplot(X,15,1)
colormap(gray)
```

❑

10.3 Dynamic Graphics

We now present some methods for dynamic graphics. These allow us to interact with our plots to uncover structure, remove outliers, locate groups, etc. Specifically, we cover labeling observations of interest, deleting points,

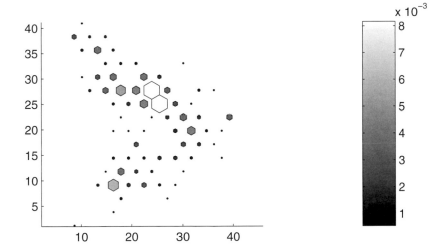

FIGURE 10.6
The top plot is a scatterplot with hexagonal bins. This would be an alternative to the plot shown in Figure 10.3. The bottom plot is for the same data, but the color of the symbols encodes the value of the probability density at that bin. The density is obtained in a manner similar to the bivariate histogram. (**SEE COLOR INSERT.**)

finding and displaying subsets of our data, linking and brushing. As with the tour methods discussed in Chapter 4, it is difficult to convey these ideas via static graphs on the pages of this book; so the reader is encouraged to try these methods to understand the techniques.

10.3.1 Identification of Data

One can identify points in a plot in several ways. First, we can add labels to certain data points of interest or we can highlight observations by using some other plotting symbol or color [Becker, Cleveland, and Wilks, 1987].

We label points in our plot by adding some text that identifies the observation. Showing all labels is often not possible because overplotting occurs, so nothing can be distinguished. Thus, a way to selectively add labels to our plot is a useful capability. One can look at accomplishing this in two ways. The user can click on a point (or select several points) on the plot, and the labels are added. Or, the user can select an observation (or several) from a list, at which point the observations are labeled. We explore both of these methods in the next example.

We might also have the need to interactively delete points because they could be outliers, and they make it difficult to view other observations. For example, we might have an extreme value that pushes all of the remaining data into one small region of the plot, making it difficult to resolve the bulk of the observations.

Instead of labeling individual cases or observations, we might have the need to highlight the points that belong to some subset of our data. For example, with the document clustering data, we could view the scatterplot of each topic separately or highlight the points in one class with different color and symbol type. These options allow direct comparison on the same scale, but overlap and overplotting can hinder our understanding. We can also plot these groups separately in panels, as we do in scatterplot matrices.

A dynamic method for visualizing subsets of the data that is reminiscent of data tours is called *alternagraphics* [Tukey, 1973]. This type of tour cycles through the subsets of the data (e.g., classes), showing each one separately in its own scatterplot. The cycle pauses at each plot for some fixed time step, so it is important that all plots be on a common scale for ease of comparison. Alternatively, one could show all of the data throughout the cycle, and highlight each subset at each step using different color and/or symbol. We leave the implementation of this idea as an exercise to the reader.

Example 10.5

The Statistics Toolbox provides a function for labeling data on a plot called **gname**. The user can invoke this function using an input argument containing strings for the case names. This must be in the form of a string matrix. An alternative syntax is to call **gname** without any input argument,

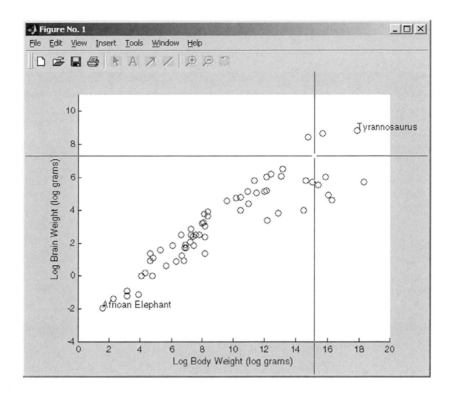

FIGURE 10.7
We constructed a scatterplot of the **animal** data. We then call the **gname** function using the animal names (converted to a string matrix). When **gname** is invoked, the figure window becomes active and a crosshair appears. The user clicks near a point and the observation is labeled. Instead of using a crosshair, one can use a bounding box to label enclosed points. See Example 10.5 for the MATLAB commands.

in which case observations are labeled with their case number. Once the function is called, the figure window becomes active, and a set of crosshairs appears. The user can click near the observation to be labeled, and the label will appear. This continues until the return key is pressed. Instead of using the crosshairs on individual points, one can use a bounding box (click on the plot, hold the left button down and drag), and enclosed points are identified. We use the animal data to show how to use **gname**. This data set contains the brain weights and body weights of several animal types [Crile and Quiring, 1940].

```
load animal
% Plot BrainWeight against BodyWeight.
scatter(log(BodyWeight),log(BrainWeight))
xlabel('Log Body Weight (log grams)')
ylabel('Log Brain Weight (log grams)')
```

```
% Change the axis to provide more room.
axis([0 20 -4 11])
% Need to convert animal names to string matrix.
% Input argument must be string matrix.
cases = char(AnimalName);
gname(cases)
```

The scatterplot with two labeled observations is shown in Figure 10.7. We provide an alternative function to **gname** that allows one to perform many of the identification operations discussed previously. The function is called **scattergui**, and it requires two input arguments. The first is the $n \times 2$ matrix to be plotted; the default symbol is blue dots. The user can right click somewhere in the axes (but not on a point) to bring up the shortcut menu. Several options are available, such as selecting subsets of data or cases for identification and deleting points. See the **help** on this function for more information on its use. The MATLAB code given below shows how to call this function.

```
% Now let's look at scattergui using the BPM data.
load L1bpm
% Reduce the dimensionality using Isomap.
options.dims = 1:10;     % These are for ISOMAP.
options.display = 0;
[Yiso, Riso, Eiso] = isomap(L1bpm, 'k', 7, options);
% Get the data out.
X = Yiso.coords{2}';
scattergui(X,classlab)
% Right click on the axes, and a list box comes up.
% Select one of the classes to highlight.
```

When the user clicks on the **Select Class** menu option, a list box comes up with the various classes available for highlighting. We chose class 6, and the result is shown in Figure 10.8, where we see the class 6 data displayed as red x's.
❑

   While we illustrated and implemented these ideas using a 2–D scatterplot, they carry over easily into other types of graphs, such as parallel coordinates or Andrews' curves (Section 10.6).

10.3.2 Linking

The idea behind *linking* is to make connections between multiple views of our data with the goal of providing information about the data as a whole. An early idea for linking observations was proposed by Diaconis and Friedman [1980]. They proposed drawing a line connecting the same observation in two scatterplots. Another idea is one we've seen before: use

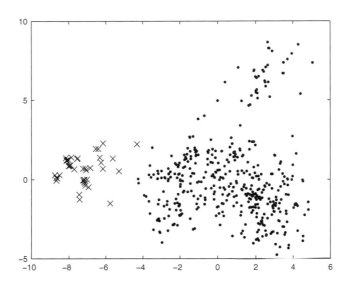

FIGURE 10.8
This shows the BPM data reduced to 2–D using ISOMAP and displayed using **scattergui**. We selected class 6 for highlighting as red x's via the shortcut menu, which is available by right-clicking inside the axes. (**SEE COLOR INSERT.**)

different colors and/or symbols for the same observations in all of the scatterplot panels. Finally, we could manually select points by drawing a polygon around the desired subset of points and subsequently highlighting these in all scatterplots.

We have already seen one way of linking views via the grand tour or a partial linking of scatterplot graphs in a scatterplot matrix. While we can apply these ideas to linking observations in all open plots (scatterplots, histograms, dendrograms, etc.), we restrict our attention to linking observations in panels of a scatterplot matrix.

Example 10.6
In this example, we show how to do linking in a brute-force way (i.e., non-interactive) using the **plotmatrix** function. We return to the **oronsay** data, but we use different variables. This creates an initial scatterplot matrix, as we've seen before.

```
load oronsay
X = [oronsay(:,7),oronsay(:,8),oronsay(:,9),];
% Get the initial plot.
% We need some of the handles to the subplots.
[H,AX,BigAx,P,PAx] = plotmatrix(X,'o');
Hdots = findobj('type','line');
set(Hdots,'markersize',3)
```

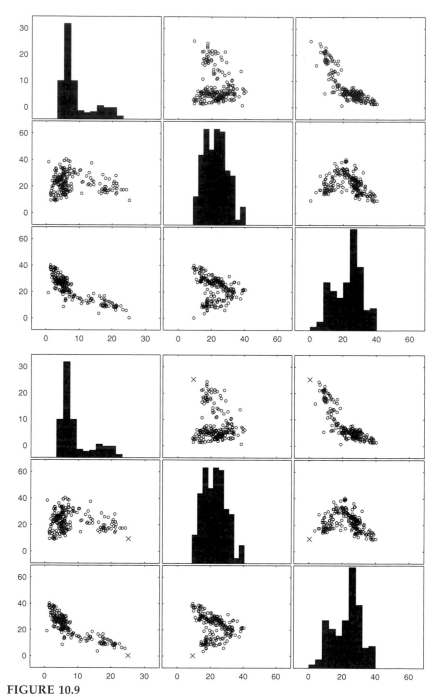

FIGURE 10.9
We have a default scatterplot matrix for three variables of the **oronsay** data (columns 7 through 9) in the top figure. We link observation 71 in all panels and display it using the 'x' symbol. This is shown in the bottom figure.

We called the **plotmatrix** function with output arguments that contain some handle information that we can use next to display linked points with different symbols. The scatterplot matrix is shown in Figure 10.9 (top). The following code shows how to highlight observation 71 in all scatterplots.

```
% The matrix AX contains the handles to the axes.
% Loop through these and change observation 71 to a
% different marker.
% Get the point that will be linked.
linkpt = X(71,:);
% Remove it from the other matrix.
X(71,:) = [];
% Now change in all of the plots.
for i = 1:2
    for j = (i+1):3
        % Change that observation to 'x'.
        axes(AX(i,j))
        cla, axis manual
        line('xdata',linkpt(j),'ydata',linkpt(i),...
            'markersize',5,'marker','x')
        line('xdata',X(:,j),'ydata',X(:,i),...
            'markersize',3,'marker','o',...
            'linestyle','none')
        axes(AX(j,i))
        cla, axis manual
        line('xdata',linkpt(i),'ydata',linkpt(j),...
            'markersize',5,'marker','x')
        line('xdata',X(:,i),'ydata',X(:,j),...
            'markersize',3,'marker','o',...
            'linestyle','none')
    end
end
```

This plot is given in Figure 10.9 (bottom). This plot would be much easier to do using the **gplotmatrix** function in the Statistics Toolbox, but we thought showing it this way would help motivate the need for interactive graphical techniques. The next example presents a function that allows one to interactively highlight points in a scatterplot panel and link them to the rest of the plots by plotting in a different color.
❑

10.3.3 Brushing

Brushing was first described by Becker and Cleveland [1987] in the context of scatterplots, and it encompassed a set of dynamic graphical methods for visualizing and understanding multivariate data. One of its main uses is to interactively link data between scatterplots of the data. A brush consists of a square or rectangle created in one plot. The brush can be of default size and shape (rectangle or square), or it can be constructed interactively (e.g., creating a bounding box with the mouse). The brush is under the control of the user; the user can click on the brush and drag it within the plot.

Several brushing operations are described by Becker and Cleveland. These include highlight, delete, and label. When the user drags the brush over observations in the plot, then the operation is carried out on corresponding points in all scatterplots. The outcome of the delete and label operations is obvious. In the highlight mode, brushed observations are shown with a different symbol and/or a different color.

Three brushing modes when using the highlighting operation are also available. The first is the *transient* paint mode. In this case, only those points that are in the current brush are highlighted. As observations move outside the scope of the brush, they are no longer highlighted. The *lasting* mode is the opposite; once points are brushed, they stay brushed. Finally, we can use the *undo* mode to remove the highlighting.

Example 10.7

We wrote a function called **brushscatter** that implements the highlighting operation and the three modes discussed above. The basic syntax is shown below, where we use the **oronsay** data as in Example 10.3.

```
% Use the same oronsay columns as in Example 10.3
load oronsay
X = [oronsay(:,8),oronsay(:,9),oronsay(:,10)];
% Get the labels for these.
clabs = labcol(8:10);
% Call the function - the labels are optional.
brushscatter(X,clabs)
```

The scatterplot matrix is shown in Figure 10.10, where we see some of the points have been brushed using the brush in the second panel of row one. The brush is in the transient mode, so only the points inside the brush are highlighted in all scatterplots. Note that the axes labels are not used on the scatterplots to maximize the use of the display space. However, we provide the range of the variables in the corners of the diagonal boxes. This is how they were implemented in the early literature. Several options (e.g., three modes, deleting the brush and resetting the plots to their original form) are available by right-clicking on one of the diagonal plots – the ones with the variable names. Brushes can be constructed in any of the scatterplot panels by creating a bounding box in the usual manner. A default brush is not implemented in this function.

❑

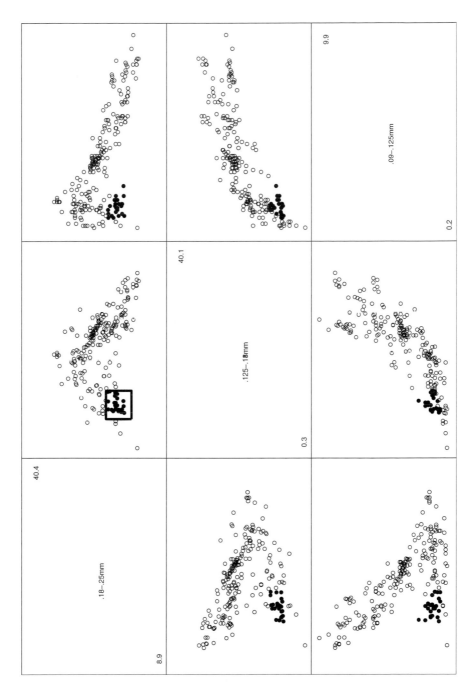

FIGURE 10.10
This is the scatterplot matrix with brushing and linking. This mode is transient, where only points inside the brush are highlighted. Corresponding points are highlighted in all scatterplots. (**SEE COLOR INSERT.**)

The main MATLAB package provides tools for brushing and linking plots. One can brush data that are displayed in 2–D and 3–D graphs, as well as surface plots. Not all plots can be brushed; see the **help** documentation on the **brush** function for a list.

One can enable brushing in several ways. One is by calling the **brush** function. There is also a button on the **Figure** or the Variable Editor (see the **Desktop** menu) toolbars. Finally, one could select **Brush** in the **Figure** window **Tools** menu.

The **brush** option enables the user to interact with the plot, similar to zooming or plot editing. However, unlike these modes, it allows the user to manipulate the data by interactively selecting, removing, or replacing individual data values. Once the data are brushed, one can perform the following tasks from the **Tools** menu or short-cut menus (right-clicking any brushed data point):

- Remove all brushed or unbrushed observations
- Replace the brushed data points with a constant value or NaNs
- Paste the brushed values to the command window
- Create a variable that contains the brushed values

Along with brushing, MATLAB includes the ability to link data in multiple plots and the workspace. This is enabled by the function **linkdata** or a button in the **Figure** window toolbar. An information bar appears that identifies data sources for the graphs and includes an editing option. When linking is enabled, the data displayed in the plot is connected with the workspace and other graphs that use the data. Any subsequent changes to the data made in the workspace or a graph (brushing, removing, or changing observations) are reflected in any linked object that uses the data as a source. This includes the Variable Editor, plots, and the workspace.

10.4 Coplots

As we have seen in earlier chapters, we sometimes need to understand how a response variable depends on one or more predictor variables. We could explore this by estimating a function that represents the relationship and then visualizing it using lines, surfaces, or contours. We will not delve into this option any further in this text. Instead, we present *coplots* for showing slices of relationships for given values of another variable. We look only at the three variable cases (one is conditional). The reader is referred to Cleveland [1993] for an extension to coplots with two conditional variables.

The idea behind coplots is to arrange subplots of one dependent variable against the independent variable. These subplots can be scatterplots, with or without smooths, or some other graphic indicating the relationship between them. Each subplot displays the relationship for a range of data over a given interval of the second variable.

The subplots are called *dependence panels*, and they are arranged in a matrix-like layout. The *given panel* is at the top, and this shows the interval of values for each subplot. The usual arrangement of the coplots is left to right and bottom to top.[1] Note that the intervals of the given variable have two main properties [Becker and Cleveland, 1991]. First, we want to have approximately the same number of observations in each interval. Second, we want the overlap to be the same in successive intervals. We require the dependence intervals to be large enough so there are enough points for effects to be seen and relationships to be estimated (via smoothing or some other procedure). On the other hand, if the length of the interval is too big, then we might get a false view of the relationship. Cleveland [1993] presents an equal-count algorithm for selecting the intervals, which is used in the **coplot** function illustrated in the next example.

Example 10.8

We turn to Cleveland [1993] and the Data Visualization Toolbox function[2] called **coplot**. We updated the function to make it compatible with later versions of MATLAB. These data contain three variables: abrasion loss, tensile strength, and hardness. Abrasion loss is a response variable, and the others are predictors, and we would like to understand how abrasion loss depends on the factors. The following MATLAB code constructs the coplot in Figure 10.11. Note that the conditioning variable must be in the first column of the input matrix.

```
load abrasion
% Get the data into one matrix.
% We are conditioning on hardness.
X = [hardness(:) tensile(:) abrasion(:)];
labels = {'Hardness'; 'Tensile Strength';...
          'Abrasion Loss'};
% Set up the parameters for the coplot.
% These are the parameters for the intervals.
np = 6;      % Number of given intervals.
overlap = 3/4;  % Amount of interval overlap.
intervalParams = [np overlap];
%   Parameters for loess curve:
alpha = 3/4;
lambda = 1;
robustFlag = 0;
fitParams = [alpha lambda robustFlag];
```

---

[1] This is backwards from MATLAB's usual way of numbering subplots: left to right and top to bottom.

[2] Also see the Data Visualization Toolbox M-file **book_4_3.m**.

```
% Call the function.
coplot(X,labels,intervalParams,fitParams)
```

The coplot is shown in Figure 10.11. A loess smooth is fit to each of the subsets of data, based on the conditioning variable. We see by the curves that most of them have a similar general shape – decreasing left to right with a possible slight increase at the end. However, the loess curve in the top row, third panel shows a different pattern – a slight increase at first, followed by a downward trend. The reader is asked to explore abrasion loss against hardness, with the tensile strength serving as the conditioning variable. We note that other plots could be used in the panels, such as scatterplots alone, histograms, line plots, etc., but these are not implemented at this time.
❑

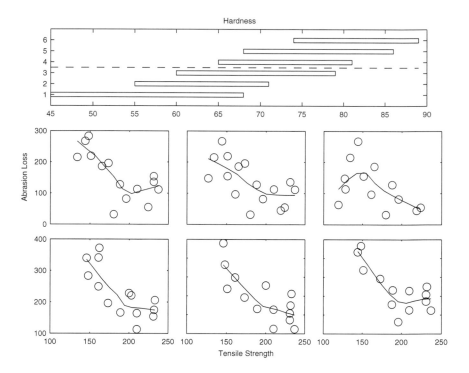

FIGURE 10.11

This is a coplot of abrasion loss against tensile strength, with the hardness as the given variable. The loess curves follow a similar pattern, except for the one in the upper row, third column.

10.5 Dot Charts

A *dot chart* is a visualization method that is somewhat different than others presented in this chapter in that it is typically used with smaller data sets that have labels. Further, it is not used for multivariate data in the sense that we have been looking at so far. However, we think it is a useful way to graphically summarize interesting statistics describing our data, and thus, it is a part of EDA. We first describe the basic dot chart and several variations, such as using scale breaks, error bars, and adding labels for the sample cumulative distribution function. This is followed by a multiway dot chart, where individual dot charts are laid out in panels according to some categorical variable.

10.5.1 Basic Dot Chart

A data analyst could use a dot chart as a replacement for bar charts [Cleveland, 1984]. They are sometimes called *dot plots* [Cleveland, 1993], but should not be confused with the other type of dot plot sometimes seen in introductory statistics books.[3] An example of a dot chart is shown in Figure 10.12. The labels are shown along the left vertical axis, and the value of the datum is given by the placement of a dot (or circle) along the horizontal axis. Dotted lines connecting the dots with the labels help the viewer connect the observation with the label. If there are just a few observations, then the lines can be omitted.

If the data are ordered, then the dots are a visual summary of the sample cumulative distribution function. Cleveland [1984] suggests that this be made explicit by specifying the sample cumulative distribution function on the right vertical axis of the graph. Thus, we can use the following label at the *i*-th order statistic:

$$\frac{(i - 0.5)}{n}.$$

Another useful addition to the dot chart is to convey a sense of the variation via error bars. This is something that is easily shown on dot charts, but is difficult to show on bar charts. As described earlier in dynamic graphics, we might have an outlying observation(s) that makes the remaining values difficult to understand. Before, we suggested just deleting the point, but in some cases, we need to keep all points. So, we can use a break in the scale.

---

[3] This type of dot plot has a horizontal line covering the range of the data. A dot is placed above each observation, in a vertical stack. This dot plot is reminiscent of a histogram or a bar chart in its final appearance.

Scale breaks are sometimes highlighted via hash marks along the axis, but these are not very noticeable and false impressions of the data can result. Cleveland recommends a full scale break, which is illustrated in the exercises.

Example 10.9

We provide a function for constructing dot charts. In this example, we show how to use some of the options in the basic **dotchart** function. First, we load the **oronsay** data and find the means and the standard deviations. We will plot the means as dots in the dot chart.

```
load oronsay
% Find the means and standard deviations of
% each column.
mus = mean(oronsay);
stds = std(oronsay);
% Construct the dotchart using lines to the dots.
dotchart(mus,labcol)
% Change the axes limits.
axis([-1 25 0 13])
```

The dot chart is shown in Figure 10.12 (top), where we see the dotted lines extending only to the solid dot representing the mean for that variable. We can add error bars using the following code:

```
% Now try error bar option.
dotchart(mus,labcol,'e',stds)
```

The resulting dot chart is shown in Figure 10.12 (bottom), where we have the error bars extending from +/– one standard deviation either side of the mean. Note that this provides an immediate visual impression of some summary statistics for the variables.
❑

10.5.2 Multiway Dot Chart

With *multiway data*, we have observations that have more than one categorical variable, and for which we have at least one quantitative variable that we would like to explore. The representation of the data can be laid out into panels, where each panel contains a dot chart and each row of the dot chart represents a level. To illustrate this visualization technique, we use an example from Cleveland [1993].

A census of farm animals in 26 countries was conducted in 1987 to study air pollution arising from the feces and urine of livestock [Buijsman, Maas, and Asman, 1987]. These countries include those in Europe and the former Soviet Union. A log (base 10) transformation was applied to the data to improve the resolution on the graph.

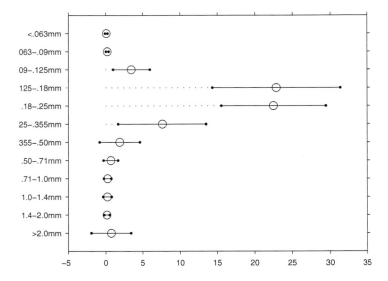

FIGURE 10.12
A regular dot chart is shown in the top panel. The dots represented the average weight for each of the sieve sizes. The dot chart at the bottom shows the same information with error bars added. These bars extend from +/− 1 standard deviation either side of the mean.

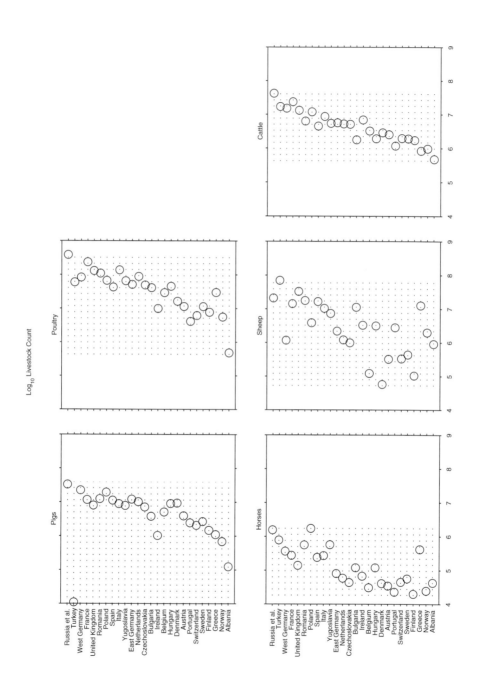

FIGURE 10.13
This is the multiway dot charts for livestock counts for various animals (shown in the panels) and countries (rows of the dot chart). Note that the dotted lines now indicate the range of the data rather than connecting values to the label, as before.

The multiway dot chart is shown in Figure 10.13, where each of the panels corresponds to livestock type. The quantitative variable associated with these is the number of livestock, for each of the different levels (categorical variable for country). We could also plot these using the countries for each panel and animal type for the levels. This will be explored in the exercises.

Note that the order of the countries (or levels) in Figure 10.13 is not in alphabetical order. It is sometimes more informative to order the data based on some summary statistic. In this case, the median of the five counts was used, and they increase from bottom to top. That is, Albania has the smallest median and Russia, et al. has the largest median. We can also order the panels in a similar manner. The median for horses is the smallest over the five animal types, and the median for poultry is the largest. From these charts, we can obtain an idea of the differences of these values in the various countries. For example, Turkey has very few pigs; poultry seems to be the more numerous type across most of the countries; and the number of cattle seems to be fairly constant among the levels. On the other hand, there is more variability in the counts of horses and sheep, with horses being the least common animal type.

Example 10.10

We provide a function in the toolbox to construct a multiway dot chart called **multiwayplot**. We apply it below to the **oronsay** data, where we use the beach/dune/midden classification for the categorical variable. Thus, we will have three panels showing the average particle weight for each classification and sieve size.

```
load oronsay
% Get the means according to each midden class.
% The beachdune variable contains class
% labels for midden (0), beach (1), and
% dune (2). Get the means for each group.
ind = find(beachdune==0);
middenmus = mean(oronsay(ind,:));
ind = find(beachdune==1);
beachmus = mean(oronsay(ind,:));
ind = find(beachdune==2);
dunemus = mean(oronsay(ind,:));
X = [middenmus(:), beachmus(:), dunemus(:)];
% Get the labels for the groups and axes.
bdlabs = {'Midden'; 'Beach'; 'Dune'};
labx = 'Average Particle Weight';
% Get the location information for the plots.
sublocs{1} = [1,3];
sublocs{2} = [1 2 3];
multiwayplot(X,labcol,labx,bdlabs,sublocs)
```

The plot given in Figure 10.14 shows that the larger sieve sizes (and the two smallest) have approximately the same average weight, while the average weight in the two sieves from 0.125mm to 0.25mm are different among the classes. Note that the horizontal axes have the same limits for easier comparison.

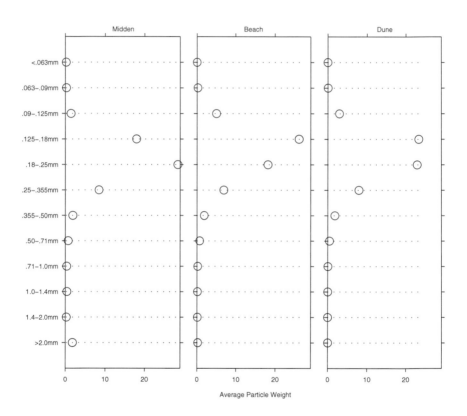

FIGURE 10.14
This is the multiway plot described in Example 10.10. Here we have the dot charts for the average particle weight, given that they come from the beach, dune, or midden.

10.6 Plotting Points as Curves

In this section, we present two methods for visualizing high-dimensional data: parallel coordinate plots and Andrews' curves. These methods are not without their problems, as we discuss shortly, but they are an efficient way of visually representing multi-dimensional relationships.

10.6.1 Parallel Coordinate Plots

In the Cartesian coordinate system, the axes are orthogonal, so the most we can view is three dimensions projected onto a computer screen or paper. If instead we draw the axes parallel to each other, then we can view many axes on the same 2–D display. This technique was developed by Wegman [1986] as a way of viewing and analyzing multi-dimensional data and was introduced by Inselberg [1985] in the context of computational geometry and computer vision.

A parallel coordinate plot for p-dimensional data is constructed by drawing p lines parallel to each other. We draw p copies of the real line representing the coordinate axes for $x_1, x_2, ..., x_p$. The lines are the same distance apart and are perpendicular to the Cartesian y axis. Additionally, they all have the same positive orientation as the Cartesian x axis, as illustrated in Figure 10.15. Some versions of parallel coordinates draw the parallel axes perpendicular to the Cartesian x axis.

Let's look at the following 4–D point:

$$\mathbf{c} = \begin{bmatrix} 1 \\ 3 \\ 7 \\ 2 \end{bmatrix}$$

This is shown in Figure 10.15, where we see that the point is a polygonal line with vertices at $(c_i, i-1)$, $i = 1,..., p$ in Cartesian coordinates on the x_i parallel axis. Thus, a point in Cartesian coordinates is represented in parallel coordinates as a series of connected line segments.

We can plot observations in parallel coordinates with colors designating what class they belong to or use some other line style to indicate group membership. The parallel coordinate display can also be used to determine the following: a) class separation in a given coordinate, b) correlation between pairs of variables (explored in the exercises), and c) clustering or groups. We could also include a categorical variable indicating the class or group label as one of the parallel axes. This helps identify groups, when color is used for each category and n is large.

Example 10.11

We are going to use a subset of the BPM data from Example 10.5 to show how to use the parallel coordinate function **csparallel**. We plot topics 6 and 9 in parallel coordinates, using different colors and linestyles.

```
load example104
% This loads up the reduced BPM features using ISOMAP.
% Use the 3-D data.
```

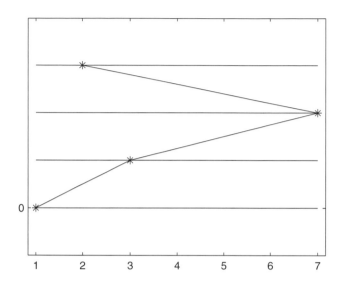

FIGURE 10.15

This shows the parallel coordinate representation for the 4–D point $c^T = (1, 3, 7, 2)$. The reader should note that the parallel axes in subsequent plots will have the parallel axes ordered from top to bottom: $x_1, x_2, ..., x_p$.

```
X = Yiso.coords{3}';
% Find the observations for the two topics: 6 and 9.
ind6 = find(classlab == 6);
ind9 = find(classlab == 9);
% Put the data into one matrix.
x = [X(ind6,:);X(ind9,:)];
% Use the csparallel function from the Computational
% Statistics Toolbox.4
% Construct the plot.
csparallel(x)
```

The parallel coordinates plot is illustrated in Figure 10.16. Several features should be noted regarding this plot. First, there is evidence of two groups in dimensions one and three. These two variables should be good features to use in classification and clustering. Second, the topics seem to overlap in dimension two, so this is not such a useful feature. Finally, although we did not use different colors or line styles to make this more apparent, it appears that observations within a topic have similar line shapes, and they are different from those in the other topic. Example 10.15 should make this point clearer.

❏

---

[4] This is available for download; see Appendix B for information.

FIGURE 10.16
In this figure we have the parallel coordinates plot for the BPM data in Example 10.11. There is evidence for two groups in dimensions one and three. We see considerable overlap in dimension two.

In parallel coordinate plots, the order of the variables makes a difference. Adjacent parallel axes provide some insights about the relationship between consecutive variables. To see other pairwise relationships, we must permute the order of the parallel axes. Wegman [1990] provides a systematic way of finding all permutations such that all adjacencies in the parallel coordinate display will be visited. These tours will be covered at the end of the chapter.

10.6.2 Andrews' Curves

Andrews' curves [Andrews, 1972] were developed as a method for visualizing multi-dimensional data by mapping each observation onto a function. This is similar to star plots in that each observation or sample point is represented by a glyph, except that in this case the glyph is a curve. The Andrews' function is defined as

$$f_x(t) = x_1/\sqrt{2} + x_2\sin t + x_3\cos t + x_4\sin 2t + x_5\cos 2t + \dots, \qquad (10.1)$$

where the range of t is given by $-\pi \le t \le \pi$. We see by Equation 10.1 that each observation is projected onto a set of orthogonal basis functions represented by sines and cosines, and that each sample point is now represented by a curve. Because of this definition, the Andrews' functions produce infinitely many projections onto the basis vectors over the range of t. We now illustrate the MATLAB code to obtain Andrews' curves.

Example 10.12

We use a small data set to show how to get Andrews' curves. The data we have are the following observations:

$$\mathbf{x}_1 = (2, 6, 4)$$
$$\mathbf{x}_2 = (5, 7, 3)$$
$$\mathbf{x}_3 = (1, 8, 9).$$

Using Equation 10.1, we construct three curves, one corresponding to each data point. The Andrews' curves for the data are:

$$f_{\mathbf{x}_1}(t) = 2 \div \sqrt{2} + 6\sin t + 4\cos t$$
$$f_{\mathbf{x}_2}(t) = 5 \div \sqrt{2} + 7\sin t + 3\cos t$$
$$f_{\mathbf{x}_3}(t) = 1 \div \sqrt{2} + 8\sin t + 9\cos t.$$

We can plot these three functions in MATLAB using the following commands.

```
% Get the domain.
t = linspace(-pi,pi);
% Evaluate function values for each observation.
f1 = 2/sqrt(2)+6*sin(t)+4*cos(t);
f2 = 5/sqrt(2)+7*sin(t)+3*cos(t);
f3 = 1/sqrt(2)+8*sin(t)+9*cos(t);
plot(t,f1,'-.',t,f2,':',t,f3,'--')
legend('F1','F2','F3')
xlabel('t')
```

The Andrews' curves for these data are shown in Figure 10.17.
❑

It has been shown [Andrews, 1972; Embrechts and Herzberg, 1991] that because of the mathematical properties of the trigonometric functions, the Andrews' functions preserve means, distances (up to a constant), and variances. One consequence of this is that observations that are close together should produce Andrews' curves that are also closer together. Thus, one use of these curves is to look for clustering of the data points.

Embrechts and Herzberg [1991] discuss how other projections could be constructed and used with Andrews' curves. One possibility is to set one of the variables to zero and re-plot the curve. If the resulting plots remain nearly the same, then that variable has *low discriminating power*. If the curves are

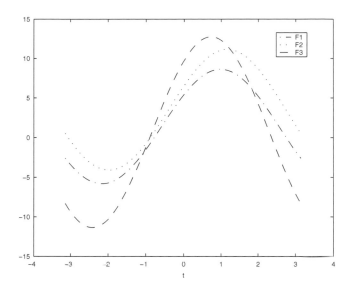

FIGURE 10.17
Andrews' curves for the three data points in Example 10.12.

greatly affected by setting the variable to zero, then it carries a lot of information. Of course, other types of projections (such as nonlinear dimensionality reduction, PCA, SVD, etc.) can be constructed and the results viewed using Andrews' curves.

Andrews' curves are dependent on the order of the variables. Lower frequency terms exert more influence on the shape of the curves, so reordering the variables and viewing the resulting plot might provide insights about the data. By lower frequency terms, we mean those that are first in the sum given in Equation 10.1. Embrechts and Herzberg [1991] also suggest that the data be rescaled so they are centered at the origin and have covariance equal to the identity matrix. Andrews' curves can be extended by using orthogonal bases other than sines and cosines. For example, Embrechts and Herzberg illustrate Andrews' curves using Legendre polynomials and Chebychev polynomials.

Example 10.13

We now construct the Andrews' curves for the data from Example 10.11 using a function called **csandrews**. The following code yields the plot in Figure 10.18. Not surprisingly, we see features similar to what we saw in parallel coordinates. The curves for each group have similar shapes, although topic 6 is less coherent. Also, the curves for topic 6 are quite a bit different than those for topic 9. One of the disadvantages of the Andrews' curves over the parallel coordinates is that we do not see information about the separate variables as we do in parallel coordinates.

```
load example104
% This loads up the reduced BPM features using ISOMAP.
% Use the 3-D data.
X = Yiso.coords{3}';
% Find the observations for the two topics: 6 and 9.
ind6 = find(classlab == 6);
ind9 = find(classlab == 9);
% This function is from the Comp Statistics Toolbox.
% Construct the plot for topic 6.
csandrews(X(ind6,:),'-','r')
hold on
% Construct the plot for topic 9.
csandrews(X(ind9,:),':','g')
```
❑

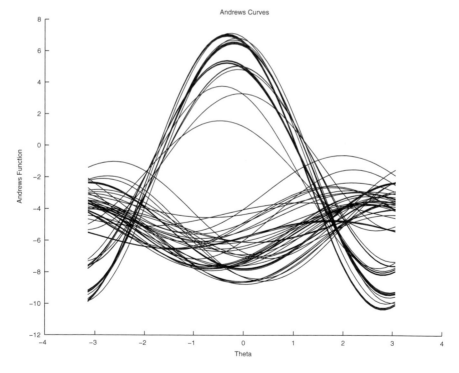

FIGURE 10.18

This is the Andrews' curves version of Figure 10.16. Similar curve shapes for each topic and different overall shapes indicate two groups in the data. The solid line is topic 6, and the dashed line is topic 9.

The Statistics Toolbox has functions for constructing parallel coordinate plots and Andrews' curves plots. The **parallelcoords** function constructs horizontal parallel coordinate plots. Options include grouping (plotting lines using different colors based on class membership), standardizing (PCA or z-score), and plotting only the median and/or other quantiles. The function to construct Andrews' curves is called **andrewsplot**. It has the same options found in **parallelcoords**.

10.6.3 Andrews' Images

When we have very large data sets, then using Andrews' curves to visualize our data is not very useful because of overplotting. An alternative way to visualize Andrews' curves is to plot them as an image. This approach combines the concept of the data image and Andrews' curves.

We first find the values of the Andrews' function (Equation 10.1) for each observation over the domain $-\pi \le t \le \pi$. These values then become a row of a matrix **A**, so that each row of **A** corresponds to one observation that has been transformed using the Andrews' function. Let's say that we divide our domain so we have twenty values of t from $-\pi$ to π. Then the matrix **A** is a matrix with dimension $n \times 20$.

The next step is to visualize the matrix **A** as a data image. This means that the height of the curves (or the values of the entries in **A**) is indicated by their color. As with the data image, it sometimes helps to re-order the rows of the matrix **A** according to some clustering. We will explore this in the next example.

Example 10.14

We are going to use the **iris** data to illustrate the construction of Andrews' images. As usual, we first load the data and put it into a data matrix.

```
load iris
data = [setosa;versicolor;virginica];
```

Recall that in unsupervised learning or clustering applications, we do not know whether the observations can be grouped in some sensible way or how many groups might be there. We would like to use a visualization method like Andrews' images to help us answer these questions visually. The current ordering of the rows in the **data** matrix includes the group information, so we are going to re-order the rows of the data matrix.

```
% Let's re-order the rows of the data
% matrix to make it more interesting.
data = data(randperm(150),:);
```

The **csandrews** function we used in a previous example plots the Andrews' curves and does not return the values of the curve for the observations. So, we wrote a function called **andrewsvals** that returns the matrix **A**. We call this function as the next step in our procedure.

```
% Now get the values of the curves as
% rows of a matrix A.
[A,t] = andrewsvals(data);
```

We now create the Andrews' image and show it in the top of Figure 10.19.

```
% Create the Andrews' image.
imagesc(A)
```

As expected, we do not see a lot of structure in the Andrews' image because we re-order the rows. However, we can now see individual curves instead of overplotting. As we did with the data images, we will impose an ordering on the rows of the Andrews' image that is based on clustering. This can be used to help us visualize the output of the clustering method. In the example below, we use agglomerative clustering, but any clustering method can also be used to order the observations.

```
% Cluster the raw data and then order the rows
% of A using the dendrogram leaves.
ydist = pdist(data);
Z = linkage(ydist,'complete');
subplot(1,2,1)
[H, T, perm] = dendrogram(Z,0,'orientation','left');
axis off
subplot(1,2,2)
imagesc(flipud(y(perm,:)));
```

Our clusters are now readily visible in the Andrews' image shown in the bottom panel of Figure 10.19, and the clusters obtained from agglomerative clustering are seen in the values of the Andrews functions.
❑

10.6.4 More Plot Matrices

So far, we have discussed the use of Andrews' functions and parallel coordinate plots for locating groups and understanding structure in multivariate data. When we know the true groups or categories in our data set, then we can use different line styles and color to visually separate them on the plots. However, we might have a large sample size and/or many groups, making it difficult to explore the data set.

We borrow from the scatterplot matrix and panel graphs (e.g., multiway dot charts, coplots) concepts and apply these to Andrews' curves and parallel coordinate plots. We simply plot each group separately in its own subplot, using either Andrews' curves or parallel coordinates. Common scales are used for all plots to allow direct comparison of the groups.

FIGURE 10.19
The top panel shows an Andrews' image of the **iris** data, where the rows have been randomly re-ordered. Each transformed observation can be seen, but there is not a lot of structure visible The bottom panel displays the Andrews' image after the rows have been re-ordered based on the leaves of the dendrogram shown to the left. We also added a colorbar on the right that indicates the true class membership of the observations. (**SEE COLOR INSERT.**)

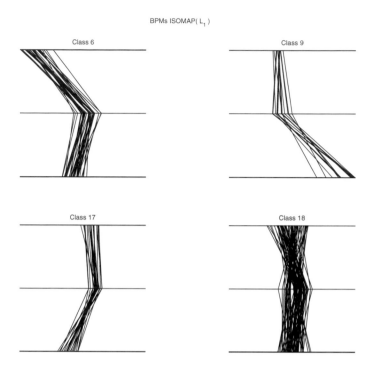

FIGURE 10.20
This shows the plot matrix of parallel coordinate plots for the BPM data in Example 10.15.

Example 10.15

Continuing with the same BPM data, we now add two more topics (17 and 18) and plot each topic separately in its own subplot.

```
load example104
% This loads up the reduced BPM features using ISOMAP.
% Use the 3-D data.
X = Yiso.coords{3}';
% Find the observations for the topics.
inds = ismember(classlab,[6 9 17 18]);
% This function comes with the text:
plotmatrixpara(X(inds,:),classlab(inds),[],...
    'BPMs ISOMAP( L_1 )')
```

See Figure 10.20 for this plot. Using these three features, we have some confidence that we would be able to discriminate between the three topics: 6, 9, and 17. Each has differently shaped line segments, and the line segments are similar within each group. However, topic 18 is a different story. The lines in this topic indicate that we would have some trouble distinguishing topics 17 and 18. Also, the topic 18 lines are not coherent; they seem to have different shapes within the topic. We can do something similar with Andrews' curves. The code for this is given below, and the plot is given in Figure 10.21. Analysis of this plot is left as an exercise to the reader.

```
plotmatrixandr(X(inds,:),classlab(inds))
```

The functions **plotmatrixandr** and **plotmatrixpara** are provided with the EDA Toolbox.

❑

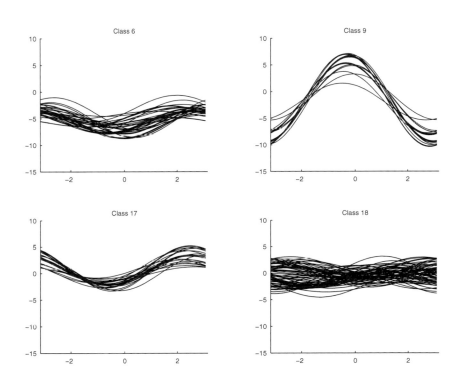

FIGURE 10.21
Here we have the plot matrix with Andrews' curves for the data in Example 10.15.

10.7 Data Tours Revisited

In Chapter 4, we discussed the basic ideas behind tours and motion graphics, but we only used 2–D scatterplots for displaying the data. Some of these ideas are easily extended to higher dimensional representations of the data. In this section, we discuss how the grand tour can be used with scatterplot matrices and parallel coordinates, as well as a different type of animated graphics called permutation tours.

10.7.1 Grand Tour

Wegman [1991] and Wegman and Solka [2002] describe the grand tour in k dimensions, where $k \leq p$. The basic procedure outlined in Chapter 4 remains the same, but we replace the manifold of two-planes with a manifold of k-planes. Thus, we would use

$$\mathbf{A}_K = \mathbf{Q}_K \mathbf{E}_{1, \ldots, k}$$

to project the data, where the columns of $\mathbf{E}_{1, \ldots, k}$ contain the first k basis vectors.

The other change we must make is in how we display the data to the user; 2–D scatterplots can no longer be used. Now that we have some ways to visualize multivariate data, we can combine these with the grand tour. For example, we could use a 3–D scatterplot if $k = 3$. For this or higher values of k, we might use the scatterplot matrix display, parallel coordinates, or Andrews' curves. We illustrate this in the next example.

Example 10.16

To go on a grand tour in k dimensions, use the function called **kdimtour**. This implements the torus grand tour, as described in Chapter 4, but now the display can either be parallel coordinates or Andrews' curves. The user can specify the maximum number of iterations, the type of display, and the number of dimensions $k \leq p$. We tour the topic 6 data from the previous examples and show the tour after a few iterations in Figure 10.22 (top).

```
% We show the tour at several iterations.
% Parallel coordinates will be used here.
% Andrews' curves are left as an exercise.
% We see that this shows some grouping of
% the lines - all seem to follow the same 'structure'.
ind6 = find(classlab==6);
x = X(ind6,:);
% Default tour is parallel coordinates.
% We have 10 iterations and k = 3.
kdimtour(x,10,3)
```

If the lines stay as a cohesive group for most of the tour, then that is an indication of true groups in the data, because the grouping is evident under various rotations/projections. Let's see what happens if we go further along in the tour.

```
% Now at the 90th iteration.
```

```
% We see that the grouping falls apart.
kdimtour(x,90,3)
```

The lines stay together until the end of this tour (90 iterations), at which time, they become rather incoherent. This plot is given in Figure 10.22 (bottom). ❏

10.7.2 Permutation Tour

One of the criticisms of the parallel coordinate plots and Andrews' curves is the dependency on the order of the variables. In the case of parallel coordinate displays, the position of the axes is important in that the relationships between pairwise axes are readily visible. In other words, the relationship between variables on nonadjacent axes is difficult to understand and compare. With Andrews' curves, the variables that are placed first carry more weight in the resulting curve. Thus, it would be useful to have a type of tour we call a *permutation tour*, where we look at replotting the points based on reordering the variables or axes.

A permutation tour can be one of two types: either a full tour of all possible permutations or a partial one. In the first case, we have $p!$ permutations or possible steps in our tour, but this yields many duplicate adjacencies in the case of parallel coordinates. We describe a much shorter permutation tour first described in Wegman [1990] and later in Wegman and Solka [2002]. This could also be used with Andrews' curves, but a full permutation tour might be more useful in this case, since knowing what variables (or axes) are adjacent is not an issue.

To illustrate this procedure, we first draw a graph, where each vertex corresponds to an axis. We place edges between the vertices to indicate that the two axes (vertices) are adjacent. We use the zig-zag pattern shown in Figure 10.23 (for $p = 6$) to obtain the smallest set of orderings such that every possible adjacency is present.

We can obtain these sequences in the following manner. We start with the sequence

$$p_{k+1} = (p_k + (-1)^{k+1}k) \bmod p \qquad k = 1, 2, \ldots, p-1, \qquad (10.2)$$

where we have $p_1 = 1$. In this use of the mod function, it is understood that $0 \bmod p = p \bmod p = p$. To get all of the zig-zag patterns, we apply the following to the sequence given above

$$p_k^{(j+1)} = (p_k^j + 1) \bmod p \qquad j = 1, 2, \ldots, \left\lfloor \frac{p-1}{2} \right\rfloor, \qquad (10.3)$$

where $\lfloor \bullet \rfloor$ is the greatest integer function. To get started, we let $p_k^1 = p_k$.

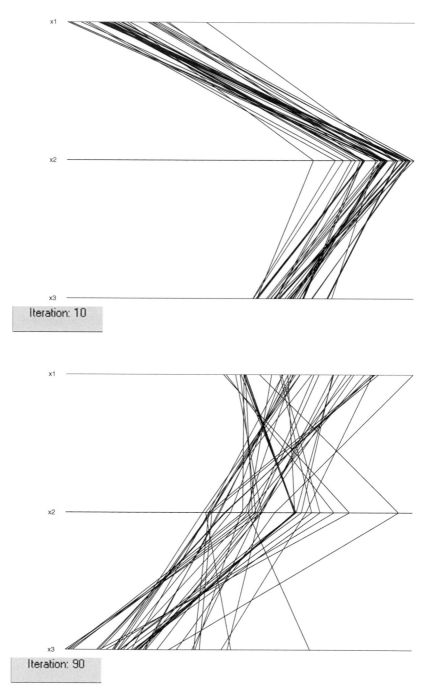

FIGURE 10.22
The top parallel coordinates plot is the 10-*th* iteration in our grand tour. The bottom plot is later on in the tour.

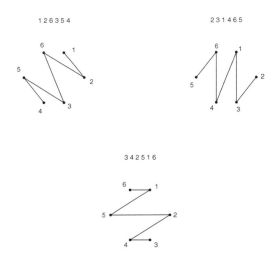

FIGURE 10.23
This figure shows the minimal number of permutations needed to obtain all the adjacencies for $p = 6$.

For the case of p even, the definitions given in Equations 10.2 and 10.3 yield an extra sequence, so some redundant adjacencies are obtained. In the case of p odd, the extra sequence is needed to generate all adjacencies, but again some redundant adjacencies will occur. To illustrate these ideas, we show in the next example how the sequences in Figure 10.23 are obtained using this formulation.

Example 10.17

We start by getting the first sequence given in Equation 10.2. We must alter the results from the MATLAB function **mod** to include the correct result for the case of $0 \mod p$ and $p \mod p$.

```
p = 6;
N = ceil((p-1)/2);
% Get the first sequence.
P(1) = 1;
for k = 1:(p-1)
    tmp(k) = (P(k) + (-1)^(k+1)*k);
    P(k+1) = mod(tmp(k),p);
end
% To match our definition of 'mod':
P(find(P==0)) = p;
```

We find the rest of the permutations by applying Equation 10.3 to the previous sequence.

```
for j = 1:N;
    P(j+1,:) = mod(P(j,:)+1,p);
    ind = find(P(j+1,:)==0);
    P(j+1,ind) = p;
end
```

We now apply this idea to parallel coordinates and Andrews' curves. Use the function we wrote called **permtourparallel** for a parallel coordinate permutation tour, based on Wegman's minimum permutation scheme. The syntax is

permtourparallel(X)

Note that this is very different from the grand tour in Example 10.16. Here we are just swapping adjacent axes; we are not rotating the data as in the grand tour. The plot freezes at each iteration of the tour; the user must hit any key to continue. This allows one to examine the plot for structure before moving on. We also include a function for Andrews' curves permutation tours. The user can either do Wegman's minimal tour or the full permutation tour (i.e., all possible permutations). The syntax for this function is

permtourandrews(X)

for a full permutation tour. To run the Wegman minimal permutation tour, use

permtourandrews(X,flag)

The input argument **flag** can be anything. As stated before, the full tour with Andrews' curves is more informative, because we are not concerned about adjacencies, as we are with parallel coordinates.
❑

10.8 Biplots

The *biplot* was originally developed by Gabriel [1971] as a graphic display of matrices with rank two and showed how it could be applied to convey the results of principal component analysis. Others have gone on to discuss how biplots are an enhanced version of scatterplots, since they superimpose a representation of the variables as vectors in the display, allowing the analyst to study the relationships between the data set and the original variables. Just like scatterplots, they can be used to search for patterns and interesting structure.

The biplot can be used with most of the transformation or dimensionality reduction methods that were described in Chapter 2. This includes factor analysis and nonnegative matrix factorization, in addition to principal component analysis (PCA). We will present the basic 2–D biplot as applied to PCA in the following discussion. However, it should be noted that the ideas generalize to 3–D and more [Young, Valero-Mora, and Friendly, 2006].

In the case of PCA, the two axes of a biplot usually represent the principal components that correspond to the largest eigenvalues. The principal component scores (i.e., the transformed observations) are displayed as points, and the variables of the original data are shown in the plot as vectors. In this way, we are seeing two views of the data – the observations as points and the variables as vectors.

The points in a biplot are the same as a scatterplot, so they can be interpreted in a similar way. Points that are close together in a biplot have similar values with respect to the axes in the display.

Each vector is usually aligned in the direction that is most strongly related to that variable, and the length of the vector conveys the magnitude of that relationship. Long vectors contribute more to the interpretation or meaning (with respect to the original variables) of the displayed component axes. Additionally, vectors that point in the same direction have similar responses or meaning, while vectors pointing in the opposite direction indicate a negative relationship between the two variables [Young, Valero-Mora, and Friendly, 2006].

The example given below uses the **biplot** function that is available in the Statistics Toolbox to illustrate the ideas we have just discussed. We apply it to the results of nonnegative matrix factorization to show that it can be used in different types of transformations.

Example 10.18

We use the **oronsay** data set in this example, where we will reduce the data to 2–D using nonnegative matrix factorization. This makes sense with these data, since all elements of the data matrix are nonnegative.

```
load oronsay
% Do nonnegative matrix factorization to
% reduce to 2-D
[W,H] = nnmf(oronsay,2);
```

Recall from Chapters 2 and 5 that the **W** matrix represents the transformed variables in 2–D (in this example), and the rows of the **H** matrix contain the coefficients of the 12 original variables in the **oronsay** data set. Next, we use the **W** matrix to create a scatterplot, where the marker styles and colors indicate the sampling site.

```
% Construct a scatterplot with characters given
% by marker style. This uses a function gscatter
% that is in the Statistics Toolbox.
gscatter(W(:,1),W(:,2),midden,'rgb','xo*')
hold on
```

The first argument to the **biplot** function is the coefficient matrix that comes from principal component analysis (**princomp**, **pcacov**), factor analysis (**factoran**), or nonnegative matrix factorization (**nnmf**).

```
biplot(max(W(:))*H','VarLabels',labcol)
hold off
axis([0 45 0 40])
```

We can see in this plot that one of the long vectors points through the cluster of observations that correspond to class 1 (*Cnoc Coig*). This indicates that this variable (0.125 - 0.18 mm) is important for that cluster of points.
❑

FIGURE 10.24
This shows the scatterplot (points) and biplot (vectors) of the **oronsay** data set that was explored in Example 10.18. The horizontal and vertical axes represent the two dimensions obtained via nonnegative matrix factorization transformation. The placement of the points is obtained from this transformation, and the vectors show how their associated variables contribute to the plane formed by the axes.

We could have used the transpose of **H** alone for the biplot shown in the previous example, but the vectors need to be scaled to have the same magnitude as the transformed data in **W** otherwise the vectors would be too small. Note that the **biplot** function will also construct 3–D plots, if the coefficient matrix contains three columns.

The MATLAB **biplot** function uses a sign convention that forces the element with the largest magnitude in each column of the coefficient matrix to be positive. This could cause some of the vectors to be displayed in the opposite direction, but it does not change the interpretation of the plot.

10.9 Summary and Further Reading

We start off by recommending some books that describe scientific and statistical visualization in general. One of the earliest ones in this area is *Semiology of Graphics: Diagrams, Networks, Maps* by Jacques Bertin [1983], originally published in French in 1967. This book discusses rules and properties of a graphic system and provides many examples. Edward Tufte wrote several books on visualization and graphics. His first book, *The Visual Display of Quantitative Information* [Tufte, 1983], shows how to depict numbers. The second in the series is called *Envisioning Information* [Tufte, 1990], and illustrates how to deal with pictures of nouns (e.g., maps, aerial photographs, weather data). The third book is entitled *Visual Explanations* [Tufte, 1997], and it discusses how to illustrate pictures of verbs (e.g., dynamic data, information that changes over time). His latest book is *Beautiful Evidence* [2006]. Among other things, these books by Tufte also provide many examples of good graphics and bad graphics. For more examples of graphical mistakes, we highly recommend the book by Wainer [1997]. Wainer discusses this subject in a way that is accessible to the general reader. Wainer also published a book called *Graphic Discovery* [2005] that details some of the history of graphical displays in a very thought provoking and entertaining way.

We referenced this book several times already in the text, but the reader should consult Cleveland's *Visualizing Data* [1993] for more information and examples on univariate and multivariate data. He includes extensive discussions on visualization tools, the relationship of visualization to classical statistical methods, and even some of the cognitive aspects of data visualization and perception. Another excellent resource on graphics for data analysis is Chambers et al. [1983]. For books on visualizing categorical data, see Blasius and Greenacre [1998] and Friendly [2000]. We mention Wilkinson's *The Grammar of Graphics* [1999] for those who are interested in applying statistical and scientific visualization in a distributed computing environment. This book provides a foundation for quantitative graphics and is based on a Java graphics library. While we did not cover issues related to the exploration of spatial data, we think it is worthwhile to point the reader to a book by Carr and Pickle [2010] that presents many techniques for EDA by linking statistical information to small maps. Unwin, Theus, and Hofmann [2006] wrote a book that addresses issues with visualizing massive data sets, and they offer many techniques to explore them.

Many papers have been written on visualization for the purposes of exploratory data analysis and data mining. One recent one by Wegman

[2003] discusses techniques and strategies for visual data mining on high-dimensional and large data sets. Wegman and Carr [1993] present many visualization techniques, such as stereo plots, mesh plots, parallel coordinates, and more. Other survey articles include Anscombe [1973], Weihs and Schmidli [1990], Young et al. [1993], and McLeod and Provost [2001]. Carr et al. [1987] include other scatterplot methods besides hexagonal binning, such as sunflower plots. For an entertaining discussion of various plotting methods, such as scatterplots, pie charts, line charts, etc., and their inventors, see Friendly and Wainer [2004].

An extensive discussion of brushing scatterplots can be found in the paper by Becker and Cleveland [1987]. They show several brushing techniques for linking single points and clusters, conditioning on a single variable (a type of coplot), conditioning on two variables, and subsetting with categorical variables. Papers that provide nice reviews of these methods and others for dynamic graphics are Becker and Cleveland [1991], Buja et al. [1991], Stuetzle [1987], Swayne, Cook, and Buja [1991], and Becker, Cleveland, and Wilks [1987]. Theus and Urbanek [2009] wrote a text about interactive graphics and show how this can be used for analyzing data. It contains many examples and uses free software called Mondrian (see Appendix B). Another excellent book on interactive graphics is by Cook and Swayne [2007]. They use R and GGobi (both are free) to illustrate the concepts.

For more information on dot charts and scaling, see Cleveland [1984]. He compares dot charts with bar charts, shows why the dot charts are a useful graphical method in EDA, and presents grouped dot charts. He also has an excellent discussion of the importance of a meaningful baseline with dot charts. If there is a zero on the scale or another baseline value, then the dotted lines should end at the dot. If this is not the case, then the dotted lines should continue across the graph.

Parallel coordinate techniques were expanded upon and described in a statistical setting by Wegman [1990]. Wegman [1990] also gave a rigorous explanation of the properties of parallel coordinates as a projective transformation and illustrated the duality properties between the parallel coordinate representation and the Cartesian orthogonal coordinate representation. Extensions to parallel coordinates that address the problem of over-plotting include saturation and brushing [Wegman and Luo, 1997; Wegman, 2003] and conveying aggregated information [Fua et al., 1999].

In the previous pages, we mentioned the papers by Andrews [1972] and Embrechts and Herzberg [1991] that describe Andrews' curves plots and their extensions. Additional information on these plots can be found in Jackson [1991] and Jolliffe [1986].

An excellent resource for details on the theory of biplots and its many variations is Gower and Hand [1996]. They describe a geometric approach that shows how they can be used with components analysis, correspondence analysis, and canonical variate analysis. There is also a very interesting discussion of biplots in Young, Valero-Mora, and Friendly [2006], where they connect them with higher-dimensional versions, as well as spinning and orbiting plots. A process for constructing interactive biplots is presented in Udina [2005]. He describes a data structure that is applicable for several types of biplots, as well as options for implementing it in web-based environments. He reviews some of the software for creating biplots that is currently available and presents an implementation in XLISP-STAT. A good resource for applications of biplots in genetics and agriculture is Yan and Kang [2002].

Exercises

10.1   Write a MATLAB function that will construct a profile plot, where each observation is a bar chart with the height of the bar corresponding to the value of the variable for that data point. The functions **subplot** and **bar** might be useful. Try your function on the **cereal** data. Compare to the other glyph plots.

10.2   The MATLAB Statistics Toolbox has a function that will create glyph plots called **glyphplot**. Do a **help** on this function and recreate Figures 10.1 and 10.2.

10.3   Do a help on **gscatter** and repeat Example 10.1.

10.4   Repeat Example 10.1 using **plot** and **plot3**, where you specify the plot symbol to get a scatterplot.

10.5   Construct a scatterplot matrix using all variables of the **oronsay** data set. Analyze the results and compare to Figure 10.5. Construct a grouped scatterplot matrix using just a subset of the most interesting variables and both classifications. Do you see evidence of groups?

10.6   Repeat Example 10.4 and vary the number of bins. Compare your results.

10.7   Write a MATLAB function that will construct a scatterplot matrix using hexagonal bins.

10.8   Load the **hamster** data and construct a scatterplot matrix. There is an unusual observation in the scatterplot of spleen (variable 5) weight against liver (variable 3) weight. Use linking to find and highlight that observation in all plots. By looking at the plots, decide whether this is because the spleen is enlarged or the liver is underdeveloped [Becker and Cleveland, 1991].

10.9   Construct a dot chart using the **brainweight** data. Analyze your results.

10.10 Implement Tukey's alternagraphics in MATLAB. This function should cycle through given subsets of data and display them in a scatterplot. Use it on the **oronsay** data.

10.11 Randomly generate a set of 2–D ($n = 20$) data that have a correlation coefficient of 1, and generate another set of 2–D data that have a correlation coefficient of -1. Construct parallel coordinate plots of each and discuss your results.

10.12 Generate a set of bivariate data ($n = 20$), with one group in the first dimension (first column of **X**) and two clusters in the second dimension (i.e., second column of **X**). Construct a parallel coordinates plot. Now generate a set of bivariate data that has two groups in both dimensions and graph them in parallel coordinates. Comment on the results.

10.13 Use the **playfair** data and construct a dot chart with the cdf (cumulative distribution) option. Do a **help** and try some of the other options in the **dotchart** function.

10.14 Using the **livestock** data, construct a multiway dot chart as in Figure 10.13. Use the countries for the panels and animal type for the levels. Analyze the results.

10.15 Repeat Example 10.8 for the **abrasion** loss data using tensile strength as the conditioning variable. Discuss your results.

10.16 Construct coplots using the **ethanol** and **software** data sets. Analyze your results.

10.17 Embrechts and Herzberg [1991] present the idea of projections with the Andrews' curves, as discussed in Section 10.6.2. Implement this in MATLAB and apply it to the **iris** data to find any variables that have low discriminating power.

10.18 Try the **scattergui** function with the following data sets. Note that you will have to either reduce the dimensionality or pick two dimensions to use.
   a. **skulls**
   b. **sparrow**
   c. **oronsay** (both classifications)
   d. BPM data sets
   e. **spam**
   f. gene expression data sets

10.19 Repeat Example 10.6 using **gplotmatrix**.

10.20 Use the **gplotmatrix** function with the following data sets.
   a. **iris**
   b. **oronsay** (both classifications)
   c. BPM data sets

10.21 Run the permutation tour with Andrews' curves and parallel coordinates. Use the data in Example 10.16.

10.22 Find the adjacencies as outlined in Example 10.17 for a permutation tour with $p = 7$. Is the last sequence needed to obtain all adjacencies? Are some of them redundant?

10.23 Analyze the Andrews' curves shown in Figure 10.21.

10.24 Apply the **permtourandrews, kdimtour** and **permtourparallel** to the following data sets. Reduce the dimensionality first, if needed.
   a. **environmental**

    b. **oronsay**

    c. **iris**

    d. **posse** data sets

    e. **skulls**

    f. BPM data sets

    g. **pollen**

    h. gene expression data sets

10.25 The MATLAB Statistics Toolbox has functions for Andrews' curves and parallel coordinate plots. Do a **help** on these for information on how to use them and the available options. Explore their capabilities using the **oronsay** data. The functions are called **parallelcoords** and **andrewsplot.**

10.26 Do a **help** on the **scatter** function. Construct a 2–D scatterplot of the data in Example 10.1, using the argument that controls the symbol size to make the circles smaller. Compare your plot with Figure 10.3 (top) and discuss the issue of overplotting.

10.27 Apply nonnegative matrix factorization and principal component analysis to the **iris** data. Display a biplot and comment on your results.

10.28 Construct 2–D scatterplots with marginal histograms of the following data sets. Reduce the dimensionality to 2–D using an appropriate method from previous chapters. (Hint: use the **scatterhist** function for the plot.)

    a. **environmental**

    b. **oronsay**

    c. **iris**

    d. **skulls**

    e. BPM data sets

Chapter 11

Visualizing Categorical Data

In this chapter, we discuss methods for exploration and visualization of categorical data. A *categorical variable* is one that can take on values from a discrete set of categories, levels, or factors. Alternatively, the type of data represented by a categorical variable can be assigned to distinct groups or categories. The categories could have an order associated with them, in which case they are called *ordinal*. Alternatively, they could be unordered and are known as *nominal* variables.

Some examples of categorical data are given here:

- Marital status could take on values *Married, Divorced, Widowed, Never Married.*
- Gender would have categories *Male* or *Female.*
- Race might have groups *Caucasian, Black,* and *Hispanic.*
- Numeric ages could be assigned to categories "0 to 19", "20 to 29", "30 to 39", etc.

Note that there is an inherent order for the *age* variable, whereas the other examples have no intrinsic or natural order.

Friendly [2000] points out that categorical data can be in one of two forms: case form or frequency form. If we have the data for each individual sample unit, then this is considered to be *case* data. We often encounter case data when we have data sets comprised of a mixture of continuous and categorical variables measured on each sample unit. For example, we might record *Gender, Race,* and *Blood Pressure*, where the first two are considered to be categorical.

If the categorical data have been tabulated or aggregated, then it is said to be in *frequency form*. This type of data is obtained by counting the number of units that fall into each category. In this case, the data are usually represented in a tabular format.

Friendly and Meyer [2015] also make a distinction between *frequency data* and *count data*. The number of children in a particular household would be a count, and the number of independent households with a given number of children is considered frequency data.

In what follows, we first describe some common discrete distributions and how to estimate their parameters. We then look at ways to visualize the distribution shapes of univariate categorical data. We conclude with a discussion of ways to visualize and explore tabular data.

11.1 Discrete Distributions

Probability distributions can have either continuous or discrete variables. We covered some aspects of continuous distributions in previous chapters. For example, we discussed finite mixture probability distributions in Chapter 6 and kernel density estimation in Chapter 9.

A discrete random variable is one that takes on a value in a countable (finite or infinite) set of numbers. These are typically counts of some characteristic. For instance, we could be recording the number of laptop computers owned by a company or the number of typographical errors in a document.

With discrete variables, we have a *probability mass function*, rather than a density function. The probability mass function assigns a probability to the event that the random variable will take on specific discrete values in the domain. Two of the most common discrete distributions are the binomial and the Poisson [Martinez and Martinez, 2015].

11.1.1 Binomial Distribution

Let's say we have an experiment whose outcome can be labeled as one of two values. We usually refer to these as a 'success' or a 'failure.' If we let the value of $X = 1$ denote a successful outcome and $X = 0$ represent a failure, then we can write the probability mass function as

$$P(X = 0) = 1 - p$$
$$P(X = 1) = p \, , \tag{11.1}$$

where p ($0 \le p \le 1$) represents the probability of a successful outcome. Note that we are using the words 'success' and 'failure' to represent one of two states and not their usual dictionary meaning. A random variable following this mass function is known as a *Bernoulli random variable* [Agresti, 2007].

We can repeat the experiment for n trials, where each trial is independent and results in a success with probability p. If a variable X denotes the number of successes in the n trials, then we say it follows a *binomial distribution* with parameters n and p. The binomial probability mass function is

$$P(X = x) = \binom{n}{x} p^x (1-p)^{n-x} \; ; x = 0, 1, 2, ..., n \; ; 0 \le p \le 1. \quad (11.2)$$

The mean and variance of a binomial distribution are given by

$$E[X] = np,$$

and

$$Var(X) = np(1-p).$$

Say we have a set of data in tabular or frequency form, where each entry represents the number of observations n_x with a specific outcome x for $x = 0, 1, 2, ...$. We might be interested in understanding the process that generated the data. To do this, we have to determine which discrete probability distribution best fits the data [Friendly, 2000]. This will be the main focus in this first section.

TABLE 11.1

Frequency Distribution for the Number of Females in a Queue of Size $n = 10$ [Hoaglin and Tukey, 1985]

Number of Females (x)	Number of Queues (n_x)	Expected Number Under Binomial
0	1	0.332
1	3	2.552
2	4	8.843
3	23	18.155
4	25	24.461
5	19	22.599
6	18	14.499
7	5	6.379
8	1	1.842
9	1	0.315
10	0	0.024

An example of data in this format is given in Table 11.1. Hoaglin and Tukey [1985] provide a frequency distribution representing the number of females in 100 queues of length 10. It is reasonable to assume these data might follow a binomial distribution. To assess this assumption, we can first estimate the success probability p, and then see how well the data fit with the binomial. If we know n (the number of trials), then we can find an estimate of p using

$$\hat{p} = \frac{1}{n \times N}\sum_x xn_x, \qquad (11.3)$$

where

$$N = \sum_x n_x.$$

So, in Table 11.1, we see that $N = \sum_x n_x = 100$. Using Equation 11.3, we get an estimated success probability of

$$\hat{p} = \frac{1}{10 \times 100}[0 \times 1 + 1 \times 3 + \dots + 9 \times 1 + 10 \times 0] = \frac{1}{1000} \times 435 = 0.435.$$

The expected number of queues (of size $n = 10$) with x women is found by first estimating the binomial probabilities for each value of x using $\hat{p} = 0.435$ in Equation 11.2 and then multiplying these estimated binomial probabilities by $N = 100$. These values are shown in the third column of Table 11.1, where we see that the binomial distribution seems to be a reasonable fit. The reader is asked to explore this further in the exercises.

11.1.2 Poisson Distribution

The Poisson distribution is often used to model data representing counts of something, and it can be used to approximate a binomial distribution under certain conditions [Ross, 2014]. The probability mass function for the Poisson is

$$P(X = x) = e^{-\lambda}\frac{\lambda^x}{x!}; \qquad x = 0, 1, 2, \dots, \qquad (11.4)$$

where $x!$ denotes the factorial of x. The factorial of a positive integer x is

$$x! \equiv x(x - 1) \times \dots \times 2 \times 1.$$

For instance, $5! = 5 \times 4 \times 3 \times 2 \times 1 = 120$. The special case of $0!$ has a value of $0! = 1$.

The mean and variance of a Poisson random variable are the same and are given by

$$E[X] = Var(X) = \lambda.$$

We estimate the value of λ as the sample mean:

$$\hat{\lambda} = \frac{1}{N}\sum_{i=1}^{N} x_i, \tag{11.5}$$

where N in Equation 11.5 is the sample size or number of observed counts.

Example 11.1

The Statistics Toolbox has functions to generate random variables for several discrete distributions. We will use the **binornd** and **poissrnd** functions to obtain samples for the binomial and Poisson distributions.

```
% Generate binomial random variables.
% We will use a success probability of p = 0.3
% and n = 50 trials.
p = 0.3;
n = 50;
% The following generates 100 binomial
% random variables.
rb = binornd(n,p,1,100);

% Generate 100 Poisson random variables.
% We will use lambda = 4.
lambda = 4;
rp = poissrnd(lambda,1,100);
```

We can view the distribution shapes as histograms using the following steps.

```
% Construct histograms.
subplot(2,1,1)
hist(rb)
title('Binomial, n = 50, p = 0.3')
subplot(2,1,2)
hist(rp)
title('Poisson, \lambda = 4')
```

These are both illustrated in Figure 11.1. We now show how to estimate the parameters of the two distributions. For the binomial random variables, we first find the mean and then divide by n. We could also use a function **binofit**, which is part of the Statistics Toolbox.

```
% Next, we estimate the probability of success
% for the binomial data.
% We take the mean of the random variables and
% divide by n.
rb_bar = mean(rb)/n;
```

```
% We can also find this using the binofit function.
rb_bar2 = binofit(sum(rb), n*length(rb));
```

We obtain the same answer of 0.303 in both cases. Note that the **binofit** function works differently than what one would normally expect. It does not treat the input vector **rb** as the set of random variables. Therefore, we had to do some calculations on the input arguments. See the MATLAB documentation for more information.

```
% Now for the Poisson random variables.
rp_bar = mean(rp);
% We can also find this using the poissfit function.
rp_bar2 = poissfit(rp);
```

As shown above, there is a function called **poissfit** that will estimate the parameter λ. The input argument to the **poissfit** function is interpreted as a set of random variables, as one would expect. We get the same estimate in both cases, which is 3.67. Note that our estimates are fairly close to the true values of $p = 0.3$ and $\lambda = 4.0$.

❑

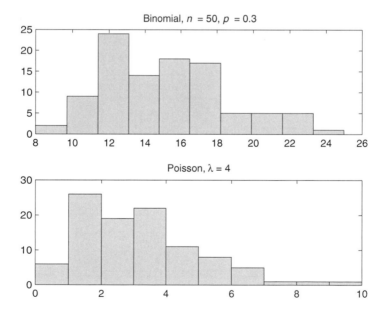

FIGURE 11.1

The top plot is a histogram of the binomial random variables we generated in Example 11.1. These are binomially distributed with parameters $n = 50$ and $p = 0.3$. Each of the random variables represents the number of successes in n trials with success probability p. The lower plot shows a histogram of the generated Poisson random variables with $\lambda = 4$.

11.2 Exploring Distribution Shapes

In this section, we present approaches for visualizing and understanding how discrete and categorical data are distributed. First, we look at plots that are similar to the quantile plots for continuous data, where we compare graphically the observed distribution of discrete random variables to theoretical ones, such as the Poisson and the binomial. We then discuss John Tukey's [1977] hanging rootogram, which can be used to compare fitted and observed frequencies.

11.2.1 Poissonness Plot

As we discussed at the beginning of this chapter, discrete or categorical data are sometimes whole number values representing the number of times something occurs for individual sample units. For example, these might be the number of traffic fatalities in an accident, the number of tablets someone owns, or the number of errors in a computer program. We sometimes have these in the form of a frequency distribution that lists the possible count values (e.g., 0, 1, 2, ...) and the number of observations that are equal to the count values. See Table 11.1 for an example.

We denote the counts as x, with $x = 0, 1, ..., L$. Thus, L is the maximum observed value for our discrete variable or counts in the data set. We are interested in all counts between 0 and L. Thus, the total number of observations in the sample is

$$N = \sum_{x=0}^{L} n_x,$$

where n_x represents the number of observations that are equal to the count x.

A ***Poissonness plot*** [Hoaglin and Tukey, 1985] is constructed by plotting the count values x on the horizontal axis and

$$\varphi(n_x) = \ln(x!n_x/N) \tag{11.6}$$

on the vertical axis. These are plotted as symbols, similar to the quantile plot. If a Poisson distribution is a reasonable model for the data, then this should follow a straight line. Systematic curvature in the plot would indicate that these data are not consistent with a Poisson distribution. The values for $\varphi(n_x)$ tend to have more variability when n_x is small, so Hoaglin and Tukey [1985] suggest plotting a special symbol or a "1" to highlight these points [Martinez and Martinez, 2015].

Hoaglin and Tukey [1985] suggest a modified Poissonness plot that is obtained by adjusting the n_x to account for the variability of the individual values. They propose the following change:

$$n_x^* = \begin{cases} n_x - 0.67 - 0.8n_x/N; & n_x \geq 2 \\ 1/e; & n_x = 1 \\ \text{undefined}; & n_x = 0. \end{cases} \tag{11.7}$$

The main effect of the modified plot is to adjust those data points with small counts that are in keeping with the other observations. Thus, if a point displayed as a "1" in a *modified* Poissonness plot seems different from the rest of the data, then it should be investigated.

Example 11.2

This example is taken from Hoaglin and Tukey [1985] and Friendly [2000]. In the late 1700s, Alexander Hamilton, John Jay, and James Madison wrote a series of 77 essays under the title of *The Federalist*. These were published in newspapers under a pseudonym. Most analysts believe that John Jay wrote 5 essays, Alexander Hamilton wrote 43, Madison wrote 14, and 3 were jointly written by Hamilton and Madison. Hamilton and Madison later claimed that they each wrote the remaining 12 papers. To verify this claim, Mosteller and Wallace [1964] used statistical methods based on the frequency of certain words contained in blocks of text. Table 11.2 gives the frequency distribution for the word *may* in papers that were known to be written by Madison.

TABLE 11.2

Frequency distribution of the word *may* in essays known to be written by James Madison.[a]

Number of Occurrences of the Word *may* (x)	Number of Blocks (n_x)
0	156
1	63
2	29
3	8
4	4
5	1
6	1

[a] The n_x represent the number of blocks of text that contained x occurrences of the word *may* [Hoaglin and Tukey, 1985].

The following MATLAB code will construct a basic Poissonness plot.

```
x = 0:6;   % vector of counts
n_x = [156 63 29 8 4 1 1];
N = sum(n_x);
% Get vector of factorials.
fact = factorial(x);

% Get phi(n_x) for plotting.
phix = log(fact.*n_x/N);
% Find the counts that are equal to 1.
% Plot these with the symbol 1.
% Plot rest with a symbol.
ind = find(n_x~=1);
plot(x(ind),phix(ind),'o')
ind = find(n_x==1);
if ~isempty(ind)
    text(x(ind),phix(ind),'1')
end
% Add some white space to see better.
axis([-0.5 max(x)+1 min(phix)-1 max(phix)+1])
title('Basic Poissonness Plot')
xlabel('Number of Occurrences - \it x')
ylabel('\it \phi (n_x)')
```

This is shown in Figure 11.2. We see some significant curvature in the plot, indicating the Poisson might not be a good model. We provide a function called **poissplot** that will create both versions of the Poissonness plot—basic and modified.

```
% Construct a modified plot.
% Put any value for the third argument to
% get a modified Poissonness plot.
poissplot(x,n_x,1)
title('Basic Poissonness Plot')
```

This is shown at the bottom of Figure 11.2, and we see that the curvature is slightly less. However, the Poisson still seems to not fit the data well.
❑

11.2.2 Binomialness Plot

A *binomialness plot* is obtained by plotting x along the horizontal axis and plotting

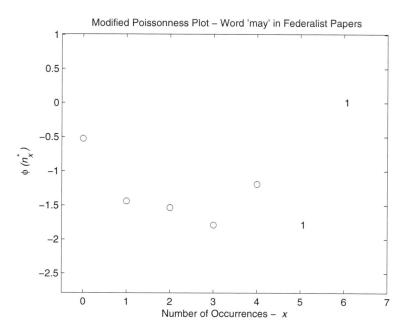

FIGURE 11.2
The top plot is the basic Poissonness plot for the frequency of the word *may* in the disputed essays. There is some curvature indicating the Poisson might not be a good model. The lower plot is the modified version that accounts for the variability of the individual values. The curvature is reduced in this plot, but the Poisson still does not seem to be a good fit.

$$\varphi(n_x^*) = \ln\left\{ \frac{n_x^*}{N \times \binom{n}{x}} \right\},$$ (11.8)

along the vertical axis. Recall that n represents the number of trials, and n_x^* is given by Equation 11.7. As with the Poissonness plot, we are looking for an approximate linear relationship between x and $\varphi(n_x^*)$.

Example 11.3

We will use the binomial random variables generated in the first example. First, we have to convert these from case form to frequency form. We can use the **tabulate** function in MATLAB to aggregate the data.

```
% Let's use the data generated in Example 11.1.
% We can get the counts for the binomial random
% variables using the tabulate function in MATLAB.
tbl = tabulate(rb);
x = tbl(:,1);
n_x = tbl(:,2);
```

We provide a function called **binoplot** with the EDA Toolbox. This will create a binomialness plot for a given number of trials n. In Example 11.1, we generated binomial random variables with $n = 50$ and $p = 0.3$.

```
% We supply a function in the EDA Toolbox to
% construct a binomialness plot.
n = 50;
binoplot(n,x,n_x)
title('Binomialness Plot')
```

This is shown in Figure 11.3, and we see that it follows a straight line. So, the binomial distribution seems to be a good fit for these data.
❏

11.2.3 Extensions of the Poissonness Plot

We provide the theoretical motivation for constructing a Poissonness plot in this section [Hoaglin, 1980; Hoaglin et al., 1985; Friendly, 2000], and we present an extension of it called a *leveled Poissonness plot*. We conclude with a discussion of confidence intervals for observations in a Poissonness plot.

Given a fixed Poisson parameter λ, we can equate the expected frequency and the observed frequency n_x, as shown here

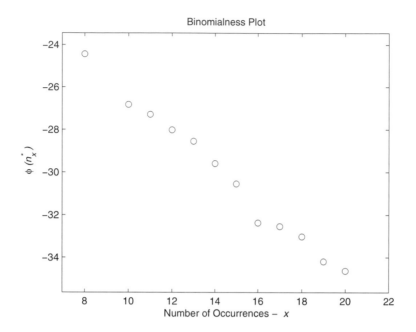

FIGURE 11.3
This is the binomialness plot for the binomial random variables generated in Example 11.1.
The points fall roughly on a straight line, so the binomial is a good model for these data.

$$n_x = Np_x = N\frac{e^{-\lambda}\lambda^x}{x!}; \qquad x = 0, 1, 2, \dots.$$

Taking the log of both sides, we get

$$\log(n_x) = \log N - \lambda + x\log\lambda - \log x!. \tag{11.9}$$

We can rearrange Equation 11.9 to produce the following

$$\log\left(\frac{x!n_x}{N}\right) = -\lambda + (\log\lambda)x. \tag{11.10}$$

The relationship in Equation 11.10 is a straight line with intercept given by $\beta_0 = -\lambda$, and a slope of $\beta_1 = \log\lambda$. It is this linear relationship that we are visualizing in a Poissonness plot. The quantity on the left is $\varphi(n_x)$ (see Equation 11.6). It is also known as the ***count metameter*** [Hoaglin et al., 1985].

We could fit a straight line to the data using $\varphi(n_x)$ as a function of x. If the points seem to follow the linear model, then we could use the slope of the fitted line to estimate λ

$$\hat{\lambda}_{PP} = \exp(\hat{\beta}_1).$$

Hoaglin et al. [1985] discuss how to level the Poissonness plot. This results in a reference curve given by a horizontal line, which makes it easier to judge the fit. We first find an initial estimate of the parameter λ, which we will denote as λ_0. Then, we produce a leveled Poissonness plot by displaying $\varphi(n_x) + [\lambda_0 - x\log(\lambda_0)]$ as a function of x. The line fitted to this relationship will have

$$\beta_0 = \lambda_0 - \lambda$$
$$\beta_1 = \log(\lambda/\lambda_0).$$

We could get an initial estimate of λ using Equation 11.5 or the slope of the fitted line in a Poissonness plot (Equation 11.10). If the Poisson process fits the data, then the leveled Poissonness plot should be nearly horizontal.

A 95% confidence interval for the modified count metameter $\varphi(n_x^*)$ can be used to better understand any deviations from a straight line. Hoaglin et al. [1985] derived confidence intervals of the form center ± half–length:

$$\varphi(n_x^*) \pm h_x. \tag{11.11}$$

The half-length h_x is given by

$$h_k = 1.96 \frac{\sqrt{1 - \hat{p}}}{\sqrt{n_x - (0.47 + 0.25\hat{p})}\sqrt{n_x}}; \qquad n_x^* \geq 2,$$

where $\hat{p} = n_x/N$.

The approximate half-lengths for $n_x^* = 1$ are a special case, and we have two of them—one to get the lower endpoint and one for the upper. The lower half-interval is 2.677, and the upper half–interval is given by

$$2.717 - \frac{2.3}{N}.$$

Note that the confidence interval is not symmetric for the case $n_x^* = 1$.

Example 11.4

In Table 11.3, we reproduce a data set from Rutherford and Geiger [1910]. They published a data set of polonium decay counts corresponding to the number of scintillations in $1/8$–minute intervals. Given that the data are counts, we could explore whether the data come from a Poisson distribution.

TABLE 11.3

Scintillations from Radioactive Decay of Polonium.[a]

Number of Scintillations (x)	Number of Intervals (n_x)
0	57
1	203
2	383
3	525
4	532
5	408
6	273
7	139
8	45
9	27
10	10
11	4
12	0
13	1
14	1
N = 2608	

[a] Hoaglin et al. [1985]

We use the modified Poissonness plot with confidence intervals to see if there is a good fit. We are going to first process the counts where $n_x^* \geq 2$.

```
% The following is the polonium data for nx > 1.
x = 0:11;
nx = [57 203 383 525 532 408 273 139 45 27 10 4];
N = 2608;
% Convert to the modified counts.
nx_mod = nx - 0.67 - .8*nx/N;
% Get the count metameter.
phix = log(factorial(x).*nx_mod/N);

% Get the half-intervals for nx > 1.
% Note that we use the unmodified n_x.
p_hat = nx/N;
h = 1.96*(sqrt(1-p_hat))./sqrt(nx - ...
    (0.47+0.25*p_hat).*sqrt(nx))
```

We now obtain the values for counts equal to 1.

```
% Get the modified counts for nx = 1.
nx_mod1 = 1/2.718268237*ones(1,2);
phix1 = log(factorial([13,14]).*nx_mod1/N);
```

```
% Those counts equal to 1 get different half-intervals.
h1_lo = 2.677*ones(1,2);
h1_hi = (2.717 - 2.3/N)*ones(1,2);
```

We can put the half–intervals into vectors for use in the MATLAB **errorbar** function.

```
% Now, get the half-intervals of the CIs.
Low = [h, h1_lo];
Up = [h, h1_hi];

% Put into vectors for plotting.
X = [x, 13, 14];
Y = [phix, phix1];
errorbar(X,Y,Low,Up,'.')
```

We fit a straight line to the data and use the 95% confidence intervals to assess any significant deviations.

```
% We can fit a straight line and add to the plot.
pf = polyfit(X,Y,1);
pv = polyval(pf,X);
hold on
plot(X,pv)
hold off

xlabel('Number of Occurrences - \it x')
ylabel('\phi (\itn^*_x )')
title('Modified Poissonness Plot - Polonium Data')
```

The plot is shown in the top of Figure 11.4. It fits quite well, except for $x = 8$. We now construct a leveled version of the confidence–interval Poissonness plot by subtracting

$$x \log \lambda_0 - \lambda_0$$

from both endpoints of the intervals. First, we need an estimate λ_0. We could use the logarithm of the slope of the fitted line, which is the first element of **pf**. Using this value, we get 3.98. Or, we could use the mean, as shown here.

```
% Now, construct a leveled confidence-interval plot.
% We need an estimate of the parameter.
% This uses the mean of the data.
lam0 = sum([nx,1,1].*[x,13,14])/N;
```

This yields a value of 3.87, which we will use to get the leveled version. The following is the code to get the leveled Poissonness plot.

```
% Find adjustment and subtract.
Del = X*log(lam0) - lam0;
errorbar(X,Y-Del,EndL-Del,EndU-Del,'.')
hold on
plot([0 14],[0, 0])
hold off

xlabel('Number of Occurrences - \it x')
ylabel('\phi (\itn^*_x ) + \lambda_0 - \itx
   \rmlog(\lambda_0 )'))
title('Leveled Poissonness Plot - Polonium Data')
```

The resulting plot is shown in the bottom of Figure 11.4. We again see that the value at $x = 8$ is lower than the horizontal reference line. It is also much easier to judge the fit using a horizontal reference line.
❏

11.2.4 Hanging Rootogram

Another way to explore the distribution shape of a data set is to fit a distribution and then plot the observed data, along with the fitted values. We then look for differences between them to assess the model fit [Friendly, 2000].

When we have discrete data, our plot would show the observed data x as bars of equal width, and the heights corresponding to the frequency of x. The fitted values are drawn as asterisks, which are connected by a smooth line. The deviation between the fitted and observed frequencies gives us a visual sense of the goodness of fit.

However, Tukey [1977] points out the problems with this approach. First, the largest frequencies drive the scale, making it difficult to see deviations in random variables with the smaller frequencies. It also puts a burden on the viewer to determine the differences between the observed (bar height) and the fitted values (points on a smooth curve).

Tukey's solution is the *hanging rootogram*. This moves the top of the bars to the fitted value, so it is hanging from the curve. This means we can now assess the differences between the fitted and observed against a horizontal line at zero.

To address the issue of large frequencies dominating the plot, the hanging rootogram shows the square root of the frequencies on the vertical axis. This causes smaller frequencies to be emphasized and become more apparent. In the next example, we show how to construct a hanging rootogram for *The Federalist* data.

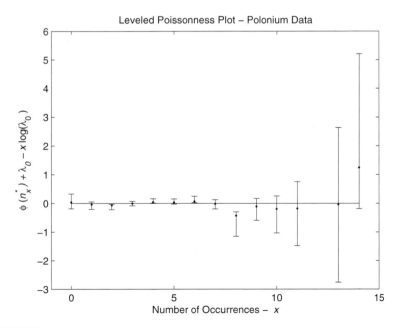

FIGURE 11.4

The top panel shows the modified Poissonness plot with confidence intervals for the polonium data from Table 11.3. A Poisson distribution seems to fit the data well, except for the value $x = 8$. The bottom panel illustrates the leveled version of the plot. It is easier to evaluate the points with respect to a horizontal reference curve.

Example 11.5

We will illustrate the hanging rootogram in this example using the frequency distribution of the word *may* in *The Federalist* essays known to be written by James Madison (see Table 11.2). Previously, we used the Poissonness plot to determine if the data follow a Poisson distribution, and it did not appear to be a good fit. Let's see how the distribution compares with a fitted Poisson. We first need to fit the data to a Poisson, as shown here.

```
% Use the Federalist data.
% Fit to a Poisson.
x = 0:6;
n_x = [156 63 29 8 4 1 1];
lambda = sum(x.*n_x)/sum(n_x);
```

Recall that the Poisson has just one parameter λ, and the estimate we get for these data is $\hat{\lambda} = 0.6565$. Next, we have to get the probability mass function for the random variables based on the fitted distribution.

```
% Get the probability mass function for x.
Yfit = poisspdf(x,lambda);
% Get the expected number of occurrences.
Expf = Yfit*sum(n_x);
```

In Figure 11.5 (top), we plot the observed frequencies as the height of the bars, and the fitted frequencies as points, which are connected by a smooth curve. This is the code to create the plot.

```
% Plot the bar chart and fitted curve.
b = bar(x,n_x,'FaceColor',[.9, .9, .9]);
hold on
vq = min(x):.2:max(x);
I = interp1(x,Expf,vq,'spline');
plot(vq,I,'k-',x,Expf,'k*')
hold off
xlabel('Number of Occurrences')
ylabel('Frequency')
```

The higher observed frequencies occur at lower values of x, making it hard to distinguish the deviations in the larger values $x = 3, 4, 5, 6$. It is also difficult to determine the differences between the observed and fitted value. A hanging rootogram is shown in the lower part of Figure 11.5. This code will produce the plot.

```
% Here is the hanging rootogram.
% Take square root of the frequencies.
ExpfS = sqrt(Expf); % Expected freqs
Is = interp1(x,ExpfS,vq,'spline');
n_xS = sqrt(n_x);   % Observed freqs
```

```
% Start the plot with the fitted curve.
plot(vq,Is,'k-')
hold on
for i = 1:length(x)
    % Display one of the rectangles and
    % hang them from the fitted curve.
    del = 0.15;
    w = 2*del;
    h = n_xS(i);
    rectangle('Position',...
        [x(i)-del, ExpfS(i)-n_xS(i), w, h],...
        'FaceColor',[.9, .9, .9]);
end

% Plot the observed values as points.
h = plot(x,ExpfS,'ko')
set(h,'MarkerFaceColor','k');
% Plot a horizontal line at y = 0;
ax = axis;
plot(ax(1:2),[0 0])
hold off
xlabel('Number of Occurrences')
% This statement produces the square root symbol.
ylabel('\surd{Frequency}')
```

It is much easier to see the deviations between the fitted and observed values in the hanging rootogram, especially at the smaller frequencies.
❑

11.3 Contingency Tables

When analyzing a data set, we are interested in exploring any relationships between our variables. When the variables are categorical, then they are often aggregated and the frequencies are displayed in the form of a *contingency table*. A contingency table contains the number of observations classified by two or more variables.

As we have discussed throughout this book, one of the objectives of EDA is to use visualization to find structure and associations that are not apparent by looking at the raw data, which in this case is a table of counts. We focus on two-way contingency tables in this section following Friendly [2000] and Meyer, et al. [2008]. Additional references for visualizing higher-dimensional tables are given at the end of the chapter. In this section, we first provide details on analyzing two–way tables. Next, we discuss plots that display tiles

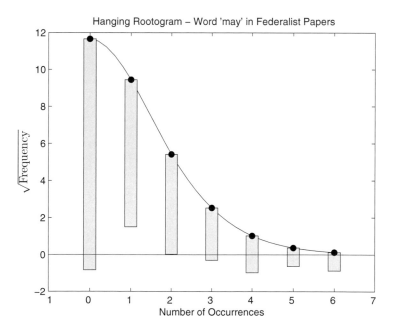

FIGURE 11.5

The upper plot shows observed frequencies as a bar height. A smooth curve shows the data fit to a Poisson distribution. The lower plot is the hanging rootogram, where the square root of the frequencies are shown on the vertical axis, and the bars are hanging from the fitted curve. We can assess the deviations between fitted and observed using the reference line at 0.

with sizes proportional to frequencies or some other statistic associated with the cells in the tables. These include bar, spine, mosaic, and sieve plots. We conclude the section with a description of log odds plots for 2×2 tables.

11.3.1 Background

There are two associations we might consider when exploring a two-way table. These are (1) the independence of the variables and (2) the distribution of one variable for given levels of the other. We first fix our notation for tables and provide some definitions.

A two-way contingency table has I rows and J columns. The cell counts or frequencies are denoted as n_{ij}, for $i = 1, ..., I$ and $j = 1, ..., J$. The row and column *marginals* are simply their summation. These sums are

$$n_{i+} = \sum_{j=1}^{J} n_{ij}, \tag{11.12}$$

where we add all of the cell frequencies across the row, and

$$n_{+j} = \sum_{i=1}^{I} n_{ij}, \tag{11.13}$$

where the counts in the columns are added. The total number of observations is given by

$$n = n_{++} = \sum_{i=1}^{I} n_{i+} = \sum_{j=1}^{J} n_{+j}. \tag{11.14}$$

The plus sign (+) in the subscript means that we add all the values for that variable.

TABLE 11.4

Hospital Data – Number of Schizophrenic Patients

Visit Frequency	Length of Stay (years)			Total
	2 – 9	10 – 19	20 +	
Regular	43	16	3	62
Less than monthly	6	11	10	27
Never	9	18	16	43
Total	58	45	29	132

An example of a two-way contingency table is given in Table 11.4. These data were originally published in Wing [1962] and are analyzed in Meyer, et al. [2008]. See Example 11.6 for more details on the data set. We have two variables—*Length of Stay* and *Visit Frequency*. The row totals are

$$n_{1+} = 62 \qquad n_{2+} = 27 \qquad n_{3+} = 43,$$

and the column totals are

$$n_{+1} = 58 \qquad n_{+2} = 45 \qquad n_{+3} = 29.$$

Adding up either the row or column totals gives $n = 132$ observations.

Assuming an underlying distribution with theoretical cell probabilities π_{ij}, we can write the null hypothesis of independence as

$$H_0: \quad \pi_{ij} = \pi_{i+}\pi_{+j}.$$

In other words, the cell probabilities are just the product of the marginals (see Equations 11.12 and 11.13).

We will need to determine the **expected** cell frequencies for plots presented later in this section. Based on the model of independence, the expected cell frequencies are calculated using the following

$$\hat{n}_{ij} = n_{i+}n_{+j}/n.$$

The table with expected values for the hospital data is shown in Table 11.5.

TABLE 11.5

Hospital Data – Expected Number of Schizophrenic Patients (under independence)

Visit Frequency	Length of Stay (years)			Total
	2 – 9	10 – 19	20 +	
Regular	27.24	21.14	13.62	62
Less than monthly	11.86	9.20	5.93	27
Never	18.89	14.66	9.45	43
Total	58	45	29	132

Later in the text we use sieve plots and mosaic plots to visually compare the observed and expected frequencies. One way to assess the difference analytically is to use **Pearson residuals**. These residuals are given by

$$r_{ij} = \frac{n_{ij} - \hat{n}_{ij}}{\sqrt{\hat{n}_{ij}}}. \tag{11.15}$$

These are also known as the **standardized raw residuals** [Meyer, et al., 2008]. These residuals are in units of standard deviations, and they measure the departure of each cell from independence. If the residual for a cell is greater than 2 or less than -2, then the departure from independence is statistically significant at the 95% level.

TABLE 11.6

Hospital Data – Pearson Residuals for Number of Schizophrenic Patients (under independence)

	Length of Stay (years)		
Visit Frequency	$2-9$	$10-19$	$20+$
Regular	3.02	−1.12	−2.88
Less than monthly	−1.70	0.59	1.67
Never	−2.28	0.87	2.13

The standardized raw residuals for the hospital visitors data are shown in Table 11.6. We see that some cells have residuals greater than 2 and less than -2. For instance, more patients staying in the hospital from 2 to 9 years get visited regularly than would be expected under the null hypothesis. On the other hand, too few patients staying 20 years or longer get visited regularly than would be expected under the null hypothesis of independence.

If we see evidence for a lack of independence in our plots, then we might perform a statistical hypothesis test. The Chi-square test of independence is often used for this purpose. The test statistic is given by

$$\chi^2 = \sum_{i,j} r_{ij}^2, \tag{11.16}$$

which is just the sum of the squared Pearson residuals. This test statistic has an asymptotic distribution that is Chi-square with $(I-1)(J-1)$ degrees of freedom when the null hypothesis of independence is true. We mention this statistic here because each squared Pearson residual is the contribution of the ij-th cell to the test statistic.

11.3.2 Bar Plots

A common plot used to visualize the frequencies in categorical data is the **bar plot**, where we display the observed values as the heights of bars. All of the bars have the same width, and the bars typically have white space between them, unlike a histogram. For example, this is what we used in the previous section to show the observed frequencies. We can also have grouped bar plots, as we demonstrate in the next example.

Example 11.6

We are going to use the hospital data from Meyer, et al. [2008] to illustrate this type of plot. The data originally came from a study on the frequency of visits to 132 schizophrenic patients in a mental hospital [Wing, 1962]. These data are given in Table 11.4 and are in the form of a two–way contingency table. The two variables are *Length of Stay* (years) and *Visit Frequency*, and we have three categories of each variable. Looking at the frequencies in each cell, we can see that hospital visits tend to decrease in frequency as patients are hospitalized longer. This relationship is more apparent when we standardize the table columns, as shown here.

```
% Enter the data.
Hosp = [43 16 3;6 11 10; 9 18 16];

% Standardize along the columns.
CSum = sum(Hosp);
HospCS = Hosp./repmat(CSum,3,1);

HospCS =
    0.7414      0.3556      0.1034
    0.1034      0.2444      0.3448
    0.1552      0.4000      0.5517
```

The **HospCS** matrix contains the hospital data standardized to the column margin. We can clearly see that *Regular Visit Frequency* decreases as *Length of Stay* increases. A contingency table is often viewed in bar plots. We will address other ways to visualize this type of data in the next section. We first create a bar plot for each column, as shown here.

```
% Create a separate bar plot for each column.
subplot(1,3,1),bar(Hosp(:,1),...
    'Facecolor',[.8,.8,.8])
ax = axis;
XYrotalabel(60, 0, gca,...
    1:3,{'Reg','<Month','Never'},...
    [],[]);
y=ylabel('Number of Patients');
set(y, 'Units', 'Normalized',...
    'Position', [-0.15, 0.5, 0]);
title('2-9 Years')

subplot(1,3,2),bar(Hosp(:,2),...
    'Facecolor',[.8,.8,.8])
axis(ax)
XYrotalabel(60, 0, gca,...
    1:3,{'Reg','<Month','Never'},...
    [],[]);
```

```
title('10-19 Years')

subplot(1,3,3),bar(Hosp(:,3),...
    'Facecolor',[.8,.8,.8])
axis(ax)
XYrotalabel(60, 0, gca,...
    1:3,{'Reg','<Month','Never'},...
    [],[]);
title('20+ Years')
```

These plots are shown in Figure 11.6. A grouped bar plot is often used for this type of data. The following is the code we used to create the plot in the top half of Figure 11.7.

```
% We could also do a grouped bar plot.
% If the input to bar is a matrix, then
% MATLAB groups them by rows. So, we input
% the transpose of the hospital data.
```

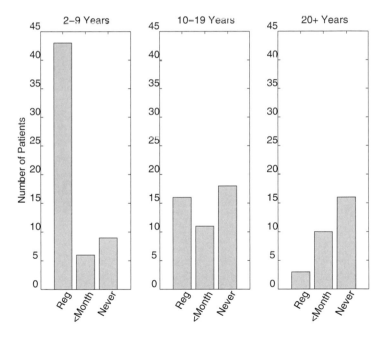

FIGURE 11.6
Here we have separate bar plots for each column of the hospital data.

```
hb = bar(Hosp');
set(hb(1),'Facecolor',[.9,.9,.9])
set(hb(2),'Facecolor',[.7,.7,.7])
set(hb(3),'Facecolor',[.5,.5,.5])
```

```
set(gca,'XTickLabel',{'2-9','10-19','20+'})
legend({'Regular','Less than monthly','Never'})
xlabel('Length of Stay')
ylabel('Number of Patients')
```

Again, we see how *Regular Visit Frequency* tends to decrease as *Length of Stay* increases. We think it is easier to judge differences along a horizontal scale. So, we can plot the bars horizontally using the **barh** function. This is given in the lower half of Figure 11.7.
❏

11.3.3 Spine Plots

Another bar–like plot is called the *spine plot* [Hummel, 1996]. The counts or frequencies are represented by the width of the bar rather than the height, as we have in the bar plot. Each bar in a spine plot has the same height, and the bars are much closer together than in the usual bar plot. We can also think of spine plots as a type of mosaic plot for just one variable, as we will see in the next section.

Example 11.7
We continue with the hospital data to illustrate spine plots. MATLAB does not have a function for spine plots, but we can use the **rectangle** function to create them, as shown here.

```
% Enter the hospital data.
Hosp = [43 16 3;6 11 10; 9 18 16];

% Each bar width is proportional to the column total.
Sx = sum(Hosp);
W = Sx/sum(Sx);

% Here is the gap between the bars.
gap = 0.01;
c = [0.9 0.9 0.9];

% Create the bars.
rectangle('Position',[0 0 W(1) 1],...
    'Facecolor',c)
rectangle('Position',[W(1)+gap,0,W(2) 1],...
    'FaceColor',c)
rectangle('Position',[W(1)+W(2)+2*gap,0,W(3) 1],...
    'FaceColor',c)

% Clean up the axes and add labels.
```

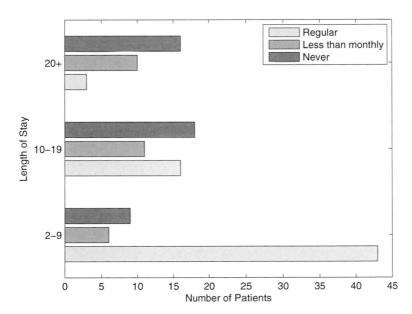

FIGURE 11.7
These are two versions of a grouped bar chart for different values of *Length of Stay* for the hospital data. We can see that regular *Visit Frequency* decreases the longer a patient stays in the hospital. In our opinion, it is easier to see this relationship in the horizontal bar chart.

```
axis tight
box on
% Remove the ticks.
set(gca,'TickLength',[0 0])
% Get the bar mid-points for the labels.
xm = W(1)/2;
xm(2) = W(1) + gap + W(2)/2;
xm(3) = W(1) + W(2) + 2*gap + W(3)/2;
% Set the labels.
set(gca,'Xtick',xm)
set(gca,'XTickLabel',{'2-9','10-19','20+'})
set(gca,'YTickLabel','')
xlabel('Length of Stay (years)')
```

The spine plot is displayed in Figure 11.8. From this plot, we see that the visit frequency for patients with long hospital stays decreases.
❑

    The spine plot in Figure 11.8 shows just one variable (*Length of Stay*), and we do not see any information about the other (*Visit Frequency*). The mosaic plot for contingency tables addresses this problem.

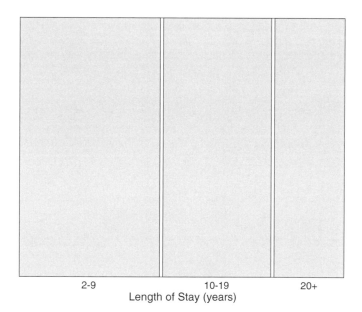

FIGURE 11.8
This shows the spine plot for the hospital data. Here, the width of the bars are proportional to the counts, while the heights of the bars are equal. This plot shows that the number of patients decreases as the *Length of Stay* increases.

11.3.4 Mosaic Plots

Mosaic plots were originally developed by Hartigan and Kleiner [1981]. They have been extended by several authors, but most notably by Friendly [1994; 1999]. *Mosaic plots* show observed cell frequencies as areas of rectangles, usually laid out in the same manner as the contingency table.

We can often think of one variable in a contingency table as explaining the other one. In the case of the hospital data, we might want to use *Length of Stay* to explain *Visit Frequency*. We need to start the mosaic plot by splitting on a variable, and the one we consider as explanatory is a good choice. As Meyer, et al. [2008] point out, whichever variable we split on first matters because it dominates the mosaic plot. We can also think of the mosaic plot as one that conditions on this first splitting variable.

The plot we get from splitting on this first variable is the same as the spine plot. However, we do not stop there. We then split each vertical bar into rectangles using the marginals of the other variable, given the first one. The result is a mosaic of tiles with areas proportional to the observed cell counts.

Example 11. 8

We return to the hospital data in this example. A function for creating mosaic plots was downloaded from MATLAB Central and adapted for our purpose [Cheng, 2014]. The code given below will construct a mosaic plot. While often not done, we can display the observed cell frequencies in each rectangle to help explore the data.

```
% Enter the hospital Data
Hosp = [43 16 3;6 11 10; 9 18 16];

% Use the columns as the conditioning variable.
% We need to re-order the rows.
data = flipud(Hosp);
[xm, ym] = mosaic_plot(data);
hold off
box on
set(gca,'tickLength',[0 0])

% Set some labels. This first function was obtained
% from MATLAB Central.
XYrotalabel(0, 60, gca,...
    xm(1,:),{'2-9','10-19','20+'},...
    ym(:,1),{'Never','< Monthly','Regular'})
y = ylabel('Visit Frequency');
set(y, 'Units', 'Normalized', ...
    'Position', [-0.1, 0.5, 0]);
x = xlabel('Length of Stay (years)')
set(x, 'Units', 'Normalized', ...
```

```
     'Position', [0.5, -0.05, 0]);

% This displays the observed frequencies.
L = cellstr(int2str(data(:)))
multi_text(xm,ym,L)
```

The mosaic plot is shown in the top panel of Figure 11.9. We see the same splits from the spine plot when we compare this to Figure 11.8. The splits in the rows and columns will align when the two variables are independent. We can clearly see that they do not align in this mosaic plot, indicating they are not independent. We can split on the other variable (*Visit Frequency*), if we use the transpose of the matrix as input to the function.

```
% Split on other variable
data = flipud(Hosp');
[xm, ym] = mosaic_plot(data);
hold off
box on
set(gca,'tickLength',[0 0])
% Set some labels.
set(gca,'Xtick',xm(1,:),...
     'XtickLabel',{'Regular','< Monthly','Never'})
set(gca,'Ytick',ym(:,1),...
     'YtickLabel',{'20+','10-19','2-9'})
xlabel('Visit Frequency')
ylabel('Length of Stay (years)');
```

This plot is shown in the lower part of Figure 11.9. In this version, we can see the same decrease in *Visit Frequency* with *Length of Stay*. Also, it is clear from the last two columns that the visitation pattern for the < *Monthly* and *Never* categories are similar. We decided not to show the observed frequencies to give an example of the more typical mosaic plot.
❑

   As we stated before, if we can consider one of the variables in a contingency table as explanatory, then we should split first on this one. The mosaic plot then shows the conditional distribution of the second (dependent) variable given the first (explanatory) one. Friendly [2000] discusses how we can shade mosaic plots according to the Pearson residuals to gain more insights. We show next how these residuals can be shown in sieve plots.

11.3.5 Sieve Plots

In Table 11.4, we showed the expected frequencies of the hospital data under the hypothesis of independence. These can be visualized in a mosaic-type plot comprised of tiles or rectangles. The heights of the tiles are proportional to the corresponding row totals (n_{i+}), while the tile widths are proportional

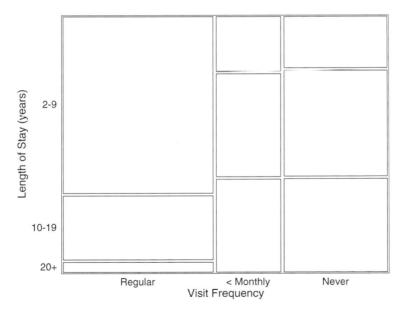

FIGURE 11.9
The first mosaic plot shows the hospital data with *Length of Stay* as the first splitting variable.
We include the observed cell frequencies. The lower mosaic plot is an example of the more
typical version, as it does not have the frequencies. In this case, we split on the *Visit Frequency*
variable. It is now apparent that the patterns for the categories < *Monthly* and *Never Visit
Frequency* and the *Length of Stay* are not too different from each other.

to the column totals (n_{+j}). Thus, we have tiles whose areas are proportional to the *expected* frequencies.

If we fill each of the rectangles with n_{ij} squares (i.e., the number of *observed* frequencies), then we have a *sieve plot*. Sieve plots are also known as *parquet diagrams* and were first published in Riedwyl and Schüpbach [1994]. This allows us to visually assess the difference between the expected frequencies given by the area of the tile and the observed frequencies given by the number of squares. The difference between the frequencies is seen as the density of shading. A tile where the number of observed frequencies exceeds the expected number will have a denser pattern of smaller squares.

It is also common practice to visualize the sign of the standardized raw residuals using a different line style and color to plot the small squares in the tiles. Cells or tiles with positive residuals have squares shown with blue solid lines. Those cells with negative residuals have squares with red dashed lines. In this way, we can see the deviations from independence using the density of the squares, color, and linestyle.

Example 11.9

The data in Table 11.7 is from Snee [1974] and Friendly [2000]. The data were recorded in an elementary statistics course taught by Professor Snee, where students gathered the data as part of a class project. The result is 592 students classified by eye and hair color. In this case, we do not have a variable that we might consider to be explanatory. Rather, we are interested in visually determining if an association might exist between hair and eye color.

TABLE 11.7

Hair Color and Eye Color Data [Snee, 1974]

Eye Color	Hair Color				Total
	Black	Brown	Red	Blond	
Green	5	29	14	16	64
Hazel	15	54	14	10	93
Blue	20	84	17	94	215
Brown	68	119	26	7	220
Total	108	286	71	127	592

We wrote a function for the EDA Toolbox called **sieve** that will produce a sieve plot. The code given below will create a sieve plot of the data in Table 11.7. First, we enter the data.

```
% This is the hair-eye color data.
data = [    5     29     14     16
        15     54     14     10
        20     84     17     94
        68    119     26      7];
```

```
% The following does a sieve plot with the
% observed frequencies as labels.
[xm,ym] = sieve(data,1);
set(gca,'tickLength',[0 0])
axis tight
```

The plot is given in Figure 11.10. Recall that a sieve plot shows rectangles with areas proportional to the *expected* frequencies under independence. Thus, the rectangles for the cells will line up (across rows and columns), as we see in the sieve plot. The following code adds the axis labels.

```
% Add some tick labels.
XYrotalabel(0, 60, gca,...
    xm(1,:),{'Black','Brown','Red','Blond'},...
    ym(:,1),{'Brown','Blue','Hazel','Green'})

% Add some axis labels.
y = ylabel('Eye Color');
set(y, 'Units', 'Normalized',...
    'Position', [-0.1, 0.5, 0]);
x = xlabel('Hair Color');
set(x, 'Units', 'Normalized',...
    'Position', [0.5, -0.05, 0]);
```

From the sieve plot, we see that blond hair and blue eyes seem to have a strong association, as does brown eyes with black and brown hair.
❑

In summary, a sieve plot has tiles with areas proportional to the expected frequencies under the model of independence. Therefore, the columns and rows of the rectangles line up. Each of the rectangles contains n_{ij} squares and are colored according to the sign of the Pearson residual. Little squares shown with red dotted lines indicate a negative residual (observed is less than expected) for the cell, and solid blue lines for the squares correspond to positive residuals (observed is greater than expected). A pattern of higher density cells along the diagonal show a linear association. The reader is asked to explore this idea in the exercises.

11.3.6 Log Odds Plot

We will use a rather well-known data set containing admissions information to the University of California, Berkeley (UCB) to explain odds ratios and the log odds plot. These data are admissions to Berkeley graduate programs, and are classified by *Gender* (*Male* or *Female*) and *Admitted* (*Yes* or *No*). These data were for admissions in 1973, and they reflect information aggregated over the six largest departments [Bickel et al., 1975; Friendly, 2000].

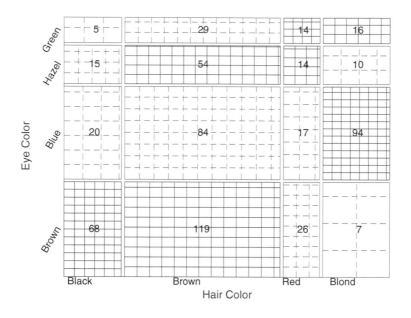

FIGURE 11.10

Here we have the sieve plot for the hair and eye color data of Table 11.7. We see a general association between hair and eye color. Under the hypothesis of independence, cells with dotted-line squares have smaller than expected frequencies, and those with solid-line squares have larger than expected cell frequencies.

This is a special type of contingency table with dimensions 2×2. In other words, we have two categories for each variable. For this type of table, we are interested in assessing any association between the two variables (i.e., gender and admission in the Berkeley data). For instance, we might be interested in determining if males are more likely to be admitted than females. The data are shown in Table 11.8.

TABLE 11.8

UC Berkeley Admissions Data [Bickel, et al., 1975]

| Gender | Admitted | | Total |
	Yes	No	
Male	1198	1493	2691
Female	557	1278	1835
Total	1755	2771	4526

A useful measure of association for a 2×2 table is the ***odds ratio***. When we have only two categories for a variable, we can designate one as a *success* and denote the probability of success as π. The ***odds*** for a success is defined as the probability of success divided by the probability of failure:

$$\text{odds} = \frac{\pi}{1 - \pi}. \tag{11.17}$$

If the probabilities are equal $\pi = 1 - \pi = 0.5$, then the ratio in Equation 11.17 is one. If the probability of success is greater than the probability of failure, then $\pi > 1 - \pi$ and the odds are greater than one. On the other hand, if the probability of failure is greater, then the odds are less than one.

Taking the log of the odds in Equation 11.17 yields a measure of association with better properties, in that it varies additively around zero. This is called the *logit* or *log odds* and is given by

$$\text{logit}(\pi) = \log(\text{odds}) = \log\left(\frac{\pi}{1 - \pi}\right). \tag{11.18}$$

Letting π_1 be the success probability for one group (e.g., row) in a 2×2 table, and π_2 be the probability of success for the other, then we can write the *odds ratio* for the two groups as

$$\text{odds ratio} = \theta = \frac{\pi_1/(1 - \pi_1)}{\pi_2/(1 - \pi_2)}. \tag{11.19}$$

Taking the log of Equation 11.19 yields the *log odds ratio*.

The sample odds ratio is found using estimates of the necessary quantities, as shown here

$$\hat{\theta} = \frac{p_1/(1 - p_1)}{p_2/(1 - p_2)} = \frac{n_{11}/n_{12}}{n_{21}/n_{22}} = \frac{n_{11}n_{22}}{n_{12}n_{21}}, \tag{11.20}$$

where p_i is an estimate for π_i. Taking the log of the sample odds ratio results in a sampling distribution that is closer to the normal distribution with a mean of $\log(\theta)$, and a standard deviation given by Equation 11.21. This is called the *asymptotic standard error* (ASE):

$$\text{ASE}(\log(\hat{\theta})) = \left\{ \frac{1}{n_{11}} + \frac{1}{n_{12}} + \frac{1}{n_{21}} + \frac{1}{n_{22}} \right\}^{1/2}. \tag{11.21}$$

The $100(1 - \alpha)\%$ confidence interval for $\log(\theta)$ can be estimated using

$$\log(\hat{\theta}) \pm z_{1 - \alpha/2} \text{ASE}(\log(\hat{\theta})). \tag{11.22}$$

We can exponentiate the end points in Equation 11.22 to get the confidence interval for the odds ratio. See Agresti [2007] for more details.

Let's calculate the sample odds ratio of the UCB admissions data in Table 11.8. We are interested in comparing the odds of getting admitted for each gender:

$$\hat{\theta} = \frac{\text{odds}(Yes|Males)}{\text{odds}(Yes|Females)} = \frac{n_{11}n_{22}}{n_{12}n_{21}} = \frac{1198 \times 1278}{1493 \times 557} = 1.84 \,.$$

This indicates that men are almost twice as likely to be admitted as women. The confidence interval for the sample log odds ratio is found as follows:

$$\log(\hat{\theta}) \pm z_{1-\alpha/2}\text{ASE}(\log(\hat{\theta}))$$

$$\log(1.84) \pm 1.96 \times \left\{ \frac{1}{1198} + \frac{1}{1493} + \frac{1}{557} + \frac{1}{1278} \right\}^{1/2}$$

$$0.6014 \pm 1.96 \times 0.0639$$

Thus, the 95% confidence interval of the log odds ratio is [0.4851, 0.7356]. We can get the corresponding interval for the odds ratio by exponentiating the end points, as shown here

$$[\exp(0.4851), \exp(0.7356)]$$
$$[1.6244, 2.0867]$$

The interval for the log odds ratio does not contain zero. So, there is evidence of an association between gender and being admitted to the university. We are going to explore this more in the next example.

Example 11.10

The UCB admissions data are often used to illustrate Simpson's paradox, and we will explore this idea using a *log odds ratio plot*. This paradox refers to the phenomenon that an association between two variables reverses sign or disappears when we divide the data into subpopulations. The UCB data in Table 11.8 are obtained by aggregating over the admissions data for six departments. Let's see what happens when we stratify on the departments. First, load the **UCBadmissions** data that contains a $2 \times 2 \times 6$ array.

```
% Load the data
load UCBadmissions
```

Just as a check, we find the aggregated UCB data by summing over the third dimension. This should give us the data shown in Table 11.8. Then, we calculate the odds ratio.

```
% We can find the aggregated UCB data
% by summing over the third dimension.
```

```
UCBtot = sum(UCB,3);
% Find the odds ratio.
oddsR = UCBtot(1,1)*UCBtot(2,2)/...
    (UCBtot(1,2)*UCBtot(2,1))
```

This gives us 1.8411, as we had previously. Next, we find the log odds ratio for each of the six departments.

```
% Get the odds ratio for each department.
OR = zeros(1,6);
for i = 1:6
    OR(i) = UCB(1,1,i)*UCB(2,2,i)/...
        (UCB(1,2,i)*UCB(2,1,i));
end
% Get the log odds ratio.
Lor = log(OR);
```

We can use the 95% confidence intervals of the log odds ratios to measure the associations in each department. The following code calculates the endpoints of the intervals. Each column is an endpoint, and each row corresponds to one of the departments.

```
% Get the 95% confidence intervals.
% Variance is given by sqrt(sum(1/nij));
UCBi = 1./UCB;
for i = 1:6
    UCBse(i) = sqrt(sum(sum(UCBi(:,:,i))));
    % Get 95% CI.
    UCBci(i,:) = [Lor(i)-1.96*UCBse(i),...
        Lor(i)+1.96*UCBse(i)];
end
```

The confidence intervals are shown here.

```
UCBci =

    -1.5670    -0.5372
    -1.0777     0.6377
    -0.1572     0.4070
    -0.3764     0.2124
    -0.1923     0.5927
    -0.7870     0.4092
```

From these, we see that only the first department has a significant association between the two variables of gender and admittance because it does not contain zero. It is interesting to note that it is in the opposite direction of what we expected. It is in favor of women being admitted! The odds ratio for this department is the first element in the **OR** vector:

```
>> OR(1)
```

```
ans =

    0.3492
```

Taking the inverse of this, we find that $1/0.3492 = 2.86$, which means that women are almost three times as likely as men to be admitted to this graduate department. These associations across strata (i.e., departments) are easy to see in a log odds ratio plot. The MATLAB code to construct one of these is shown below.

```
% Construct the odds ratio plot.
E = 1.96*UCBse;
bar(Lor,'FaceColor',[.9, .9, .9])
hold on
h = errorbar(1:length(E),Lor,E,'ro')
set(h,'MarkerFaceColor','r');
set(gca,'XtickLabel',{'A','B','C','D','E','F'})
title('Log Odds Ratio for Admittance by Department')
xlabel('Department')
ylabel('LOR(Admit/Gender)')
```

The plot is shown in Figure 11.11. The line for the first confidence interval does not cross the horizontal line at zero, indicating that it is significant. Some implicit assumptions were made when analyzing the aggregated data. We falsely assumed that *Males* and *Females* applied to the different departments equally. It turns out that women largely applied to departments that have low admission rates [Bickel, et al., 1975; Friendly, 2000].
❑

11.4 Summary and Further Reading

In this chapter, we provided a short introduction to categorical data analysis and some relevant visualization techniques. We started off by describing how we can assess the distributions of categorical data using the Poissonness plot, the binomialness plot, and the hanging rootogram. Friendly [2000] presents *ord plots*, which can be used to find a likely theoretical probability distribution for a discrete data set [Ord, 1967]. The ord plot can be used with the Poisson, binomial, negative binomial, and logarithmic series.

We can get plots similar to the Poissonness and binomialness plots for other discrete distributions. Each of these plots is obtained by displaying the appropriate count metameter $\varphi(n_x)$ as a function of x. If the fit is a good one for that distribution, then the points should fall on a straight line. Friendly

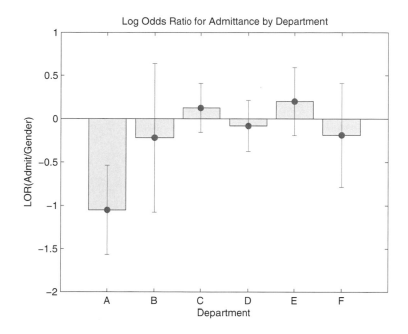

FIGURE 11.11

This is the log odds ratio plot for the UC Berkeley admissions data. There is a significant association between gender and admissions in department A because the vertical line corresponding to the 95% confidence interval does not cross zero. It turns out that women are 2.86 times as likely as men to be admitted to this department.

[2000] provides count metameters for the negative binomial, geometric, and logarithmic series distributions.

We then turned our attention to visualizing cell frequencies in contingency tables. We showed how to visualize univariate cell counts using bar plots and spine plots. Mosaic plots and sieve plots were used to understand both of the variables in two-way tables. We concluded the chapter with an illustration of the log odds ratio plot.

As mentioned in the text, tiles in the mosaic plots are sometimes colored according to the values of the Pearson residuals. There are also options for stacking and stratifying the tiles to display n–way contingency tables. For more information on these plots, see Friendly [2000].

Another way to visualize the Pearson residuals in a mosaic–type plot is the *association plot* [Meyer, et al. 2008]. This plot shows the Pearson residuals for each cell as a rectangle, where the height is proportional to the residual, and the width is proportional to the square root of the expected counts. The bar is positioned (positive or negative) from a baseline according to the sign.

Friendly [2000] describes *fourfold plots* as a way to display the odds ratios. It is like a pie chart in that it uses wedges to show frequencies. There are four quarter circles, so one is judging differences based on the radius. We can see different aspects of the data, depending on how the data are standardized.

A plot useful for viewing a special type of categorical data is called the **trilinear** or **ternary plot**. This is used for two-way contingency tables with three columns. This type of data is also known as **compositional** data. The data are shown as a point in an equilateral triangle. The location of the point indicates the relative proportions of the categories [Friendly, 2000].

We recommend Agresti [2007] for an excellent introduction to the analysis of categorical data. For a more advanced treatment of the subject, the reader should consult Agresti [2012]. We already mentioned the book by Friendly [2000] as a resource for learning more about visualizing categorical data, but the focus is on SAS implementations. Friendly and Meyer [2015] have a book on discrete data analysis and visualization using R.

Michael Friendly has a wonderful website on visualization. He includes examples of good and bad graphics, R and SAS software, papers, and online applications. Readers will probably find his online course materials to be especially helpful. The website is

http://www.datavis.ca/

Exercises

11.1 Later versions of MATLAB have a function called **histogram**. This provides more options and control over the type of histogram it produces. For example, we can obtain a valid probability density histogram using the following name–value pair in the argument list: **'Normalization', 'pdf'**. For more options, see the **help** on **histogram**. Generate random variables from the binomial and Poisson distributions, as we showed in Example 11.1. Construct histograms using the **histogram** function and compare to Figure 11.1.

11.2 Use the **randtool** in the Statistics Toolbox to generate and visualize random variables from the binomial and Poisson distributions for different parameter values.

11.3 Construct a binomialness plot of the data in Table 11.1 and assess the fit. Also try the Poissonness plot. Which one seems best?

11.4 Load the data generated in Example 11.1 (**example11_1.mat**). Create a Poissonness plot. Does this seem a reasonable fit for these data? Discuss why the Poisson distribution might be a good fit.

11.5 Generate a larger set (5,000 or more) of binomial and Poisson random variables for the distributions in Example 11.1. Estimate p and λ. Are your estimates closer to the true values?

11.6 Conduct a chi–square test of independence for the hospital data in Table 11.4. Discuss your results.

11.7 The **saxony** data set contains the number of males in $N = 6115$ Saxony families of the nineteenth century [Friendly, 2000]. Each of the families has $n = 12$ children. Construct Poissonness plots as shown in this chapter. Construct a hanging rootogram and assess the fit to a Poisson.

11.8 Create a binomialness plot using the **saxony** data and describe the fit. Also, construct a hanging rootogram and discuss your results.

11.9 Standardize the rows of the hospital data. See Example 11.6 for more help. Comment on the relationship between *Length of Stay* and *Visit Frequency*. How different are the values for *Less than Monthly* and *Never*?

11.10 Construct a grouped bar plot of the hospital data without transposing the data matrix. What patterns, if any, do you see? By which variable is it being grouped?

11.11 Create a spine plot for the rows of the hospital data by revising the code in Example 11.7.

11.12 Generate frequencies for a two-way contingency table, such that the variables are independent of each other. Use three levels for each of the variables. Construct mosaic and sieve plots and comment on the results.

11.13 The file **problem11_13** contains two data sets from Friendly [2000]. Construct sieve plots for each of them and comment on the results. What patterns do you see and what do the indicate? Calculate the row means and row totals for each data set. Discuss these results and how they relate to the patterns seen in the sieve plots.

11.14 Kendall and Stuart [1961] and Friendly [2000] have data documenting distance vision (with no glasses or contacts) for men and women in the UK from 30 to 39 years of age. The visual acuity for each eye was measured and grouped into four levels. There are two data objects in the file **vision**—one for men and one for women. Construct sieve plots for each of these data objects. Add labels as shown in Example 11. 9. What type of pattern do you see and what does it indicate? Does this seem reasonable, given what you might know about vision?

Appendix A

Proximity Measures

We provide some information on proximity measures in this appendix. *Proximity measures* are important in clustering, multi-dimensional scaling, and nonlinear dimensionality reduction methods, such as ISOMAP. The word *proximity* indicates how close objects or measurements are in space, time, or in some other way (e.g., taste, character, etc.). How we define the nearness of objects is important in our data analysis efforts and results can depend on what measure is used. There are two types of proximity measures: similarity and dissimilarity. We now describe several examples of these and include ways to transform from one to the other.

A.1 Definitions

Similarity measures indicate how alike objects are to each other. A high value means they are similar, while a small value means they are not. We denote the similarity between objects (or observations) i and j as s_{ij}. Similarities are often scaled so the maximum similarity is one (e.g., the similarity of an observation with itself is one, $s_{ii} = 1$). *Dissimilarity measures* are just the opposite. Small values mean the observations are close together and thus are alike. We denote the dissimilarity by δ_{ij}, and we have $\delta_{ii} = 0$. In both cases, we have $s_{ij} \geq 0$ and $\delta_{ij} \geq 0$.

Hartigan [1967] and Cormack [1971] provide a taxonomy of twelve proximity structures; we list just the top four of these below. As we move down in the taxonomy, constraints on the measures are relaxed. Let the observations in our data set be denoted by the set O, then the proximity measure is a real function defined on $O \times O$. The four structures S that we consider here are

S1   S defined on $O \times O$ is Euclidean distance;

S2   S defined on $O \times O$ is a metric;

S3   S defined on $O \times O$ is symmetric real-valued;

S4   S defined on $O \times O$ is real-valued.

The first structure S1 is very strict with the proximity defined as the familiar *Euclidean distance* given by

$$\delta_{ij} = \sqrt{\sum_k (x_{ik} - x_{jk})^2},$$ (A.1)

where x_{ik} is the k-th element of the i-th observation.

If we relax this and require our measure to be a metric only, then we have structure S2. A dissimilarity is a metric if the following holds:

1. $\delta_{ij} = 0$, if and only if $i = j$.
2. $\delta_{ij} = \delta_{ji}$.
3. $\delta_{ij} \leq \delta_{it} + \delta_{tj}$.

The second requirement given above means the dissimilarity is symmetric, and the third one is the triangle inequality. Removing the metric requirement produces S3, and allowing nonsymmetry yields S4. Some of the later structures in the taxonomy correspond to similarity measures.

We often represent the interpoint proximity between all objects i and j in an $n \times n$ matrix, where the ij-th element is the proximity between the i-th and j-th observation. If the proximity measure is symmetric, then we only need to provide the $n(n - 1)/2$ unique values (i.e., the upper or lower part of the *interpoint proximity matrix*). We now give examples of some commonly used proximity measures.

A.1.1 Dissimilarities

We have already defined Euclidean distance (Equation A.1), which is probably used most often. One of the problems with the Euclidean distance is its sensitivity to the scale of the variables. If one of the variables is more dispersed than the rest, then it will dominate the calculations. Thus, we recommend transforming or scaling the data as discussed in Chapter 1.

The Mahalanobis distance (squared) takes the covariance into account and is given by

$$\delta_{ij}^2 = (\mathbf{x}_i - \mathbf{x}_j)^T \Sigma^{-1} (\mathbf{x}_i - \mathbf{x}_j),$$

where Σ is the covariance matrix. Often Σ must be estimated using the data, so it can be affected by outliers.

The **Minkowski metric** defines a whole family of dissimilarities, and it is used extensively in nonlinear multidimensional scaling (see Chapter 3). It is defined by

$$\delta_{ij} = \left\{ \sum_k |x_{ik} - x_{jk}|^\lambda \right\}^{\frac{1}{\lambda}} \qquad \lambda \geq 1. \tag{A.2}$$

When the parameter $\lambda = 1$, we have the **city block metric** (sometimes called **Manhattan distance**) given by

$$\delta_{ij} = \sum_k |x_{ik} - x_{jk}|.$$

When $\lambda = 2$, then Equation A.2 becomes the Euclidean distance (Equation A.1).

The MATLAB Statistics Toolbox provides a function called **pdist** that calculates the interpoint distances between the n data points. It returns the values in a vector, but they can be converted into a square matrix using the function **squareform**. The basic syntax for the **pdist** function is

```
Y = pdist(X, distance)
```

The **distance** argument can be one of the following choices (check the **help** for any updates):

```
'euclidean'   - Euclidean distance
'seuclidean'  - Standardized Euclidean distance, each coor-
   dinate in the sum of squares is inverse weighted by the
   sample variance of that coordinate
'cityblock'   - City Block distance
'mahalanobis' - Mahalanobis distance
'minkowski'   - Minkowski distance with exponent 2
'chebychev'   - Chebychev distance (maximum coordinate dif-
   ference)
'cosine'      - One minus the cosine of the included angle
   between observations (treated as vectors)
'correlation' - One minus the sample correlation between
   observations (treated as sequences of values).
'spearman'    - One minus the sample Spearman's rank corre-
   lation between observations (treated as sequences of val-
   ues).
'hamming'     - Hamming distance, percentage of coordinates
   that differ
'jaccard'     - One minus the Jaccard coefficient, the per-
   centage of nonzero coordinates that differ
```

The **cosine, correlation,** and **jaccard** measures specified above are originally similarity measures, which have been converted to dissimilarities, two of which are covered below. It should be noted that the **minkowski** option can be used with an additional argument designating other values of the exponent.

A.1.2 Similarity Measures

A common similarity measure that can be used with real-valued vectors is the *cosine similarity measure*. This is the cosine of the angle between the two vectors and is given by

$$s_{ij} = \frac{\mathbf{x}_i^T \mathbf{x}_j}{\sqrt{\mathbf{x}_i^T \mathbf{x}_i} \sqrt{\mathbf{x}_j^T \mathbf{x}_j}} .$$

The *correlation similarity measure* between two real-valued vectors is similar to the one above, and is given by the expression

$$s_{ij} = \frac{(\mathbf{x}_i - \bar{\mathbf{x}})^T (\mathbf{x}_j - \bar{\mathbf{x}})}{\sqrt{(\mathbf{x}_i - \bar{\mathbf{x}})^T (\mathbf{x}_i - \bar{\mathbf{x}})} \sqrt{(\mathbf{x}_j - \bar{\mathbf{x}})^T (\mathbf{x}_j - \bar{\mathbf{x}})}} .$$

A.1.3 Similarity Measures for Binary Data

The proximity measures defined in the previous sections can be used with quantitative data that are continuous or discrete, but not binary. We now describe some measures for binary data.

When we have binary data, we can calculate the following frequencies:

		j-th Observation		
		1	0	
i-th Observation	1	*a*	*b*	*a + b*
	0	*c*	*d*	*c + d*
		a + c	*b + d*	*a + b + c + d*

This table shows the number of elements where the *i*-th and *j*-th observations both have a 1 (*a*), both have a 0 (*d*), etc. We use these quantities to define the following similarity measures for binary data.

First is the *Jaccard coefficient*, given by

$$s_{ij} = \frac{a}{a + b + c} .$$

Next we have the *Ochiai measure*:

$$s_{ij} = \frac{a}{\sqrt{(a+b)(a+c)}}.$$

Finally, we have the **simple matching coefficient**, that calculates the proportion of matching elements:

$$s_{ij} = \frac{a+d}{a+b+c+d}.$$

A.1.4 Dissimilarities for Probability Density Functions

In some applications, our observations might be in the form of relative frequencies or probability distributions. For example, this often happens in the case of measuring semantic similarity between two documents. Or, we might be interested in calculating a distance between two probability density functions, where one is estimated and one is the true density function. We discuss three types of measures here: Kullback-Leibler information, L_1 norm, and information radius.

Say we have two probability density functions f and g (or any function in the general case). **Kullback-Leibler** (KL) **information** measures how well function g approximates function f. The KL information is defined as

$$KL(f, g) = \int f(\mathbf{x}) \log \left\{ \frac{f(\mathbf{x})}{g(\mathbf{x})} \right\} d\mathbf{x},$$

for the continuous case. A summation is used in the discrete case, where f and g are probability mass functions. The KL information measure is sometimes called the **discrimination information**, and it measures the divergence between two functions. The higher the measure, the greater the difference between the functions.

There are two problems with the KL information measure. First, there could be cases where we get a value of infinity, which can happen often in natural language understanding applications [Manning and Schütze, 2000]. The other potential problem is that the measure is not necessarily symmetric.

The second measure we consider is the **information radius** (IRad) that overcomes these problems. It is based on the KL information and is given by

$$IRad(f, g) = KL\left[f, \frac{f+g}{2} \right] + KL\left[g, \frac{f+g}{2} \right].$$

The IRad measure tries to quantify how much information is lost if we describe two random variables that are distributed according to f and g with their average distribution. The information radius ranges from 0 for identical distributions to 2log2 for distributions that are maximally different. Here we assume that 0log0 = 0. Note that the IRad measure is symmetric and is not subject to infinite values [Manning and Schütze, 2000].

The third measure of this type that we cover is the L_1 norm. It is symmetric and well defined for arbitrary probability density functions f and g (or any function in the general case). We can think of it as a measure of the expected proportion of events that are going to be different between distributions f and g. This norm is defined as

$$L_1(f, g) = \int |f(\mathbf{x}) - g(\mathbf{x})| d\mathbf{x} .$$

The L_1 norm has a nice property. It is bounded below by zero and bounded above by two:

$$0 \le L_1(f, g) \le 2 ,$$

when f and g are valid probability density functions.

A.2 Transformations

In many instances, we start with a similarity measure and then convert to a dissimilarity measure for further processing. There are several transformations that one can use, such as

$$\delta_{ij} = 1 - s_{ij}$$
$$\delta_{ij} = c - s_{ij} \qquad \text{for some constant } c$$
$$\delta_{ij} = \sqrt{2(1 - s_{ij})} .$$

The last one is valid only when the similarity has been scaled such that $s_{ii} = 1$. For the general case, one can use

$$\delta_{ij} = \sqrt{s_{ii} - 2s_{ij} + s_{jj}} .$$

In some cases, we might want to transform from a dissimilarity to a similarity. One can use the following to accomplish this:

$$s_{ij} = (1 + \delta_{ij})^{-1} .$$

A.3 Further Reading

Most books on clustering and multidimensional scaling have extensive discussions on proximity measures, since the measure used is fundamental to these methods. We describe a few of these here.

The clustering book by Everitt, Landau, and Leese [2001] has a chapter on this topic that includes measures for data containing both continuous and categorical variables, weighting variables, standardization, and the choice of proximity measure. For additional information from a classification point of view, we recommend Gordon [1999].

Cox and Cox [2001] has an excellent discussion of proximity measures as they pertain to multidimensional scaling. Their book also includes the complete taxonomy for proximity structures mentioned previously. Finally, Manning and Schütze [2000] provide an excellent discussion of measures of similarity that can be used when measuring the semantic distance between documents, words, etc.

There are also several survey papers on dissimilarity measures. Some useful ones are Gower [1966] and Gower and Legendre [1986]. They review properties of dissimilarity coefficients, and they emphasize their metric and Euclidean nature. For a summary of distance measures that can be used when one wants to measure the distance between two functions or statistical models, we recommend Basseville [1989]. The work described in Basseville takes the signal processing point of view. Jones and Furnas [1987] study proximity measures using a geometric analysis and apply it to several measures used in information retrieval applications. Their goal is to demonstrate the relationship between a measure and its performance.

Appendix B
Software Resources for EDA

The purpose of this appendix is to provide information on internet resources for EDA. Most of these were discussed in the body of the text, but some were not. Also, most of them are for MATLAB code, but some of them are stand-alone packages.

B.1 MATLAB Programs

In this section of Appendix B, we provide websites and references to MATLAB code that can be used for EDA. Some of the code is included with the EDA Toolbox. However, we recommend that users periodically look for the most recent versions.

Bagplots

Papers on the bagplot and related topics are available for download from this same website. We also include a **bagplot** function with the EDA Toolbox, courtesy of the authors of the LIBRA Toolbox (described later in this appendix) [Verboven and Hubert, 2005]. The use of this function is governed by the license found at

    http://wis.kuleuven.be/stat/robust/LIBRA/LIBRA-home

Computational Statistics Toolbox

This toolbox was written for the book *Computational Statistics Handbook with MATLAB, Third Edition*. It contains many useful functions, some of which were used in this book. It is available for download from

    https://www.crcpress.com/Computational-Statistics-
        Handbook-with-MATLAB-Third-Edition/Martinez-Mar-
        tinez/p/book/9781466592735

and

```
http://pi-sigma.info
```

Some of the functions from this toolbox are included with the EDA Toolbox.

Data Visualization Toolbox

The authors of this MATLAB toolbox provide functions that implement the graphical techniques described in *Visualizing Data* by William Cleveland. Software, data sets, documentation, and a tutorial are available from

```
http://www.datatool.com/prod02.htm
```

Some of these functions are included with the EDA Toolbox.

Dimensionality Reduction Toolbox

This toolbox was developed by L. J. P. van der Maaten from the Universiteit Maastricht, The Netherlands. The toolbox, papers, and data sets can be downloaded at

```
http://homepage.tudelft.nl/19j49/...
        Matlab_Toolbox_for_Dimensionality_Reduction.html
```

This toolbox has methods for linear and nonlinear dimensionality reduction, as well as functionality for estimating the intrinsic dimension of a data set.

Generative Topographic Mapping

The GTM Toolbox used to be available as separate toolbox. However, it is now part of the SOM Toolbox available here

```
http://research.ics.aalto.fi/software/somtoolbox/
```

This original GTM Toolbox is included under the GNU license with the EDA Toolbox.

Hessian Eigenmaps

The code for Hessian eigenmaps or HLLE no longer has its own website. It is included with the Dimensionality Reduction Toolbox, and the EDA Toolbox.

ISOMAP

The main website for ISOMAP is found at

```
http://web.mit.edu/cocosci/isomap/isomap.html
```

This also has links to related papers, data sets, convergence proofs, and supplemental figures. The ISOMAP functions are in the EDA Toolbox.

Locally Linear Embedding – LLE

The main website for LLE is

 `http://cs.nyu.edu/~roweis/nldr.html`

From here you can download the MATLAB code, papers, and data sets. The LLE function is part of the EDA Toolbox.

MATLAB Central

You can find lots of user-contributed code at this website:

 `www.mathworks.com/matlabcentral/`

Always look here first before you write your own functions. What you need might be here already.

Microarray Analysis – MatArray Toolbox

The link below takes you to a webpage where you can download software that implements techniques for the analysis of microarray data. This toolbox includes functions for k-means, hierarchical clustering, and others.

 `www.ulb.ac.be/medecine/iribhm/microarray/toolbox/`

Model-Based Clustering Toolbox

The code from the Model-Based Clustering Toolbox is included with the EDA Toolbox. However, for those who are interested in more information on MBC, please see

 `www.stat.washington.edu/mclust/`

This website also contains links for model-based clustering functions that will work in S-Plus and R.

Nonnegative Matrix Factorization

A toolbox for nonnegative matrix factorization for signal processing is available for download at

 `http://www.bsp.brain.riken.jp/ICALAB/nmflab.html`

This collection of MATLAB functions also implements nonnegative tensor factorization.

Robust Analysis – LIBRA

A toolbox for robust statistical analysis is available for download at

 `http://wis.kuleuven.be/stat/robust/LIBRA/LIBRA-home`

The toolbox has functions for univariate location and scale, multivariate location and covariance, regression, PCA, principal component regression, partial least squares, and robust classification [Verboven and Hubert, 2005]. Graphical tools are also included for model checking and outlier detection.

Self-Organizing Map – SOM

The website for the SOM Toolbox is

```
http://research.ics.aalto.fi/software/somtoolbox/
```

This is a new website for the toolbox, and it includes a link to the previous version and site. It contains links to documentation, theory, research, and related information. The software may be used for noncommercial purposes and is governed by the terms of the GNU General Public License. Some functions in the SOM Toolbox are contributed by users and might be governed by their own copyright. Some of the functions in the SOM Toolbox are included in the EDA Toolbox.

SiZer

This software allows one to explore data through smoothing. It contains various functions and GUIs for MATLAB. It helps answer questions about what features are really there or what might be just noise. In particular, it provides information about smooths at various levels of the smoothing parameter. For code, see:

```
http://marron.web.unc.edu/sample-page/marrons-matlab-
    software/
```

For examples and explanations of SiZer and smoothing, go to:

```
https://www.unc.edu/~marron/marron_data_anal.html
```

Statistics Toolboxes – No Cost

The following are links to statistics toolboxes:

```
www.statsci.org/matlab/statbox.html
www.maths.lth.se/matstat/stixbox/
```

Text to Matrix Generator (TMG) Toolbox

The TMG Toolbox creates new term-document matrices from documents. It also updates existing term-document matrices. It includes various term-weighting methods, normalization, and stemming. The website to download this toolbox is

```
http://scgroup20.ceid.upatras.gr:8000/tmg/
```

Topic Modeling Toolbox

Mark Steyvers wrote a Topic Modeling Toolbox for MATLAB. The following website includes code, data sets, and example scripts:

psiexp.ss.uci.edu/research/programs_data/toolbox.htm

t–SNE

There is an extensive website for *t*–SNE tools written by Laurens van der Maaten. He provides implementations of *t*–SNE for different software, such as MATLAB, R, Python, and more. Here is the website

https://lvdmaaten.github.io/tsne/

This site also has links to publications and data sets.

B.2 Other Programs for EDA

The following programs are available for download (at no charge).

Finite Mixtures - EMMIX

A FORTRAN program for fitting finite mixtures with normal and t-distribution components can be downloaded from

http://www.maths.uq.edu.au/~gjm/emmix/emmix.html

GGobi

GGobi is a data visualization system for viewing high-dimensional data, and it includes brushing, linking, plot matrices, multi-dimensional scaling, and many other capabilities. The home page for GGobi is

www.ggobi.org/

Versions are available for Windows and Linux. See the text by Cook and Swayne [2007] for information on how to use GGobi and to connect it with R.

MANET

For Macintosh users, one can use MANET (missings are now equally treated) found at:

http://www.rosuda.org/MANET/

MANET provides various graphical tools that are designed for studying multivariate features.

Model-Based Clustering

Software for model-based clustering can be downloaded at

> `http://www.stat.washington.edu/mclust/`

Versions are available for S-Plus and R. One can also obtain some of the papers and technical reports at this website, in addition to joining a mailing list.

Mondrian

This software is a general purpose package for visualization and EDA [Theus and Urbanek, 2009]. It is well-suited for large data sets, categorical data, and spatial data. All plots are linked, and many options for plotting and interacting with the graphics are available. The website for Mondrian is

> `http://www.theusrus.de/Mondrian/`

R for Statistical Computing

R is a language and environment for statistical computing. It is distributed as free software under the GNU license. It operates under Windows, UNIX systems (such as Linux), and the Macintosh. One of the benefits of the R language is the availability of many user-contributed packages. Please see the following website for more information.

> `www.r-project.org/`

B.3 EDA Toolbox

The Exploratory Data Analysis Toolbox is available for download from two sites:

> `https://www.crcpress.com/Exploratory-Data-Analysis-`
> `with-MATLAB-Third-Edition/Martinez-Martinez-`
> `Solka/p/book/9781498776066`

and

> `http://pi-sigma.info`

Please review the **readme** file for installation instructions and information on any changes. For complete functionality, you must have the MATLAB Statistics Toolbox, version 4 or higher. Please see the accompanying documentation for more information on the use of the EDA Toolbox.

Appendix C

Description of Data Sets

In this appendix, we describe the various data sets used in the book. All data sets are downloaded with the accompanying software. We provide the data sets in MATLAB binary format (MAT-files).

abrasion
These data are the results from an experiment measuring three variables for 30 rubber specimens [Davies, 1957; Cleveland and Devlin, 1988]: tensile strength, hardness, and abrasion loss. The abrasion loss is the amount of material abraded per unit of energy. Tensile strength measures the force required to break a specimen (per unit area). Hardness is the rebound height of an indenter dropped onto the specimen. Abrasion loss is measured in g/hp-hour; tensile strength is measured in kg/cm^2; and the unit for hardness is ° Shore. The intent of the experiment was to determine the relationship between abrasion loss with respect to hardness and tensile strength.

animal
This data set contains the brain weights and body weights of several types of animals [Crile and Quiring, 1940]. According to biologists, the relationship between these two variables is interesting, since the ratio of brain weight to body weight is a measure of intelligence [Becker, Cleveland, and Wilks, 1987]. The MAT-file contains variables: **AnimalName**, **BodyWeight**, and **BrainWeight**.

BPM data sets
There are several BPM data sets included with the text. We have **iradbpm**, **ochiaibpm**, **matchbpm**, and **L1bpm**. Each data file contains the interpoint distance matrix for 503 documents and an array of class labels, as described in Chapter 1. These data can be reduced using ISOMAP before applying other analyses.

calibrat
This data set reflects the relationship between radioactivity counts (**counts**) to hormone level for 14 immunoassay calibration values (**tsh**). The original

source of the data is Tiede and Pagano [1979], and we downloaded them from the website for Simonoff [1996]:

http://pages.stern.nyu.edu/~jsimonof/SmoothMeth/.

cereal
These data were obtained from ratings of eight brands of cereal [Chakrapani and Ehrenberg, 1981; Venables and Ripley, 1994]. The **cereal** file contains a matrix where each row corresponds to an observation and each column represents one of the variables or the percent agreement to statements about the cereal. The statements are: comes back to, tastes nice, popular with all the family, very easy to digest, nourishing, natural flavor, reasonably priced, a lot of food value, stays crispy in milk, helps to keep you fit, fun for children to eat. It also contains a cell array of strings (**labs**) for the type of cereal.

environmental
This file contains data comprising 111 measurements of four variables. These include **Ozone** (PPB), **SolarRadiation** (Langleys), **Temperature** (Fahrenheit), and **WindSpeed** (MPH). These were initially examined in Bruntz et al. [1974], where the intent was to study the mechanisms that might lead to air pollution.

ethanol
A single-cylinder engine was run with either ethanol or indolene. This data set contains 110 measurements of compression ratio (**Compression**), equivalence ratio (**Equivalence**), and $N0_x$ in the exhaust (**NOx**). The goal was to understand how $N0_x$ depends on the compression and equivalence ratios [Brinkman, 1981; Cleveland and Devlin, 1988].

example96
This is the data used in Example 9.6 to illustrate the box-percentile plots.

example104
This loads the data used in several examples to illustrate methods for plotting high-dimensional data. It is a subset of the BPM data that was reduced using ISOMAP as outlined in Example 10.5.[1]

faithful
This is a different version of the **geyser** data set, which is also included in the EDA Toolbox. It has fewer observations, but it has an additional variable. the first column is the eruption time of the Old Faithful geyser in minutes. The second column is the waiting time to the next eruption [Härdle, 1991].

---

[1] Note that the example numbers were changed in the second edition, but we did not change the name of this file.

forearm
These data consist of 140 measurements of the length in inches of the forearm of adult males [Hand et al., 1994; Pearson and Lee, 1903].

galaxy
The **galaxy** data set contains measurements of the velocities of the spiral galaxy NGC 7531. The array **EastWest** contains the velocities in the east-west direction, covering around 135 arc sec. The array **NorthSouth** contains the velocities in the north-south direction, covering approximately 200 arc sec. The measurements were taken at the Cerro Tololo Inter-American Observatory in July and October of 1981 [Buta, 1987].

geyser
These data represent the waiting times (in minutes) between eruptions of the Old Faithful geyser at Yellowstone National Park [Hand et al., 1994; Scott, 2015].

hamster
This data set contains measurements of organ weights for hamsters with congenital heart failure [Becker and Cleveland, 1991]. The organs are heart, kidney, liver, lung, spleen, and testes.

iris
The **iris** data were collected by Anderson [1935] and were analyzed by Fisher [1936] (and many statisticians since then!). The data set consists of 150 observations containing four measurements based on the petals and sepals of three species of iris. The three species are: *Iris setosa*, *Iris virginica*, and *Iris versicolor*. When the **iris** data file is loaded, you get three 50×4 matrices, one corresponding to each species.

leukemia
The **leukemia** data set is described in detail in Chapter 1. It measures the gene expression levels of patients with acute leukemia.

lsiex
This file contains the term-document matrix used in Examples 2.3 and 2.4.

lungA, lungB
The **lung** data set is another one that measures gene expression levels. Here the classes correspond to various types of lung cancer.

oronsay
The **oronsay** data set consists of particle size measurements. It is described in Chapter 1. The data can be classified according to the sampling site as well as the type (beach, dune, midden).

ozone
The **ozone** data set [Cleveland, 1993] contains measurements for the daily maximum ozone (ppb) concentrations at ground level from May 1 to September 30 in 1974. The measurements were taken at Yonkers, New York and Stamford, Connecticut.

playfair
This data set is described in Cleveland [1993] and Tufte [1983], and it is based on William Playfair's 1801 published displays of demographic and economic data. The **playfair** data set consists of the 22 observations representing the populations (thousands) of cities at the end of the 1700s and the diameters of the circles Playfair used to encode the population information. This MAT-file also includes a cell array containing the names of the cities.

pollen
This data set was generated for a data analysis competition at the 1986 Joint Meetings of the American Statistical Association. It contains 3848 observations, each with five fictitious variables: ridge, nub, crack, weight, and density. The data contain several interesting features and structures. See Becker et al. [1986] and Slomka [1986] for information and results on the analysis of these artificial data.

posse
The **posse** file contains several data sets generated for simulation studies in Posse [1995b]. These data sets are called **croix** (a cross), **struct2** (an L-shape), **boite** (a donut), **groupe** (four clusters), **curve** (two curved groups), and **spiral** (a spiral). Each data set has 400 observations in 8-D. These data can be used in PPEDA and other data tours.

salmon
The **salmon** data set was downloaded from the website for the book by Simonoff [1996]: **pages.stern.nyu.edu/~jsimonof/SmoothMeth/**. The MAT-file contains 28 observations in a 2–D matrix. The first column represents the size (in thousands of fish) of the annual spawning stock of Sockeye salmon along the Skeena River from 1940 to 1967. The second column represents the number of new catchable-size fish or recruits, again in thousands of fish.

scurve
This file contains data randomly generated from an S-curve manifold. See Example 3.5 for more information.

singer
This file contains several variables representing the height in inches of singers in the New York Choral Society [Cleveland, 1993; Chambers et al.,

1983]. There are four voice parts: sopranos, altos, tenors, and basses. The sopranos and altos are women, and the tenors and basses are men.

skulls

These data were taken from Cox and Cox [2001]. The data originally came from a paper by Fawcett [1901], where they detailed measurements and statistics of skulls belonging to the Naqada race in Upper Egypt. The **skulls** file contains an array called **skullsdata** for forty observations, 18 of which are female and 22 are male. The variables are greatest length, breadth, height, auricular height, circumference above the superciliary ridges, sagittal circumference, cross-circumference, upper face height, nasal breadth, nasal height, cephalic index, and ratio of height to length.

snowfall

This contains the amount of snowfall (inches) that fell in Buffalo, New York from 1910 to 1972 [Parzen, 1979].

software

This file contains data collected on software inspections. The variables are normalized by the size of the inspection (the number of pages or SLOC – single lines of code). The file **software.mat** contains the preparation time in minutes (**prepage, prepsloc**), the total work hours in minutes for the meeting (**mtgsloc**), and the number of defects found (**defpage, defsloc**). A more detailed description can be found in Chapter 1.

spam

These data were downloaded from the UCI Machine Learning Repository:

http://www.ics.uci.edu/~mlearn/MLRepository/.html

Anyone who uses E-mail understands the problem of spam, which is unsolicited E-mail, commercial or otherwise. For example, spam can be chain letters, pornography, advertisements, foreign money-making schemes, etc. This data set came from Hewlett-Packard Labs and was generated in 1999. The **spam** data set consists of 58 variables: 57 continuous and one class label. If an observation is labeled class 1, then it is considered to be spam. If it is of class 0, then it is not considered spam. The first 48 attributes represent the percentage of words in the E-mail that match some specified word corresponding to spam or not spam. There are an additional six variables that specify the percentage of characters in the E-mail that match a specified character. Others refer to attributes relating to uninterrupted sequences of capital letters. More information on the attributes is available at the above internet link. One can use these data to build classifiers that will discriminate between spam and nonspam E-emails. In this application, a low false positive rate (classifying an E-mail as spam when it is not) is very important.

sparrow

These data are taken from Manly [1994]. They represent some measurements taken on sparrows collected after a storm on February 1, 1898. Eight morphological characteristics and the weight were measured for each bird, five of which are provided in this data set. These are on female sparrows only. The variables are total length, alar extent, length of beak and head, length of humerus, and length of keel of sternum. All lengths are in millimeters. The first 21 of these birds survived, and the rest died.

swissroll

This data file contains a set of variables randomly generated from the Swiss roll manifold with a hole in it. It also has the data in the reduced space (from ISOMAP and HLLE) that was used in Example 3.5.

votfraud

These data represent the Democratic over Republican pluralities of voting machine and absentee votes for 22 Philadelphia County elections. The variable **machine** is the Democratic plurality in machine votes, and the variable **absentee** is the Democratic plurality in absentee votes [Simonoff, 1996].

yeast

The **yeast** data set is described in Chapter 1. It contains the gene expression levels over two cell cycles and five phases.

Appendix D

MATLAB® Basics

The purpose of this appendix is to provide some introductory information to help you get started using MATLAB. It has been adapted from *Statistics in MATLAB: A Primer* [Martinez and Cho, 2014]. We will describe:

- The desktop environment
- How to get help from several sources
- Ways to get your data into and out of MATLAB
- The different data types in MATLAB
- How to work with arrays
- Functions and commands
- Simple plotting functions

D.1 Desktop Environment

This section will provide information on the desktop environment. Figure D.1 shows the desktop layout for MATLAB, where we chose to view just the Command Window. We made the window smaller, which causes some of the sections—Code, Environment, and Resources, in this case—to be collapsed. Simply click on the arrow buttons for the sections to see what tools are available or make the window bigger by resizing with the mouse.

The default desktop layout includes the following panels or windows: Command Window, Current Folder, Workspace, and Command History. You can personalize your desktop layout by choosing those panels or sub-windows that you need for your work, dragging the panels to different places within the desktop, or re-sizing them.

The Command Window is the main interface to communicate with MATLAB. The window shows the MATLAB prompt, which is shown as a double arrow: >>. This is where you enter commands, functions, and other code. MATLAB will also display information in this spot in response to certain commands.

FIGURE D.1
*This is a screenshot of the main MATLAB window or desktop, where we have chosen to view only the Command Window in our layout. The desktop layout can be changed by clicking on the **Environment** section and the **Layout** option. MATLAB now uses the familiar ribbon interface common to Microsoft® applications.*

MATLAB now uses the typical ribbon interface found in recent Microsoft Windows® applications. The interface for MATLAB is comprised of tabs and sections, as highlighted in Figure D.1. The ribbon interface will be different depending on the selected tab (**Home, Plots,** or **Apps**).

We will provide more detail on the Figure window in a later section, but we mention it here briefly. Plots are displayed in a separate Figure window with its own user interface, which is comprised of the more familiar menu options and toolbar buttons. The main menu items include **File, Edit, Tools, Help,** and more. See Figure D.6 for an example.

The Workspace sub-window provides a listing of the variables that are in the current workspace, along with information about the type of variable, the size, and summary information. You can double-click on the **Variable** icon to open it in a spreadsheet-like interface. Using this additional interface, you can change elements, delete and insert rows, select and plot columns, and much more.

The Script Editor is a useful tool. This can be opened by clicking the **New Script** button on the ribbon interface. This editor has its own ribbon interface with helpful options for programming in MATLAB. The editor also has some nice features to make it easier to write error-free MATLAB code, such as tracking parentheses, suggesting the use of a semi-colon, and warnings about erroneous expressions or syntax.

D.2 Getting Help and Other Documentation

This appendix is meant to be a brief introduction to MATLAB—just enough to get you started. So, the reader is encouraged to make use of the many sources of assistance available via the online documentation, command line tools, help files, and the user community. A starting point could be to select the HELP button found on the HOME ribbon and the RESOURCES section; see Figure D.2.

FIGURE D.2
This shows the options available from the HELP button on the RESOURCES section of the HOME tab. Click the question mark to open a documentation window or click the arrow button to access more options.

There are several options to get help at the command line. They are the easiest and quickest way to get some help, especially once you become familiar with MATLAB.

You can find a shortened version of the documentation for a function by typing

$$\texttt{help } \textit{functionname}$$

at the command line. This will provide information on the syntax for the function, definitions of input/output arguments, and examples. The command

$$\texttt{doc } \textit{functionname}$$

will open the documentation for **functionname** in a separate window.

You can access MATLAB documentation in several ways. One is via the HELP button (see Figure D.2) on the HOME ribbon. This opens a window that has links to the documentation for your installed toolboxes. There is also a link to .pdf documentation on all toolboxes (scroll down to the bottom of the list on the website). Clicking on that link will take you to the Documentation Center at the MathWorks, Inc. website.

There is a vast MATLAB user community. This is a great resource for finding answers, obtaining user-written code, and following news groups. The main portal for this community is MATLAB Central, and it can be found at this link:

http://www.mathworks.com/matlabcentral/

D.3 Data Import and Export

Getting data into MATLAB is the first step in any analysis. Similarly, we might need to export data for use in other software (e.g., R) or to save the objects we created for future analysis in MATLAB. There are two main approaches: using command line functions (see Table D.1) or the interactive Import Wizard.

D.3.1 Data Import and Export in Base MATLAB®

The main functions to use for importing and exporting MATLAB specific data files (**.mat** files) via the command line are the **load** function for importing and the **save** function for exporting. These functions can also be used with an ASCII text file.

MATLAB has several options for reading in ASCII text files. If the file has numerical entries separated by white space, then the **load** (for exporting use **save -ascii**) command can be employed. If the values are separated by a different character, such as commas or tabs, then **dlmread** can be used. The character used to separate the values is inferred from the file or it can be specified as an argument. A related function called **dlmwrite** is used for exporting data in an ASCII file with different delimiters.

Most software packages for data analysis provide an option to import and export files where the data entries are separated by commas (**.csv** files). MATLAB has functions **csvread** and **csvwrite** to handle this type of file. Note that the file must contain only numeric data to use these functions.

You can import and export spreadsheet files with the **.xls** or **.xlsx** extensions using **xlsread** and **xlswrite**. The **xlsread** function will also read in OpenDocument™ spreadsheets that have the **.ods** file extension. For more information on OASIS OpenDocument formats, see

http://www.opendocumentformat.org/.

Sometimes a file can contain numerical and text data. There are two command-line options for reading in these types of files. One is the **importdata** function. This function can handle headers for ASCII files and spreadsheets, but the rest of the data should be in tabular form and has to be numeric. If you have an ASCII file with columns of non-numeric data (characters or formatted dates/times), then you can use the **textscan** function or the Import Wizard described next.

TABLE D.1

Common Data I/O Functions in Base MATLAB®

load, save	Read and write **.mat** files Read and write text files using the **-ascii** flag
dlmread, dlmwrite	Handles text files with specified delimiter
csvread, csvwrite	Use for comma delimited or **.csv** files
xlsread, xlswrite	Read and write spreadsheet files
importdata, textscan	Use for files that have a mixture of text and numbers

The Import Wizard can also be used to import data. It is started by clicking on the IMPORT DATA button located on the MATLAB desktop HOME ribbon's VARIABLE section; see Figure D.3. You can also start it by typing **uiimport** at the command line. The wizard guides you through the process of importing data from many different recognized file types. It allows you to view the contents of a file, to select the variables and the observations to import, to specify a value for unimportable cells, and more.

FIGURE D.3
Select the IMPORT DATA *button to start the Import Wizard.*

D.3.2 Data Import and Export with the Statistics Toolbox

There are some special functions for importing and exporting data in the Statistics Toolbox; see Table D.2 for a summary of the functions. These are particularly useful for statisticians and data analysts, and they include functions for importing and exporting tabular data, as well as data stored in the SAS XPORT format.

The functions **tblread** and **tblwrite** will import and export data in a tabular or matrix-like format. The data file must have the variable names on the first row, and the case names (or record identification variable) are in the first column. The data entries would start in the cell corresponding to the second row and second column (position (2, 2) in a matrix).

The basic syntax to *interactively* select a file to load is:

[data, vnames, cnames] = tblread,

where **vnames** contains the variable names (first row of the file) and **cnames** has the names of the observations (first column of the file). The variable **data** is a numeric matrix, where rows correspond to the observations, and the columns are the variables or characteristics.

Calling **tblread** with no input argument opens a window interface for selecting files. You can also specify a file name by using an input argument, as follows:

[data, vnames, cnames] = tblread(*filename*).

The following function call

tblwrite(data,vnames,cnames,*filename*,delim)

will export the data to a file with the delimiter specified by **delim**. Use **help tblwrite** at the command line for delimiter options.

Statisticians and data analysts often have to read in SAS files. The function **xptread** can be used to import files in the SAS XPORT transport format. The file can be selected interactively by leaving out the file name as the input argument or you can use the following syntax to read in a specific file:

data = xptread(*filename*).

The documentation for the Statistics Toolbox describes the functions we identified above and some additional ones for reading and exporting case names (**caseread, casewrite**). MATLAB documentation and help pages can be accessed from the desktop environment by clicking on the HELP button (on the HOME ribbon), selecting STATISTICS TOOLBOX … EXPLORATORY DATA ANALYSIS … DATA IMPORT AND EXPORT.

TABLE D.2

Data I/O Functions in the Statistics Toolbox

`tblread, tblwrite`	Data in tabular format with variable names on the first row and case names in the first column
`xptread`	SAS XPORT (transport) format files
`caseread, casewrite`	Import and export text files with one case name per line
`export`	Write a dataset array to a tab-delimited file
`tdfread`	Import a tab-delimited file that has text and numeric data

D.4 Data in MATLAB®

We now describe the basic data types in MATLAB and the Statistics Toolbox. We also discuss how to merge data sets and to create special arrays that might prove useful in data analysis. We conclude with a short introduction to object-oriented programming constructs and how they are used in MATLAB.

D.4.1 Data Objects in Base MATLAB®

One can consider two main aspects of a data object in MATLAB—the object class and what it contains. We can think of the object *class* as the type of container that holds the data. The most common ones are arrays, cell arrays, and structures. The *content* of objects in MATLAB can be numeric (e.g., double precision floating point or integers) or characters (e.g., text or strings). We now describe the common types or classes of objects in the base MATLAB software.

The word *array* is a general term encompassing scalars, vectors, matrices, and multi-dimensional arrays. All of these objects have a *dimension* associated with them. You can think of the dimension as representing the number of indexes you need to specify to access elements in the array. We will represent this dimension with the letter k.

- **Scalar**: A scalar is just a single number (or character), and it has dimension $k = 0$. You do not need to specify an address because it is just a single element.

- **Vector**: A vector is usually a column of values, but MATLAB also has row vectors. A vector has dimension $k = 1$ because you have to provide one value to address an element in the column (or row) vector.
- **Matrix**: A matrix is an object that has rows and columns—like a table. To extract or access an element in a matrix, you have to specify what row it is in and also the column. Thus, the dimension of a matrix is $k = 2$.
- **Multi-dimensional array**: A multi-dimensional array has dimension $k > 2$. For example, we could think of a three-dimensional array being organized in pages (the third dimension), where each page contains a matrix. So, to access an element in such an array, we need to provide the row, column, and page number.

As a data analyst, you will usually import data using **load** or some other method we described previously. However, you will likely also need to type in or construct arrays for testing code, entering parameters, or getting the arrays into the right form for calling functions. We now cover some of the ways to build small arrays.

Commas or spaces concatenate elements (or other arrays) as columns. In other words, it puts them together as a row. Thus, the following MATLAB code will produce a row vector:

$$x = [1, 4, 5]$$

Or, we can concatenate two column vectors **a** and **b** to create one matrix, as shown here:

$$Y = [a \ b]$$

The semi-colon will stack elements as rows. So, we would obtain a column vector from this command:

$$z = [1; 4; 5]$$

As another example, we could put three row vectors together to get a matrix, as shown next:

$$Y = [a; b; c]$$

It is important to note that the building blocks of your arrays have to be conformal in terms of the number of elements in the sub-arrays when using the comma or semi-colon to merge data sets. Otherwise, you will get an error.

We might need to generate a regular sequence of values when working in MATLAB. We can accomplish this by using the colon. We get a sequence of values from one to ten with the following syntax:

$$x = 1:10$$

Other step sizes can be used, too. For instance, this will give us a sequence from one to ten in steps of 0.5:

$$x = 1:0.5:10$$

and this will yield a decreasing sequence:

$$x = 10:-1:1$$

We can create an array of all zeros or all ones. Arrays of this type are often used in data analysis. Here is how we can create a 3×3 matrix of zeros:

$$Z = zeros(3,3)$$

This next function call will produce a multidimensional array with $k = 3$:

$$O = ones(2,4,3)$$

The array **o** has three pages (third dimension), and each page has a matrix with two rows and four columns.

There is a special type of array in MATLAB called the *empty array*. An empty array is one that contains no elements. Thus, it does not have any dimensions associated with it, as we mentioned above with other arrays. The empty array is designated by closed square brackets, as shown here: **[]**. It can be used to delete elements of an array at the command line, and it is sometimes returned in response to function calls and logical expressions.

Here is an example of the first case, where we show how to delete an element from a vector **x**.

```
% Create a vector x.
x = [2, 4, 6];
% Delete the second element.
x(2) = [];
% Display the vector x.
disp(x)
        2       6
```

A *cell array* is a useful data object, especially when working with strings. The elements of a cell array are called *cells*. Each cell provides a flexible container for our data because it can hold data of any type—even other cell arrays. Furthermore, each element of the cell array can have a different size.

The cell array has an overall structure that is similar to basic numeric or character data arrays covered previously, and as such, they have to be conformal in their overall arrangement. For example, the cells are arranged in rows, columns, pages, etc., as with arrays. If we have a cell array with two rows, then each of its rows has to have the same number of cells. However, the content of the cells can be different in terms of data type and size.

We can create an empty cell array using the function **cell**, as shown here, where we set up a $2 \times 4 \times 3$ array of cells. Each of the cells in the following cell array object is empty:

```
cell_array = cell(2,4,3)
```

You can also construct a cell array and fill it with data, as shown in the code given next. Note that curly braces are used to denote a cell array.

```
% Create a cell array, where one cell contains
% numbers and another cell element is a string.
cell_array2 = {[1,2], 'This is a string'};
```

Like cell arrays, *structures* allow one to combine dissimilar data into a single variable. The basic syntax to create a structure is

```
S = struct('field1',data1,'field2',data2,...).
```

Note that the structure can have one or more fields, along with the associated data values. Let's use this to create a small structure.

```
% Create a structure called author with four fields.
author = struct(...
    'name',{{'Wendy','Angel'}},...
    'area',{{'Clustering','Visualization'}},...
    'deg',{{'PhD','PhD'}});
```

A dot notation is used to extract or access a field. Suppose we want to get all of the names in our **author** structure, then we can use

```
all_names = author.name
```

to get all of the entries in the **name** field. We will discuss how to access individual data elements and records in the next section.

A *table* is a type of data object in the base MATLAB software. The table object is like cell arrays and structures. We can combine data of different data types in one object. However, it has a table-like format that is familiar to statisticians and data analysts. The rows of a table object would contain the observations or cases, and the columns correspond to the characteristics or features.

We can use the **table** function to create a table object employing variables that are in the workspace. The basic syntax is

```
% Create a table using two objects.
mytab = table(var1,var2)
```

You can import a file as a table object using the **readtable** function. This function works with delimited text files (**.txt**, **.dat**, or **.csv**). It will also read in an Excel spreadsheet file with **.xls** or **.xlsx** extensions.

D.4.2 Accessing Data Elements

In this section, we demonstrate how you can identify elements of arrays, cell objects, and structures. This is useful in data analysis because we often need to analyze subsets of our data or to create new data sets by combining others. Table D.3 provides some examples of how to access elements of arrays. These can be numeric, string, or cell arrays. In the case of cell arrays, the notation is used to access the cell elements, but not the *contents* of the cells. Curly braces **{ }** are used to get to the data that are inside the cells. For example, **A{1,1}** would give us the contents of the cell, which would have a numeric or character type. Whereas, **A(1,1)** is the cell, and it has a type (or class) of cell.

These two notations can be combined to access part of the contents of a cell. To get the first two elements of the vector contents of cell **A(1,1)**, we can use

$$A\{1,1\}(1:2)$$

The curly braces in **A{1,1}** tells MATLAB to go inside the cell in position **(1,1)**, and the **(1:2)** points to elements 1 and 2 inside the cell.

TABLE D.3

Examples of Accessing Elements of Arrays

Notation	Usage
a(i)	Access the *i*th element (cell) of a row or column vector array (cell array)
a(3:5)	Access elements 3 through 5 of a vector or cell array
A(:,i)	Access the *i*th column of a matrix or cell array. In this case, the colon in the row dimension tells MATLAB to access all rows.
A(i,:)	Access the *i*th row of a matrix or cell array. The colon tells MATLAB to gather all of the columns.
A(2:4,1:2)	Access the elements in the second, third, and fourth rows and the first two columns
A(1,3,4)	Access the element in the first row, third column on the fourth entry of dimension 3 (sometimes called the page).

Recall that we can access entire fields in a structure using the dot notation. We can extract partial content from the fields by using the techniques we described for numeric and cell arrays, as illustrated next.

```
% Display Wendy's degree.
author.deg(1)

ans = 'PhD'
```

The **ans** object above is actually a one-cell array. To get the contents of the cell (as a string object), we use the curly braces, as shown here.

```
author.deg{1}
```

```
ans = PhD
```

The **ans** result is now a **char** array. The dot notation is used to access the field. We then specify the elements using the notation for arrays (Table D.3).

The techniques for manipulating subsets of data in table objects are similar to structures and arrays, but they have some additional options because of the column or variable names. This shows how to create a sub-table with the first three records and all variables.

```
% Get a sub-table by extracting the first 3 rows.
newtab = mytab(1:3,:)
```

We are able to extract a column of the table using the dot notation that we had with structures.

```
% Get second variable in the table we
% created in the previous section.
vt = mytab.var2;
```

A *dataset array* is a special data object that is included in the Statistics Toolbox, and it can be used to store variables of different data types. As an example, you can combine numeric, logical, character, and categorical data in one array. Each row of the dataset array corresponds to an observation, and each column corresponds to a variable. Therefore, each column has to have elements that are of the same data type. However, the individual columns can be different. For instance, one column can be numeric, and another can be text.

A dataset array can be created using variables that exist in the workspace or as a result of importing data. The function **dataset** is used in either case, as shown here. The basic syntax for creating a dataset array from various file types is shown next.

```
% Create from a tab-delimited text file.
ds = dataset('File','filename.txt')
```

```
% Create from a .csv file.
ds = dataset('File','filename.csv','Delimiter',',')
```

```
% Create from an Excel file.
ds = dataset('XLSFile','filename.xlsx')
```

There are several options for creating a dataset array from variables in the workspace. They are listed here.

```
% Create by combining three different variables.
ds = dataset(var1, var2, var3);

% Create by converting a numeric matrix called data.
ds = mat2dataset(data);
```

A dataset array has its own set of defined operations, and you cannot operate on this type of array in the same manner as a numeric array. We will discuss this idea in more detail shortly, when we cover object-oriented programming.

The dataset array might be removed in future versions of MATLAB, but it was still available in version 2015a. Because of this, it is recommended that you use the **table** object in base MATLAB instead of a dataset array.

D.4.3 Object-Oriented Programming

Certain aspects of MATLAB are *object-oriented*, which is a programming approach based on three main ideas:

1. **Classes and objects**: A *class* is a description or definition of a programming construct. An *object* is a specific instance of a class.
2. **Properties**: These are aspects of the object that can be manipulated or extracted.
3. **Methods**: These are behaviors, operations, or functions that are defined for the class.

The benefit of object-oriented programming is that the computer code for the class and the methods are defined once, and the same method can be applied to different instances of the class without worrying about the details.

Every data object has a class associated with it. There is a function called **class** that will return the class of an object. This can be very helpful when trying to understand how to access elements and to perform other data analytic tasks. You will encounter some special object classes throughout this book. There are several instances of unique classes that are defined in the Statistics Toolbox. Some examples of these include probability distributions, models, and trees.

D.5 Workspace and Syntax

In this section, we cover some additional topics you might find helpful when using MATLAB. These include command line functions for managing your workspace and files, punctuation, arithmetic operators, and functions.

D.5.1 File and Workspace Management

You can enter MATLAB expressions interactively at the command line or save them in an M-file. This special MATLAB file is used for saving scripts or writing functions. We described the Script Editor in an earlier section, which is a very handy tool for writing and saving MATLAB code.

As stated previously, we will not be discussing how to write your own programs or functions, but you might find it helpful to write *script* M-files. These script files are just text files with the **.m** extension, and they contain any expressions or commands you want to execute. Thus, it is important to know some commands for file management. There are lots of options for directory and file management on the desktop. In a previous section, we briefly mentioned some interactive tools to interface with MATLAB. Table D.4 provides some commands to list, view, and delete files.

TABLE D.4

File Management Commands

Command	Usage
dir, ls	Shows the files in the present directory
delete *filename*	Deletes *filename*
pwd	Shows the present directory
cd *dir*	Changes the directory. There is also a pop-up menu and button on the desktop that allows the user to change directory, as we show here.
edit *filename*	Brings up *filename* in the editor
type *filename*	Displays the contents of the file in the command window
which *filename*	Displays the path to *filename*. This can help determine whether a file is part of base MATLAB.
what	Lists the **.m** files and **.mat** files that are in the current directory

Variables created in a session (and not deleted) live in the MATLAB *workspace*. You can recall the variable at any time by typing in the variable name with no punctuation at the end. Note that variable names in MATLAB are case sensitive, so **Temp**, **temp**, and **TEMP** are different variables.

As with file management, there are several tools on the desktop to help you manage your workspace. For example, there is a **Variable** section on the

TABLE D.5

Commands for Workspace Management

Command	Usage
`who`	Lists all variables in the workspace.
`whos`	Lists all variables in the workspace along with the size in bytes, array dimensions, and object type.
`clear`	Removes all variables from the workspace.
`clear x y`	Removes variables **x** and **y** from the workspace.

desktop ribbon interface that allows you to create a variable, open current variables in a spreadsheet-like interface, save the workspace, and clear it. Some commands to use for workspace management are given in Table D.5.

D.5.2 Syntax in MATLAB®

Punctuation and syntax are important in any programming language. If you get this wrong, then you will either get errors or your results will not be what you expect. Some of the common punctuation characters used in MATLAB are described in Table D.6.

MATLAB has the usual arithmetic operators for addition, subtraction, multiplication, division, and exponentiation. These are designated by $+$, $-$, $*$, $/$, $\wedge$, respectively. It is important to remember that MATLAB interprets these operators in a linear algebra sense and uses the corresponding operation definition for arrays, also.

For example, if we multiply two matrices **A** and **B**, then they must be dimensionally correct. In other words, the number of columns of **A** must equal the number of rows of **B**. Similarly, adding and subtracting arrays requires the same number of elements and configuration for the arrays. Thus, you can add or subtract *row* vectors with the same number of elements, but you will get an error of you try to add or subtract a row and column vector, even if they have the same number of elements. See any linear algebra book for more information on these concepts and how the operations are defined with arrays [Strang, 1993].

Addition and subtraction operations are defined element-wise on arrays, as we have in linear algebra. In some cases, we might find it useful to perform other element-by-element operations on arrays. For instance, we might want to square each element of an array or multiply two arrays element-wise. To do this, we change the notation for the multiplication, division, and exponentiation operators by adding a period before the operator. As an example, we could square each element of **A**, as follows:

TABLE D.6

List of MATLAB® Punctuation

Punctuation	Usage
%	A percent sign denotes a comment line. Information after the **%** is ignored.
,	When used to separate commands on a single line, a comma tells MATLAB to display the results of the preceding command. When used to combine elements or arrays, a comma or a blank space groups elements along a row. A comma also has other uses, including separating function arguments and array subscripts.
;	When used after a line of input or between commands on a single line, a semicolon tells MATLAB not to display the results of the preceding command. When used to combine elements or arrays, a semicolon stacks them in a column.
...	Three periods denote the continuation of a statement onto the next line.
:	The colon specifies a range of numbers. For example, **1:10** means the numbers 1 through 10. A colon in an array dimension accesses all elements in that dimension.

A.^2

A summary of these element-by-element operators are given in Table D.7.

MATLAB follows the usual order of operations we are familiar with from mathematics and computer programming. The precedence can be changed by using parentheses.

TABLE D.7

List of Element-by-Element Operators

Operator	Usage
.*	Multiply element by element
./	Divide element by element
.^	Raise each element to a power

D.5.3 Functions in MATLAB®

MATLAB is a powerful computing environment, but one can also view it as a programming language. Most computer programming languages have mini-programs; these are called *functions* in MATLAB.

In most cases, there are two different ways to call or invoke functions in MATLAB: *function syntax* or *command syntax*, as we describe next. What type of syntax to use is up to the user and depends on what you need to accomplish. For example, you would typically use the function syntax option when the output from the function is needed for other tasks.

Function syntax

The function syntax approach works with input arguments and/or output variables. The basic syntax includes the function name and is followed by arguments enclosed in parentheses. Here is an illustration of the syntax:

$$functionname(arg1, \ldots, argk)$$

The statement above does not return any output from the function to the current workspace. The output of the function can be assigned to one or more output variables enclosed in square brackets:

$$[out1, \ldots, outm] = functionname(arg1, \ldots, argk)$$

You do not need the brackets, if you have only one output variable. The number of inputs and outputs you use depends on the definition of the function and what you want to accomplish. Always look at the **help** pages for a function to get information on the definitions of the arguments and the possible outputs. There are many more options for calling functions than what we describe in this book.

Command Syntax

The main difference between command and function syntax is how you designate the input arguments. With command syntax, you specify the function name followed by arguments separated by spaces. There are no parentheses with command syntax. The basic form of a command syntax is shown here:

$$functionname\ arg1\ \ldots\ arg2$$

The other main difference with command syntax pertains to the outputs from the function. You cannot obtain any output values with commands; you must use function syntax for that purpose.

D.6 Basic Plot Functions

In this section, we discuss the main functions for plotting in two and three dimensions. We also describe some useful auxiliary functions to add content to the graph. Type **help graph2d** or **help graph3d** for a list of plotting functions in the base MATLAB software. Table D.8 contains a list of common plotting functions in base MATLAB. We will examine several of these functions starting off with **plot** for creating 2–D graphics.

TABLE D.8

List of Plotting Functions in Base MATLAB®

area	Plot curve and fill in the area
bar, bar3	2–D and 3–D bar plots
contour, contour3	Display isolines of a surface
errorbar	Plot error bars with curve
hist	Histogram
image	Plot an image
pie, pie3	Pie charts
plot, plot3	2–D and 3–D lines
plotmatrix	Matrix of scatterplots
scatter, scatter3	2–D and 3–D scatterplots
stem, stem3	Stem plot for discrete data

D.6.1 Plotting 2D Data

The main function for creating a 2D plot is called **plot**. When the function **plot** is called, it opens a new Figure window. It scales the axes to fit the limits of the data, and it plots the points, as specified by the arguments to the function. The default is to plot the points and connect them with straight lines. If a Figure window is already available, then it produces the plot in the current Figure window, replacing what is there.

The main syntax for **plot** is

$$\texttt{plot(x,y,'color\_linestyle\_marker')}$$

where **x** and **y** are vectors of the same size. The **x** values correspond to the horizontal axis, and the **y** values are represented on the vertical axis.

Several pairs of vectors (for the horizontal and vertical axes) can be provided to **plot**. MATLAB will plot the values given by the pairs of vectors on the same set of axes in the Figure window. If just one vector is provided as an argument, then the function plots the values in the vector against the index $1 \ldots n$, where n is the length of the vector. For example, the following command plots two curves

$$plot(x1,y1,x2,y2)$$

The first curve plots **y1** against **x1**, and the second shows **y2** versus the values in **x2**.

Many arguments can be used with the **plot** function giving the user a lot of control over the appearance of the graph. Most of them require the use of MATLAB's Handle Graphics® system, which is beyond the scope of this introduction. The interested reader is referred to Marchand and Holland [2003] for more details on Handle Graphics.

Type **help plot** at the command line to see what can be done with **plot**. We present some of the basic options here. The default line style in MATLAB is a solid line, but there are other options as listed in Table D.9. The first column in the table contains the notation for the line style that is used in the **plot** function. For example, one would use the notation

$$plot(x,y,':')$$

to create a dotted line with the default marker style and color. Note that the specification of the line style is given within single quotes (denoting a string) and is placed immediately after the vectors containing the observations to be graphed with the line style.

We can also specify different colors and marker (or point) styles within the single quotes. The predefined colors are given in Table D.10. There are thirteen marker styles, and they are listed in Table D.11. They include circles, asterisks, stars, x-marks, diamonds, squares, triangles, and more.

The following is a brief list of common plotting tasks:

- Solid green line with no markers or plotting with just points:

$$plot(x,y,'g'), \ plot(x,y,'.')$$

- Dashed blue line with points shown as an asterisk:

$$plot(x,y,'b--*')$$

- Two lines with different colors and line styles (see Figure D.4):

$$plot(x,y,'r:',x2,y2,'k.-o')$$

TABLE D.9

Line Styles for Plots

Notation	Line Type
–	solid line
:	dotted line
–.	dash-dot line
– –	dashed line

TABLE D.10

Line Colors for Plots

Notation	Color
b	blue
g	green
r	red
c	cyan
m	magenta
y	yellow
k	black
w	white

TABLE D.11

Marker Styles for Plots

Notation	Marker Style
.	point
o	circle
x	x-mark
+	plus
*	star
s	square
d	diamond
v	down triangle
^	up triangle
<	left triangle
>	right triangle
p	pentagram
h	hexagram

It is always good practice to include labels for all axes and a title for the plot. You can add these to your plot using the functions **xlabel**, **ylabel**, and **title**. The basic input argument for these functions is a text string:

xlabel('*text*'),ylabel('*text*'),title('*text*')

In the bulleted list above, we showed how to plot two sets of **x** and **y** pairs, and this can be extended to plot any number of lines on one plot. There is another mechanism to plot multiple lines that can be useful in loops and other situations. The command to use is **hold on**, which tells MATLAB to apply subsequent graphing commands to the current plot. To unfreeze the current plot, use the command **hold off**.

We sometimes want to plot different lines (or any other type of graph) in a Figure window, but we want each one on their own set of axes. We can do this through the use of the **subplot** function, which creates a matrix of plots in a Figure window. The basic syntax is

subplot(m, n, p)

This produces **m** rows and **n** columns of plots in one Figure window. The third argument denotes what plot is active. Any plotting commands after this function call are applied to the **p**-th plot. The axes are numbered from top to bottom and left to right.

D.6.2 Plotting 3D Data

We can plot three variables in MATLAB using the **plot3** function, which works similarly to **plot**. In this case, we have to specify three arguments that correspond to the three axes. The basic syntax is

plot3(x,y,z)

where **x**, **y**, and **z** are vectors with the same number of elements. The function **plot3** will graph a line in 3D through the points with coordinates given by the elements of the vectors. There is a **zlabel** function to add an axes label to the third dimension.

Sometimes, one of our variables is a response or dependent variable, and the other two are predictors or independent variables. Notationally, this situation is given by the relationship

$$z = f(x, y).$$

The z values define a surface given by points above the x–y plane.

We can plot this type of relationship in MATLAB using the **mesh** or **surf** functions. Straight lines are used to connect adjacent points on the surface. The **mesh** function shows the surface as a wireframe with colored lines, where the color is proportional to the height of the surface. The **surf** function fills in the surface facets with color.

The **surf** and **mesh** functions require three matrix arguments, as shown here

surf(X,Y,Z), mesh(X,Y,Z)

The **X** and **Y** matrices contain repeated rows and columns corresponding to the domain of the function. If you do not have these already, then you can generate the matrices using a function called **meshgrid**. This function takes two vectors **x** and **y** that specify the domains, and it creates the matrices for constructing the surface. The **x** vector is copied as rows of the **X** matrix, and the vector **y** is copied as columns of the **Y** matrix.

As stated previously, the color is mapped to the height of the surface using the default color map. The definition of the color map can be changed using the function **colormap**. See the **help** on **graph3d** for a list of built-in color maps. You can also add a bar to your graph that illustrates the scale associated with the colors by using the command **colorbar**.

MATLAB conveys 3D surfaces on a 2D screen, but this is just one view of the surface. Sometimes interesting structures are hidden from view. We can change the view interactively via a toolbar button in the Figure window, as shown in Figure D.4. We can also use the **view** function on the command line, as shown here

view(azimuth, elevation)

The first argument **azimuth** defines the horizontal rotation. The second argument **elevation** corresponds to the vertical rotation. Both of these are given in degrees. For example, we are looking directly overhead (2D view) if the azimuth is equal to zero, and the elevation is given by 90 degrees.

Sometimes, it is easier to rotate the surface interactively to find a good view. We can use the R<small>OTATE</small> **3D** button in the Figure window, as shown in Figure D.4. When the button is pushed, the cursor changes to a curved arrow. At this point, you can click in the plot area and rotate it while holding the left mouse button. The current azimuth and elevation is given in the lower left corner, as you rotate the axes.

FIGURE D.4
Click on the R<small>OTATE</small> **3D** *button to rotate 3D plots. The current azimuth and elevation is indicated in the lower left corner as the axes rotate.*

D.6.3 Scatterplots

The scatterplot is one of the main tools a statistician should use before doing any analysis of the data or modeling. A *scatterplot* is a plot of one variable against another, where (x, y) pairs are plotted as points. The points are not connected by lines. This type of plot can be used to explore the distribution of multivariate data, to assess the relationship between two variables, or to look for groups and other structure. These were also discussed in Chapter 5, but we provide additional information here for completeness.

We can easily use the functions **plot** and **plot3** to create 2D and 3D scatterplots. We just specify the desired marker or symbol, as shown here:

plot(x,y,'o'), plot3(x,y,z,'*')

The call to **plot** shows the markers as open circles, and the call to **plot3** uses the asterisk as the plotting symbol.

The **plot** and **plot3** functions are best for the case where you are using only one or two marker styles and/or symbol colors. MATLAB has two special functions for scatterplots that provide more control over the symbols used in the plot. These are called **scatter** and **scatter3**. The syntax is

$$\texttt{scatter(x,y,s,c), scatter3(x,y,z,s,c)}$$

The first two (or three for **scatter3**) arguments represent the coordinates for the points. The optional arguments **s** (marker size) and **c** (marker color) allow one to control the appearance of the symbols. These inputs can be a single value, in which case, they are applied to all markers. Alternatively, one could assign a color and/or size to each point.

The default symbol is an open circle. An additional input argument specifying the marker (see **help** on **plot** for a list) can be used to get a different plotting symbol. If we assign a different size to each of the open circles, possibly corresponding to some other variable, then this is known as a *bubble plot*.

D.6.4 Scatterplot Matrix

We often have data with many variables or dimensions. In this case, we can use the *scatterplot matrix* to look at all 2D scatterplots, which gives us an idea of the pair-wise relationships or distributions in the data. A scatterplot matrix is a matrix of plots, where each one is a 2D scatterplot.

MATLAB provides a function called **plotmatrix** that takes care of the plotting commands for us, and we do not have to worry about multiple uses of **subplot**. The syntax for **plotmatrix** is:

$$\texttt{plotmatrix(X), plotmatrix(X,Y)}$$

The first of these plots the columns of **X** as scatterplots, with histograms of the columns on the diagonal. This is the version used most often by statisticians. The second plots the columns of **Y** against the columns of **X**.

D.6.5 GUIs for Graphics

MATLAB has several graphical user interfaces (GUIs) to make plotting data easier. We describe the main tools in this section, including simple edits using menu options in the Figure window, the plotting tools interface, and the PLOTS tab on the desktop ribbon.

We often want to add simple graphics objects to our plot, such as labels, titles, arrows, rectangles, circles, and more. We can do most of these tasks via the command line, but it can be frustrating trying to get them to appear just the way we want them to. We can use the INSERT menu on the Figure window, as shown in Figure D.5, to help with these tasks. Just select the desired option, and interactively add the object to the plot.

Perhaps the most comprehensive GUIs for working with graphics are the plotting tools. This is an interactive set of tools that work similarly to the

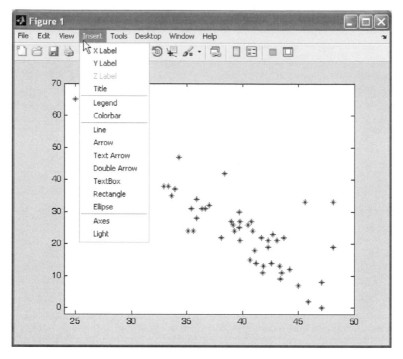

FIGURE D.5
Select the **INSERT** *menu on the Figure window to add objects to your plot.*

main MATLAB desktop environment. In other words, they can be part of an expanded Figure window, or they can be undocked by clicking the downward arrow in the upper right corner of the tool. Look at **help** on **plottools** for more information and examples of using these tools.

The plotting tool GUIs consist of three panels or editors, as listed here:

- **Property Editor**: This provides access to some of the properties of the graphics objects in a figure. This includes the Figure window, the axes, line objects, and text. The editor can be started by using the command **propertyeditor**. It can also be opened via the TOOLS > EDIT PLOT menu item in a Figure window. Double-click on a highlighted graphics object to open the editor.
- **Figure Palette**: This tool allows the user to add and position axes, plot variables from the workspace, and annotate the graph. The command **figurepalette** will open the editor.
- **Plot Browser**: The browser is used to add data to the plot and to control the visibility of existing objects, such as the axes. This can be opened using the command **plotbrowser**.

The plotting tools can be opened in different ways, and we specified some of them in the list given above. One option is to use the command

plottools. This will add panels to an existing Figure window, or it will open a new Figure window and the tools, if one is not open already. The most recently used plotting tools are opened for viewing.

One could also click on the highlighted toolbar button shown in Figure D.6. The button on the right shows the plotting tools, and the one on the left closes them. Finally, the tools can be accessed using the VIEW menu on the Figure window. Clicking on the desired tool will toggle it on or off.

CLOSE PLOT TOOLS   OPEN PLOT TOOLS

FIGURE D.6
This shows the toolbar buttons for a Figure window. The buttons on the right will open and close the interactive plotting tools.

An example of a Figure window with the plotting tools open is given in Figure D.7. Only the Property Editor is opened because that was the last tool we used. Select other editors using the VIEW menu.

Another GUI option to create plots is available via the PLOTS tab on the main MATLAB ribbon interface. The Workspace browser has to be open because this is used to select variables for plotting. Click on the desired variables in the browser, while holding the CTRL key. The variables will appear in the left section of the PLOTS tab.

Next, select the type of plot by clicking the corresponding picture. There is a downward arrow button that provides access to a complete gallery of plots. It will show only those plots that are appropriate for the types of variables selected for plotting. See Figure D.8 for an example.

D.7 Summary and Further Reading

MATLAB has more graphing functions as part of the base software. The main ones were described in this chapter, and we provide an expanded list in Table D.8. Use the **help** *functionname* at the command line for information on how to use them and to learn about related functions. Table D.12 has a list of auxiliary functions for enhancing your plots.

We already recommended the Marchand and Holland book [2003] for more information on Handle Graphics. This text also has some useful tips and ideas for creating plots and building GUIs. You should always consult

FIGURE D.7

Here is an example of a Figure window with the potting tools open. Only the Property Editor is viewed because this is the last one that was used. Select other editors using the **VIEW** *menu.*

FIGURE D.8

This is a portion of the **PLOTS** *ribbon in the main MATLAB desktop environment. Click the* **PLOTS** *tab to access it. Variables for plotting are selected in the Workspace browser, and they appear in the left section of the ribbon. Click on one of the icons in the right section of the* **PLOTS** *ribbon to create the plot. More options are available using the downward arrow button in the plots section.*

TABLE D.12

List of Auxiliary Plotting Functions in Base MATLAB®

`axis`	Change axes scales and appearance
`box`	Draw a box around the axes
`grid`	Add grid lines at the tick marks
`gtext`	Add text interactively
`hidden`	Remove hidden lines in **mesh** plots
`hold`	Hold the current axes
`legend`	Insert a legend
`plotedit`	Tools for annotation and editing
`rotate`	Rotate using given angles
`subplot`	Include multiple axes in figure window
`text`	Insert text at locations
`title`	Put a title on the plot
`xlabel, ylabel, zlabel`	Label the axes
`view`	Specify the view for a 3–D plot
`zoom`	Zoom in and out of the plot

the MATLAB documentation and help files for examples. For example, there is a section called Graphics in the MATLAB documentation center. Recall that you get to the documentation by clicking the **HELP** button in the **RESOURCES** section of the **HOME** ribbon and selecting the MATLAB link.

We now provide some references to books that describe scientific and statistical visualization, in general. One of the earliest ones in this area is called the *Semiology of Graphics: Diagrams, Networks, Maps* [Bertin, 1983]. This book discusses rules and properties of graphics. For examples of graphical mistakes, we recommend the book by Wainer [1997]. Wainer also published a book called *Graphic Discovery: A Trout in the Milk and Other Visual Adventures* [2004] detailing some of the history of graphical displays in a very thought provoking and entertaining way. The book *Visualizing Data* [Cleveland, 1993] includes descriptions of visualization tools, the relationship of visualization to classical statistical methods, and some of the cognitive aspects of data visualization and perception. Another excellent resource on graphics for data analysis is Chambers et al. [1983]. Finally, we highly recommend Naomi Robbins' [2013] book called *Creating More Effective Graphs*. This text provides a wonderful introduction on ways to convey data correctly and effectively.

There is a *Graphics* section in the online documentation for base MATLAB. Recall that you can access this documentation via the **HELP** button on the **RESOURCES** tab. The Statistics Toolbox documentation has a chapter called *Exploratory Data Analysis*, under which is a section on *Statistical Visualization*. This has details about univariate and multivariate plots.

References

Agresti, A. 2007. *An Introduction to Categorical Data Analysis, Second Edition*, New York: John Wiley & Sons.

Agresti, A. 2012. *Categorical Data Analysis, Third Edition*, New York: John Wiley & Sons.

Ahalt, A., A. K. Krishnamurthy, P. Chen, and D. E. Melton. 1990. "Competitive learning algorithms for vector quantization," *Neural Networks*, **3**:277–290.

Alter, O., P. O. Brown, and D. Botstein. 2000. "Singular value decomposition for genome–wide expression data processing and modeling," *Proceedings of the National Academy of Science*, **97**:10101–10106.

Amsaleg, L., O. Chelly, T. Furon, S. Girard, M. E. Houle, K. Kawarabayashi, M. Nett. 2015. "Estimating local intrinsic dimensionality," *Proceedings KDD 2015*, Sydney, Australia, p. 29–38.

Anderberg, M. R. 1973. *Cluster Analysis for Applications*, New York: Academic Press.

Anderson, E. 1935. "The irises of the Gaspe Peninsula," *Bulletin of the American Iris Society*, **59**:2–5.

Andrews, D. F. 1972. "Plots of high–dimensional data," *Biometrics*, **28**:125–136.

Andrews, D. F. 1974. "A robust method of multiple linear regression," *Technometrics*, **16**:523–531.

Andrews, D. F. and A. M. Herzberg. 1985. *Data: A Collection of Problems from Many Fields for the Student and Research Worker*, New York: Springer–Verlag.

Anscombe, F. J. 1973. "Graphs in statistical analysis," *The American Statistician*, **27**: 17–21.

Arbelaitz, O., I. Gurrutxaga, J. Muguerza, J. M. Perez, and I. Perona, 2013. "An extensive comparative study of cluster validity indices," *Pattern Recognition*, **46**:243–256.

Asimov, D. 1985. "The grand tour: A tool for viewing multidimensional data," *SIAM Journal of Scientific and Statistical Computing*, **6**:128–143.

Asimov, D. and A. Buja. 1994. "The grand tour via geodesic interpolation of 2–frames," in *Visual Data Exploration and Analysis, Symposium on Electronic Imaging Science and Technology, IS&T/SPIE*.

Baeza–Yates, R. and B. Ribero–Neto. 1999. *Modern Information Retrieval*, New York, NY: ACM Press.

Bailey, T. A. and R. Dubes. 1982. "Cluster validity profiles," *Pattern Recognition*, **15**:61–83.

Balasubramanian, M. and E. L. Schwartz. 2002. "The isomap algorithm and topological stability (with rejoinder)," *Science*, **295**:7.

Banfield, A. D. and A. E. Raftery. 1993. "Model–based Gaussian and non–Gaussian clustering," *Biometrics*, **49**:803–821.

Basseville, M. 1989. "Distance measures for signal processing and pattern recognition," *Signal Processing*, **18**:349–369.

Becker, R. A. and W. S. Cleveland. 1987. "Brushing scatterplots," *Technometrics*, **29**:127–142.

Becker, R. A. and W. S. Cleveland. 1991. "Viewing multivariate scattered data," *Pixel*, July/August, 36–41.

Becker, R. A., W. S. Cleveland, and A. R. Wilks. 1987. "Dynamic graphics for data analysis," *Statistical Science*, **2**:355–395.

Becker, R. A., L. Denby, R. McGill, and A. Wilks. 1986. "Datacryptanalysis: A case study," *Proceedings of the Section on Statistical Graphics*, 92–91.

Becketti, S. and W. Gould. 1987. "Rangefinder box plots," *The American Statistician*, **41**:149.

Bellman, R. E. 1961. *Adaptive Control Processes*, Princeton, NJ: Princeton University Press.

Bengio, Y. 2009. "Learning deep architectures for AI," *Foundations and Trends in Machine Learning*, **2**:1–127.

Benjamini, Y. 1988. "Opening the box of a boxplot," *The American Statistician*, **42**: 257–262.

Bennett, G. W. 1988. "Determination of anaerobic threshold," *Canadian Journal of Statistics*, **16**:307–310.

Bensmail, H., G. Celeux, A. E. Raftery, and C. P. Robert. 1997. "Inference in model–based cluster analysis," *Statistics and Computing*, **7**:1–10.

Berry, M. W. (editor) 2003. *Survey of Text Mining 1: Clustering, Classification, and Retrieval*, New York: Springer.

Berry, M. W. and M. Browne. 2005. *Understanding Search Engines: Mathematical Modeling and Text Retrieval*, 2nd Edition, Philadelphia, PA: SIAM.

Berry, M. W. and M. Castellanos (editors). 2007. *Survey of Text Mining 2: Clustering, Classification, and Retrieval*, New York: Springer.

Berry, M. W., S. T. Dumais, and G. W. O'Brien. 1995. "Using linear algebra for intelligent information retrieval," *SIAM Review*, **37**:573–595.

Berry, M. W., Z. Drmac, and E. R. Jessup. 1999. "Matrices, vector spaces, and information retrieval," *SIAM Review*, **41**:335–362.

Berry, M. W., M. Browne, A. N. Langville, V. P. Pauca, and R. J. Plemmons. 2007. "Algorithms and applications for approximate nonnegative matrix factorization," *Computational Statistics & Data Analysis*, **52**:155–173.

Bertin, J. 1983. *Semiology of Graphics: Diagrams, Networks, Maps.* Madison, WI: The University of Wisconsin Press.

Bhattacharjee, A., W. G. Richards, J. Staunton, C. Li, S. Monti, P. Vasa, C. Ladd, J. Beheshti, R. Bueno, M. Gillette, M. Loda, G. Weber, E. J. Mark, E. S. Lander, W. Wong, B. E. Johnson, T. R. Bolub, D. J. Sugarbaker, and M. Meyerson. 2001. "Classification of human lung carcinomas by mRNA expression profiling reveals distinct adenocarcinoma subclasses," *Proceedings of the National Academy of Science*, **98**:13790–13795.

Bickel, P. J., E. A. Hammel, and J. W. O'Connell. 1975. "Sex bias in graduate admissions: Data from Berkeley," *Science*, **187**:398–403.

Biernacki, C. and G. Govaert. 1997. "Using the classification likelihood to choose the number of clusters," *Computing Science and Statistics*, **29**(2):451–457.

Biernacki, C., G. Celeux, and G. Govaert. 1999. "An improvement of the NEC criterion for assessing the number of clusters in a mixture model," *Pattern Recognition Letters*, **20**:267–272.

Binder, D. A. 1978. "Bayesian cluster analysis," *Biometrika*, **65**:31–38.

Bingham, E. and H. Mannila. 2001. "Random projection in dimensionality reduction: Applications to image and text data," in *Proceedings of the Seventh ACM SIGKDD International Conference on Knowledge Discovery and Data Mining*, pp. 245–250.

Bishop, C. M., G. E. Hinton, and I. G. D. Strachan. 1997. "GTM through time," *Proceedings IEEE 5th International Conference on Artificial Neural Networks*, Cambridge, UK, 111–116.

Bishop, C. M., M. Svensén, and C. K. I. Williams. 1996. "GTM: The generative topographic mapping," Neural Computing Research Group, Technical Report NCRG/96/030.

Bishop, C. M., M. Svensén, and C. K. I. Williams. 1997a. "Magnification factors for the SOM and GTM algorithms," *Proceedings 1997 Workshop on Self-Organizing Maps*, Helsinki University of Technology, 333–338.

Bishop, C. M., M. Svensén, and C. K. I. Williams. 1997b. "Magnification factors for the GTM algorithm," *Proceedings IEEE 5th International Conference on Artificial Neural Networks*, Cambridge, UK, 64–69.

Bishop, C. M., M. Svensén, and C. K. I. Williams. 1998a. "The generative topographic mapping," *Neural Computation*, **10**:215–234.

Bishop, C. M., M. Svensén, and C. K. I. Williams. 1998b. "Developments of the generative topographic mapping," *Neurocomputing*, **21**:203–224.

Bishop, C. M. and M. E. Tipping. 1998. "A hierarchical latent variable model for data visualization," *IEEE Transactions on Pattern Analysis and Machine Intelligence*, **20**:281–293.

Blasius, J. and M. Greenacre. 1998. *Visualizing of Categorical Data*, New York: Academic Press.

Bock, H. 1985. "On some significance tests in cluster analysis," *Journal of Classification*, **2**:77–108.

Bock, H. 1996. "Probabilistic models in cluster analysis," *Computational Statistics and Data Analysis*, **23**:5–28.

Bolton, R. J. and W. J. Krzanowski. 1999. "A characterization of principal components for projection pursuit," *The American Statistician*, **53**:108–109.

Bonner, R. 1964. "On some clustering techniques," *IBM Journal of Research and Development*, **8**:22–32.

Borg, I. and P. Groenen. 1997. *Modern Multidimensional Scaling: Theory and Applications*, New York: Springer.

Boutsidis, C., A. Zouzias, and P. Drineas. 2010. "Random projections for *k*–means clustering," in *Advances in Neural Network Processing Systems*, pp. 293–306.

Bowman, A. W. and A. Azzalini. 1997. *Applied Smoothing Techniques for Data Analysis: The Kernel Approach with S–Plus Illustrations*, Oxford: Oxford University Press.

Brinkman, N. D. 1981. "Ethanol fuel – a single–cylinder engine study of efficiency and exhaust emissions," *SAE Transactions*, **90**:1410–1424.

Bruls, D. M., C. Huizing, J. J. van Wijk. 2000. "Squarified Treemaps," in: W. de Leeuw, R. van Liere (eds.), *Data Visualization 2000, Proceedings of the Joint Eurographics and IEEE TCVG Symposium on Visualization*, 33–42, New York: Springer.

Bruntz, S. M., W. S. Cleveland, B. Kleiner, and J. L. Warner. 1974. "The dependence of ambient ozone on solar radiation, wind, temperature and mixing weight," *Symposium on Atmospheric Diffusion and Air Pollution*, Boston: American Meteorological Society, 125–128.

Bruske, J. and G. Sommer. 1998. "Intrinsic dimensionality estimation with optimally topology preserving maps," *IEEE Transactions on Pattern Analysis and Machine Intelligence*, **20**:572–575.

Bucak, S. S. and B. Gunsel. 2009. "Incremental subspace learning via non–negative matrix factorization," *Pattern Recognition*, **42**:788–797.

Buckley, M. J. 1994. "Fast computation of a discretized thin–plate smoothing spline for image data," *Biometrika*, **81**:247–258.

Buijsman, E., H. F. M. Maas, and W. A. H. Asman. 1987. "Anthropogenic NH_3 Emissions in Europe," *Atmospheric Environment*, **21**:1009–1022.

Buja, A. and D. Asimov. 1986. "Grand tour methods: An outline," *Computer Science and Statistics*, **17**:63–67.

Buja, A., J. A. McDonald, J. Michalak, and W. Stuetzle. 1991. "Interactive data visualization using focusing and linking," *IEEE Visualization, Proceedings of the 2nd Conference on Visualization '91*, 156–163.

Buta, R. 1987. "The structure and dynamics of ringed galaxies, III: Surface photometry and kinematics of the ringed nonbarred spiral NGC 7531," *The Astrophysical Journal Supplement Series*, **64**:1–37.

Calinski, R. and J. Harabasz. 1974. "A dendrite method for cluster analysis," *Communications in Statistics*, **3**:1–27.

Camastra, F. 2003. "Data dimensionality estimation methods: A survey," *Pattern Recognition*, **36**:2945–2954.

Camastra, F. and A. Vinciarelli. 2002. "Estimating the intrinsic dimension of data with a fractal–based approach," *IEEE Transactions on Pattern Analysis and Machine Intelligence*, **24**:1404–1407.

Campbell, J. G., C. Fraley, F. Murtagh, and A. E. Raftery. 1997. "Linear flaw detection in woven textiles using model–based clustering," *Pattern Recognition Letters*, **18**:1539–1549.

Campbell, J. G., C. Fraley, D. Stanford, F. Murtagh, and A. E. Raftery. 1999. "Model–based methods for real–time textile fault detection," *International Journal of Imaging Systems and Technology*, **10**:339–346.

Carr, D., R. Littlefield, W. Nicholson, and J. Littlefield. 1987. "Scatterplot matrix techniques for large N," *Journal of the American Statistical Association*, **82**:424–436.

Carr, D. and L. W. Pickle. 2010. *Visualizing Data Patterns with Micromaps*, Boca Raton: CRC Press.

Carter, K. M., R. Raich, and A. P. Hero III. 2010. "On local intrinsic dimension estimation and its applications," *IEEE Transactions on Signal Processing*, **58**:650–663.

Cattell, R. B. 1966. "The scree test for the number of factors," *Journal of Multivariate Behavioral Research*, 1:245–276.

Cattell, R. B. 1978. *The Scientific Use of Factor Analysis in Behavioral and Life Sciences*, New York: Plenum Press.

Celeux, G. and G. Govaert. 1995. "Gaussian parsimonious clustering models," *Pattern Recognition*, 28:781–793.

Chakrapani, T. K. and A. S. C. Ehrenberg. 1981. "An alternative to factor analysis in marketing research – Part 2: Between group analysis," *Professional Marketing Research Society Journal*, 1:32–38.

Chambers, J. 1999. "Computing with data: Concepts and challenges," *The American Statistician*, 53:73–84.

Chambers, J. M., W. S. Cleveland, B. Kleiner, and P. A. Tukey. 1983. *Graphical Methods for Data Analysis*, Boca Raton: CRC/Chapman and Hall.

Charniak, E. 1996. *Statistical Language Learning*, Cambridge, MA: The MIT Press.

Chatfield, C. 1985. "The initial examination of data," *Journal of the Royal Statistical Society, A*, 148:214–253.

Chee, M., R. Yang, E. Hubbell, A. Berno, X. C. Huang, D. Stern, J. Winkler, D. J. Lockhart, M. S. Morris, and S. P. A. Fodor. 1996. "Accessing genetic information with high–density DNA arrays," *Science*, 274:610–614.

Cheng, S. 2014. "mosic_plot.zip," `https://www.mathworks.com/matlabcentral/fileexchange/47785-mosaic-plot-zip`.

Chernoff, H. 1973. "The use of faces to represent points in k–dimensional space graphically," *Journal of the American Statistical Association*, 68:361–368.

Cho, R. J., M. J. Campbell, E. A. Winzeler, L. Steinmetz, A. Conway, L. Wodicka, T. G. Wolfsberg, A. E. Gabrielian, D. Landsman, D. J. Lockhart, and R. W. Davis. 1998. "A genome–wide transcriptional analysis of the mitotic cell cycle," *Molecular Cell*, 2:65–73.

Cichocki, A. and R. Zdunek. 2006. "Multilayer nonnegative matrix factorization," *Electronics Letters*, 42:947–948.

Cichocki, A., R. Zdunek, and S. Amari. 2006. "New algorithms for non–negative matrix fatorization in applications to blind source separation," in *IEEE International Conference on Acoustics, Speech and Signal Processing*, Toulouse, France, pp. 621–624.

Cleveland, W. S. 1979. "Robust locally weighted regression and smoothing scatterplots," *Journal of the American Statistical Association*, 74:829–836.

Cleveland, W. S. 1984. "Graphical methods for data presentation: Full scale breaks, dot charts, and multibased logging," *The American Statistician*, 38:270–280.

Cleveland, W. S. 1993. *Visualizing Data*, New York: Hobart Press.

Cleveland, W. S. and S. J. Devlin. 1988. "Locally weighted regression: An approach to regression analysis by local fitting," *Journal of the American Statistical Association*, 83:596–610.

Cleveland, W. S., S. J. Devlin, and E. Grosse. 1988. "Regression by local fitting: Methods, properties, and computational algorithms," *Journal of Econometrics*, 37:87–114.

Cleveland, W. S. and C. Loader. 1996. "Smoothing by local regression: Principles and methods, in Härdle and Schimek (eds.), *Statistical Theory and Computational Aspects of Smoothing*, Heidelberg: Phsyica–Verlag, 10–49.

Cleveland, W. S. and R. McGill. 1984. "The many faces of a scatterplot," *Journal of the American Statistical Association*, **79**:807–822.

Cohen, A., R. Gnanadesikan, J. R. Kettenring, and J. M. Landwehr. 1977. "Methodological developments in some applications of clustering," in: *Applications of Statistics*, P. R. Krishnaiah (ed.), Amsterdam: North–Holland.

Cook, D., A. Buja, and J. Cabrera. 1993. "Projection pursuit indexes based on orthonormal function expansions," *Journal of Computational and Graphical Statistics*, **2**:225–250.

Cook, D., A. Buja, J. Cabrera, and C. Hurley. 1995. "Grand tour and projection pursuit," *Journal of Computational and Graphical Statistics*, **4**:155–172.

Cook, D. and D. F. Swayne. 2007. *Interactive and Dynamic Graphics for Data Analysis: With R and GGobi (Use R)*, New York: Springer–Verlag.

Cook, W. J., W. H. Cunningham, W. R. Pulleyblank, and A. Schrijver. 1998. *Combinatorial Optimization*, New York: John Wiley & Sons.

Cormack, R. M. 1971. "A review of classification," *Journal of the Royal Statistical Society, Series A*, **134**:321–367.

Costa, J. A., A. Girotra, and A. O. Hero. 2005. "Estimating local intrinsic dimension with k–nearest neighbor graphs," *IEEE Workshop on Statistical Signal Processing (SSP)*.

Costa, J. A. and A. O. Hero. 2004. "Geodesic entropic graphs for dimension and entropy estimation in manifold learning," *IEEE Transactions on Signal Processing*, **52**:2210–2221.

Cottrell, M., J. C. Fort, and G. Pages. 1998. "Theoretical aspects of the SOM algorithm," *Neurocomputing*, **21**:119–138.

Cox, T. F. and M. A. A. Cox. 2001. *Multidimensional Scaling, 2nd Edition*, Boca Raton: Chapman & Hall/CRC.

Craven, P. and G. Wahba. 1979. "Smoothing noisy data with spline functions," *Numerische Mathematik*, **31**:377–403.

Crawford, S. 1991. "Genetic optimization for exploratory projection pursuit," *Proceedings of the 23rd Symposium on the Interface*, **23**:318–321.

Crile, G. and D. P. Quiring. 1940. "A record of the body weight and certain organ and gland weights of 3690 animals," *Ohio Journal of Science*, **15**:219–259.

Dasgupta, A. and A. E. Raftery. 1998. "Detecting features in spatial point processes with clutter via model–based clustering," *Journal of the American Statistical Association*, **93**:294–302.

Davies, O. L. 1957. *Statistical Methods in Research and Production*, New York: Hafner Press.

Davies, D. L. and D. W. Bouldin. 1979. "A cluster separation measure," *IEEE Transactions on Pattern Analysis and Machine Intelligence*, **1**:224–227.

Day, N. E. 1969. "Estimating the components of a mixture of normal distributions," *Biometrika*, **56**:463–474.

de Boor, C. 2001. *A Practical Guide to Splines, Revised Edition*, New York: Springer–Verlag.

de Leeuw, J. 1977. "Applications of convex analysis to multidimensional scaling," in *Recent Developments in Statistics*, J. R. Barra, R. Brodeau, G. Romier & B. van Cutsem (ed.), Amsterdam, The Netherlands: North–Holland, 133–145.

Deboeck, G. and T. Kohonen. 1998. *Visual Explorations in Finance using Self–Organizing Maps*, London: Springer–Verlag.

Deerwester, S., S. T. Dumais, G. W. Furnas, T. K. Landauer, and R. Harshman. 1990. "Indexing by Latent Semantic Analysis," *Journal of the American Society for Information Science*, **41**:391–407.

Demartines, P. and J. Herault. 1997. "Curvilinear component analysis: A self–organizing neural network for nonlinear mapping of data sets," *IEEE Transactions on Neural Networks*, **8**:148–154.

Dempster, A. P., N. M. Laird, and D. B. Rubin. 1977. "Maximum likelihood from incomplete data via the EM algorithm (with discussion)," *Journal of the Royal Statistical Society: B*, **39**:1–38.

Diaconis, P. 1985. "Theories of data analysis: From magical thinking through classical statistics," in *Exploring Data Tables, Trends, and Shapes*, D. Hoaglin, F. Mosteller, and J. W. Tukey (eds.), New York: John Wiley and Sons.

Diaconis, P. and J. H. Friedman. 1980. "M and N plots," Technical Report 15, Department of Statistics, Stanford University.

Ding, W., M. H. Rohban, P. Ishwar, and V. Saligrama. 2013. "Topic discovery through data dependent and random projections," in *International Conference on Machine Learning*, **28**:471–479.

Donoho, D. L. and M. Gasko. 1992. "Breakdown properties of location estimates based on halfspace depth and projected outlyingness," *The Annals of Statistics*, **20**:1803–1827.

Donoho, D. L. and C. Grimes. 2002. Technical Report 2002–27, Department of Statistics, Stanford University.

Donoho, D. L. and C. Grimes. 2003. "Hessian eigenmaps: Locally linear embedding techniques for high–dimensional data," *Proceedings of the National Academy of Science*, **100**:5591–5596.

Draper, N. R. and H. Smith. 1981. *Applied Regression Analysis, 2nd Edition*, New York: John Wiley & Sons.

du Toit, S. H. C., A. G. W. Steyn, and R. H. Stumpf. 1986. *Graphical Exploratory Data Analysis*, New York: Springer–Verlag.

Dubes, R. and A. K. Jain. 1980. "Clustering methodologies in exploratory data analysis," *Advances in Computers, Vol. 19*, New York: Academic Press.

Duda, R. O. and P. E. Hart. 1973. *Pattern Classification and Scene Analysis*, New York: John Wiley & Sons.

Duda, R. O., P. E. Hart, and D. G. Stork. 2001. *Pattern Classification, Second Edition*, New York: John Wiley & Sons.

Dunn, J. C. 1973. "A fuzzy relative of the ISODATA process and its use in detecting compact well–separated clusters," *Journal of Cybernetics*, **3**:32–57.

Edwards, A. W. F. and L. L. Cavalli–Sforza. 1965. "A method for cluster analysis," *Biometrics*, **21**:362–375.

Efromovich, S. 1999. *Nonparametric Curve Estimation: Methods, Theory, and Applications*, New York: Springer–Verlag.

Efron, B. and R. J. Tibshirani. 1993. *An Introduction to the Bootstrap*, London: Chapman and Hall.

Embrechts, P. and A. Herzberg. 1991. "Variations of Andrews' plots," *International Statistical Review*, **59**:175–194.

Emerson, J. D. and M. A. Stoto. 1983. "Transforming Data," in *Understanding Robust and Exploratory Data Analysis*, Hoaglin, Mosteller, and Tukey (eds.), New York: John Wiley & Sons.

Erickson, Jeff. 2015. *Algorithms, etc.* `http://jeffe.cs.illinois.edu/teaching/algorithms/`

Estivill–Castro, V. 2002. "Why so many clustering algorithms – A position paper," *SIGKDD Explorations*, **4**:65–75.

Esty, W. W. and J. D. Banfield. 2003. "The box–percentile plot," *Journal of Statistical Software*, **8**, `http://www.jstatsoft.org/v08/i17`.

Everitt, B. S. 1993. *Cluster Analysis, Third Edition*, New York: Edward Arnold Publishing.

Everitt, B. S. and D. J. Hand. 1981. *Finite Mixture Distributions*, London: Chapman and Hall.

Everitt, B. S., S. Landau, and M. Leese. 2001. *Cluster Analysis, Fourth Edition*, New York: Edward Arnold Publishing.

Fan, M., H. Qiao, and B. Zhang. 2008. "Intrinsic dimension estimation of manifolds by incising balls," *Pattern Recognition*, **42**:780–787.

Fawcett, C. D. 1901. "A second study of the variation and correlation of the human skull, with special reference to the Naqada crania," *Biometrika*, **1**:408–467.

Fieller, N. R. J., E. C. Flenley, and W. Olbricht. 1992. "Statistics of particle size data," *Applied Statistics*, **41**:127–146.

Fieller, N. R. J., D. D. Gilbertson, and W. Olbricht. 1984. "A new method for environmental analysis of particle size distribution data from shoreline sediments," *Nature*, **311**:648–651.

Fienberg, S. 1979. "Graphical methods in statistics," *The American Statistician*, **33**:165–178.

Fisher, R. A. 1936. "The use of multiple measurements in taxonomic problems," *Annals of Eugenics*, **7**:179–188.

Flenley, E. C., and Olbricht, W. 1993. "Classification of Archaeological Sands by Particle Size Analysis," *Proceedings of the 16th Annual Conference of the Gesellschaft für Klassifikation e. V.*, Springer, Berlin, Heidelberg, pp. 478–489.

Flick, T., L. Jones, R. Priest, and C. Herman. 1990. "Pattern classification using projection pursuit," *Pattern Recognition*, **23**:1367–1376.

Flury, B. and H. Riedwyl. 1988. *Multivariate Statistics: A Practical Approach*, London: Chapman and Hall.

Fodor, I. K. 2002. "A survey of dimension reduction techniques," Lawrence Livermore National Laboratory Technical Report, UCRL–ID–148494.

Fogel, P., S. S. Young, D. M. Hawkins, and N. Ledirac. 2007. "Inferential robust non–negative matrix factorization analysis of microarray data," *Bioinformatics*, **23**:44–49.

Fowlkes, E. B. and C. L. Mallows. 1983. "A method for comparing two hierarchical clusterings," *Journal of the American Statistical Association*, **78**:553–584.

Fox, J. 2000. *Nonparametric Simple Regression*, Thousand Oaks, CA: Sage Publications, Inc.

Frakes, W. B. and R. Baeza–Yates. 1992. *Information Retrieval: Data Structures & Algorithms*, Prentice Hall, New Jersey.

Fraley, C. 1998. "Algorithms for model–based Gaussian hierarchical clustering," *SIAM Journal on Scientific Computing*, **20**:270–281.

Fraley, C. and A. E. Raftery. 1998. "How many clusters? Which clustering method? Answers via model–based cluster analysis," *The Computer Journal*, **41**:578–588.

Fraley, C. and A. E. Raftery. 2002. "Model–based clustering, discriminant analysis, and density estimation: MCLUST," *Journal of the American Statistical Association*, **97**:611–631.

Fraley, C. and A. E. Raftery. 2003. "Enhanced software for model–based clustering, discriminant analysis, and density estimation: MCLUST," *Journal of Classification*, **20**:263–286.

Fraley, C., A. E. Raftery, and R. Wehrens. 2005. "Incremental model–based clustering for large datasets with small clusters," *Journal of Computational and Graphical Statistics*, **14**:529–546.

Freedman, D. and P. Diaconis. 1981. "On the histogram as a density estimator: L_2 theory," *Zeitschrift fur Wahrscheinlichkeitstheorie und verwandte Gebiete*, **57**:453–476.

Fridlyand, J. and S. Dudoit. 2001. "Applications of resampling methods to estimate the number of clusters and to improve the accuracy of a clustering method," Technical Report #600, Division of Biostatistics, University of California, Berkeley.

Friedman, J. 1987. "Exploratory projection pursuit," *Journal of the American Statistical Association*, **82**:249–266.

Friedman, J. and W. Stuetzle. 1981. "Projection pursuit regression," *Journal of the American Statistical Association*, **76**:817–823.

Friedman, J. and J. Tukey. 1974. "A projection pursuit algorithm for exploratory data analysis," *IEEE Transactions on Computers*, **23**:881–889.

Friedman, J., W. Stuetzle, and A. Schroeder. 1984. "Projection pursuit density estimation," *Journal of the American Statistical Association*, **79**:599–608.

Friendly, M. 1994. "Mosaic displays for multi–way contingency tables," *Journal of the American Statistical Association*, **89**:190–200.

Friendly, M. 1999. "Extending mosaic displays: Marginal conditional, and partial views of categorical data," *Journal of Computational and Graphical Statistics*, **8**:373–395.

Friendly, M. 2000. *Visualizing Categorical Data*, Cary, NC: SAS Institute, Inc.

Friendly, M. and E. Kwan. 2003. "Effect ordering for data displays," *Computational Statistics and Data Analysis*, **43**:509–539.

Friendly, M. and D. Meyer. 2015. *Discrete Data Analysis with R: Visualizing and Modeling Techniques for Categorical and Count Data*, Boca Raton: CRC Press.

Friendly, M. and H. Wainer. 2004. "Nobody's perfect," *Chance*, **17**:48–51.

Frigge, M., D. C. Hoaglin, and B. Iglewicz. 1989. "Some implementations of the boxplot," *The American Statistician*, **43**:50–54.

Fua, Y. H., M. O. Ward, and E. A. Rundensteiner. 1999. "Hierarchical parallel coordinates for exploration of large datasets," *IEEE Visualization, Proceedings of the Conference on Visualization '99*, 43 – 50.

Fukunaga, K. 1990. *Introduction to Statistical Pattern Recognition, Second Edition*, New York: Academic Press.

Fukunaga, K. and D. R. Olsen. 1971. "An algorithm for finding intrinsic dimensionality of data," *IEEE Transactions on Computers*, **20**:176–183.

Gabriel, K. R. 1971. "The biplot graphic display of matrices with application to principal component analysis," *Biometrika*, **58**:453–467.

Garcia, D. 2009. "MATLAB functions in BiomeCardio," `http://www.biomecardio.com/matlab`.

Garcia, D. 2010. "Robust smoothing of gridded data in one and higher dimensions with missing values," *Computational Statistics and Data Analysis*, **54**:1167–1178.

Gentle, J. E. 2002. *Elements of Computational Statistics*, New York: Springer–Verlag.

Golub, G. and C. Van Loan. 1996. *Matrix Computations*, Baltimore: Johns Hopkins University Press.

Golub, T. R., D. K. Slonim, P. Tamayo, C. Huard, M. Gaasenbeek, J. P. Mesirov, H. Coller, M. L. Loh, J. R. Downing, M. A. Caligiuri, C. D. Bloomfield, and E. S. Lander. 1999. "Molecular classification of cancer: Class discovery and class prediction by gene expression monitoring," *Science*, **286**:531–537.

Good, I. J. 1983. "The philosophy of exploratory data analysis, *Philosophy of Science*, 50:283–295.

Goodfellow, I., Y. Bengio, and A. Courville. 2016. *Deep Learning*, MIT Press.

Gorban, A. N., B. Kegl, D. C. Wunsch, and A. Zinovyev. 2008. *Principal Manifolds for Data Visualization and Dimension Reduction*, New York: Springer.

Gordon, A. D. 1999. *Classification*, London: Chapman and Hall.

Gower, J. C. 1966. "Some distance properties of latent root and vector methods in multivariate analysis," *Biometrika*, **53**:325–338.

Gower, J. C. and D. J. Hand. 1996. *Biplots*, London: Chapman and Hall.

Gower, J. C. and P. Legendre. 1986. "Metric and Euclidean properties of dissimilarity coefficients," *Journal of Classification*, **3**:5–48.

Gower, J. C. and G. J. S. Ross. 1969. "Minimum Spanning Trees and Single Linkage Cluster Analysis," *Journal of the Royal Statistical Society. Series C (Applied Statistics)*, **18**:54–64.

Grassberger, P. and I. Procaccia. 1983. "Measuring the strangeness of strange attractors," *Physica*, **D9**:189–208.

Green P. J. and B. W. Silverman. 1994. *Nonparametric Regression and Generalized Linear Models: A Roughness Penalty Approach*, London: Chapman and Hall.

Griffiths, A. J. F., J. H. Miller, D. T. Suzuki, R. X. Lewontin, and W. M. Gelbart. 2000. *An Introduction to Genetic Analysis*, 7th ed., New York: Freeman.

Groenen, P. 1993. *The Majorization Approach to Multidimensional Scaling: Some Problems and Extensions*, Leiden, The Netherlands: DSWO Press.

Guillamet, D. and J. Vitria. 2002. "Nonnegative matrix factorization for face recognition," in *Fifth Catalonian Conference on Artificial Intelligence*, pp. 336–344.

Guttman, L. 1968. "A general nonmetric technique for finding the smallest coordinate space for a configuration of points," *Psychometrika*, **33**:469–506.

Halkidi, M., Y. Batistakis, and M. Vazirgiannis, 2001. "On clustering validation techniques," *Journal of Intelligent Information Systems*, **17**:107–145.

Hand, D., F. Daly, A. D. Lunn, K. J. McConway, and E. Ostrowski. 1994. *A Handbook of Small Data Sets*, London: Chapman and Hall.

Hand, D., H. Mannila, and P. Smyth. 2001. *Principles of Data Mining*, Cambridge, MA: The MIT Press.

Hanselman, D. and B. Littlefield. 2004. *Mastering MATLAB 7*, New Jersey: Prentice Hall.

Hansen, P., B. Jaumard, and B. Simeone. 1996. "Espaliers: A generalization of dendrograms," *Journal of Classification*, **13**:107–127.

Härdle, W. 1991. *Smoothing Techniques with Implementation in S*, New York, NY: Springer–Verlag.

Hartigan, J. A. 1967. "Representation of similarity measures by trees," *Journal of the American Statistical Association*, **62**:1140–1158.

Hartigan, J. A. 1975. *Clustering Algorithms*, New York: John Wiley & Sons.

Hartigan, J. A. 1985. "Statistical theory in clustering," *Journal of Classification*, **2**:63–76.

Hartigan, J. A. and B. Kleiner. 1981. "Mosaics for contingency tables," in W. F. Eddy (editor), *Computer Science and Statistics: Proceedings of the 13th Symposium on the Interface*, pp. 268–273, New York: Springer–Verlag.

Hartwig, F. and B. E. Dearing. 1979. *Exploratory Data Analysis*, Newbury Park, CA: Sage University Press.

Hastie, T. J. and C. Loader. 1993. "Local regression: Automatic kernel carpentry (with discussion)," *Statistical Science*, **8**:120–143.

Hastie, T. J. and R. J. Tibshirani. 1986. "Generalized additive models," *Statistical Science*, **1**:297–318.

Hastie, T. J. and R. J.Tibshirani. 1990. *Generalized Additive Models*, London: Chapman and Hall.

Hastie, T. J., R. J. Tibshirani, and J. Friedman. 2009. *The Elements of Statistical Learning: Data Mining, Inference and Prediction, 2nd Edition*, New York: Springer.

Hastie, T., R. Tibshirani, M. B. Eisen, A. Alizadeh, R. Levy, L. Staudt, W. C. Chan, D. Botstein and P. Brown. 2000. "Gene shaving as a method for identifying distinct sets of genes with similar expression patterns," *Genome Biology*, **1**.

Hinton, G. E. and S. T. Roweis. 2002. "Stochastic neighbor embedding," in *Advances in Neural Information Processing Systems*, p. 833–840.

Hinton, G. E. and R. R. Salakhutdinov. 2006. "Reducing the dimensionality of data with neural networks," *Science*, **313**:504–507.

Hintze, J. L. and R. D. Nelson. 1998. "Violin plots: A box plot–density trace synergism," *The American Statistician*, **52**:181–184.

Hoaglin, D. C. 1980. "A poissonness plot," *The American Statistician*, **34**:146–149.

Hoaglin, D. C. 1982. "Exploratory data analysis," in *Encyclopedia of Statistical Sciences, Volume 2*, Kotz, S. and N. L. Johnson, eds., New York: John Wiley & Sons.

Hoaglin, D. C., B. Iglewicz, and J. W. Tukey. 1986. "Performance of some resistant rules for outlier labeling," *Journal of the American Statistical Association*, **81**:991–999.

Hoaglin, D. C. and J. Tukey. 1985. "Checking the shape of discrete distributions," in *Exploring Data Tables, Trends and Shapes*, D. Hoaglin, F. Mosteller, and J. W. Tukey, eds., New York: John Wiley & Sons.

Hoaglin, D. C., F. Mosteller, and J. W. Tukey (eds.). 1983. *Understanding Robust and Exploratory Data Analysis*, New York: John Wiley & Sons.

Hoaglin, D. C., F. Mosteller, and J. W. Tukey. 1985. *Exploring Data Tables, Trends, and Shapes*, New York: John Wiley & Sons.

Hoffmann H. 2015, "violin.m – Simple violin plot using matlab default kernel density estimation," `http://www.mathworks.com/matlabcentral/fileexchange/45134-violin-plot`.

Hofmann, T. 1999a. "Probabilistic latent semantic indexing," *Proceedings of the 22nd Annual ACM Conference on Research and Development in Information Retrieval*, Berkeley, CA, 50–57 (www.cs.brown.edu/~th/papers/Hofmann–SIGIR99.pdf).

Hofmann, T. 1999b. "Probabilistic latent semantic analysis," *Proceedings of the Fifteenth Conference on Uncertainty in Artificial Intelligence, UAI'99*, Stockholm, (www.cs.brown.edu/~th/papers/Hofmann–UAI99.pdf).

Hogg, R. 1974. "Adaptive robust procedures: A partial review and some suggestions for future applications and theory (with discussion)," *Journal of the American Statistical Association*, **69**:909–927.

Huber, P. J. 1973. "Robust regression: Asymptotics, conjectures, and Monte Carlo," *Annals of Statistics*, **1**:799–821.

Huber, P. J. 1981. *Robust Statistics*, New York: John Wiley & Sons.

Huber, P. J. 1985. "Projection pursuit (with discussion)," *Annals of Statistics*, **13**:435–525.

Hubert, L. J. and P. Arabie. 1985. "Comparing partitions," *Journal of Classification*, **2**:193–218.

Hummel, J. 1996. "Linked bar charts: Analyzing categorical data graphically," *Computational Statistics*, **11**:23–33.

Hurley, C. and A. Buja. 1990. "Analyzing high–dimensional data with motion graphics," *SIAM Journal of Scientific and Statistical Computing*, **11**:1193–1211.

Hyvärinen, A. 1999a. "Survey on independent component analysis," *Neural Computing Surveys*, 2:94–128.

Hyvärinen, A. 1999b. "Fast and robust fixed–point algorithms for independent component analysis," *IEEE Transactions on Neural Networks*, **10**:626–634.

Hyvärinen, A., J. Karhunen, and E. Oja. 2001. *Independent Component Analysis*, New York: John Wiley & Sons.

Hyvärinen, A. and E. Oja. 2000. "Independent component analysis: Algorithms and applications," *Neural Networks*, **13**:411–430.

Ilc, N. 2012. "Modified Dunn's cluster validity index based on graph theory," *Przeglad Elektrotechniczny (Electrical Review)*, **2**:126–131.

Ilc, N. 2013. "Modified & Generalized Dunn's Index," MATLAB Central, `https://www.mathworks.com/matlabcentral/fileex-change/42199-modified---generalized-dunn-s-index`

Inselberg, A. 1985. "The plane with parallel coordinates," *The Visual Computer*, 1:69–91.

Jackson, J. E. 1981. "Principal components and factor analysis: Part III – What is factor analysis?" *Journal of Quality Technology*, 13:125–130.

Jackson, J. E. 1991. *A User's Guide to Principal Components*, New York: John Wiley & Sons.

Jain, A. K. and R. C. Dubes. 1988. *Algorithms for Clustering Data*, New York: Prentice Hall.

Jain, A. K. and M. Law. 2005. "Data clustering: A user's dilemma," *Lecture Notes in Computer Science*, 3776:1–10.

Jain, A. K., M. N. Murty, and P. J. Flynn. 1999. "Data clustering: A review," *ACM Computing Surveys*, 31:264–323.

Jeffreys, H. 1935. "Some tests of significance, treated by the theory of probability," *Proceedings of the Cambridge Philosophy Society*, 31:203–222.

Jeffreys, H. 1961. *Theory of Probability, Third Edition*, Oxford, U. K.: Oxford University Press.

Johnson, B. and B. Shneiderman. 1991. "Treemaps: A space–filling approach to the visualization of hierarchical information structures," *Proceedings of the 2nd International IEEE Visualization Conference*, 284–291.

Johnson, W. B. and J. Lindenstrauss. 1984. "Extensions of Lipschitz mappings into a Hilbert space," *Contemporary Mathematics*, 26:189–206.

Jolliffe, I. T. 1972. "Discarding variables in a principal component analysis I: Artificial data," *Applied Statistics*, 21:160–173.

Jolliffe, I. T. 1986. *Principal Component Analysis*, New York: Springer–Verlag.

Jonas. 2009. "Violin plots for plotting multiple distributions," MATLAB Central, `http://www.mathworks.com/matlabcentral/fileex-change/23661-violin-plots-for-plotting-multiple-distribu-tions--distributionplot-m-`

Jonas. 2012. "Plot spread points (beeswarm plot)," MATLAB Central, `http://www.mathworks.com/matlabcentral/fileex-change/37105-plot-spread-points--beeswarm-plot-`

Jones, W. P. and G. W. Furnas. 1987. "Pictures of relevance: A geometric analysis of similarity measures," *Journal of the American Society for Information Science*, 38:420–442.

Jones, M. C. and R. Sibson. 1987. "What is projection pursuit" (with discussion), *Journal of the Royal Statistical Society, Series A*, 150:1–36.

Kaiser, H. F. 1960. "The application of electronic computers to factor analysis," *Educational and Psychological Measurement*, 20:141–151.

Kampstra, P. 2008. "Beanplot: A boxpot alternative for visual comparison of distributions," *Journal of Statistical Software*, Volume 28, DOI: 10.18637/jss.v028.c01.

Kangas, J. and S. Kaski. 1998. "3043 works that have been based on the self–organizing map (SOM) method developed by Kohonen," *Technical Report A50*, Helsinki University of Technology, Laboratory of Computer and Information Science.

Kaski, S. 1997. *Data Exploration Using Self–Organizing Maps*, Ph.D. dissertation, Helsinki University of Technology.

Kaski, S., T. Honkela, K. Lagus, and T. Kohonen. 1998. "WEBSOM – Self–organizing maps of document collections," *Neurocomputing*, **21**:101–117.

Kass, R. E. and A. E. Raftery. 1995. "Bayes factors," *Journal of the American Statistical Association*, **90**:773–795.

Kaufman, L. and P. J. Rousseeuw. 1990. *Finding Groups in Data: An Introduction to Cluster Analysis*, New York: John Wiley & Sons.

Kegl, B. 2003. "Intrinsic dimension estimation based on packing numbers," in *Advances in Neural Information Processing Systems (NIPS)*, 15:833–840, Cambridge, MA: The MIT Press.

Kiang, M. Y. 2001. "Extending the Kohonen self–organizing map networks for clustering analysis," *Computational Statistics and Data Analysis*, **38**:161–180.

Kimbrell, R. E. 1988. "Searching for text? Send an N–Gram!," *Byte*, May, 297 – 312.

Kimball, B. F. 1960. "On the choice of plotting positions on probability paper," *Journal of the American Statistical Association*, **55**:546–560.

Kirby, M. 2001. *Geometric Data Analysis: An Empirical Approach to Dimensionality Reduction and the Study of Patterns*, New York: John Wiley & Sons.

Kleiner, B. and J. A. Hartigan. 1981. "Representing points in many dimensions by trees and castles," *Journal of the American Statistical Association*, **76**:260–276

Kohonen, T. 1998. "The self–organizing map," *Neurocomputing*, **21**:1–6.

Kohonen, T. 2001. *Self–Organizing Maps, Third Edition*, Berlin: Springer.

Kohonen, T., S. Kaski, K. Lagus, J. Salojarvi, T. Honkela, V. Paatero, and A. Saarela. 2000. "Self organization of a massive document collection," *IEEE Transactions on Neural Networks*, **11**:574–585.

Kotz, S. and N. L. Johnson (eds.). 1986. *Encyclopedia of Statistical Sciences*, New York: John Wiley & Sons.

Kruskal, J. B. 1964a. "Multidimensional scaling by optimizing goodness of fit to a nonmetric hypothesis," *Psychometrika*, **29**:1–27.

Kruskal, J. B. 1964b. "Nonmetric multidimensional scaling: A numerical method," *Psychometrika*, **29**:115–129.

Kruskal, J. B. and M. Wish. 1978. *Multidimensional Scaling*, Newbury Park, CA: Sage Publications, Inc.

Krzanowski, W. J. and Y. T. Lai. 1988. "A criterion for determining the number of groups in a data set using sum–of–squares clustering," *Biometrics*, **44**:23–34.

Lai, M. 2015. "Giraffe: Using deep reinforcement learning to play chess," MSc Dissertation, College of London, **http://arxiv.org/abs/1509.01549**.

Lander, E. S. 1999. "Array of hope," *Nature Genetics Supplement*, **21**:3–4.

Lang, K. 1995. "Newsweeder: Learning to filter netnews," in *Proceedings of the 12th International Conference on Machine Learning*, pp. 331–339.

Langville, A., C. Meyer, and R. Albright. 2006. "Initializations for the nonnegative matrix factorization," *Proceedings of the Twelfth ACM SIGKDD International Conference on Knowledge Discovery and Data Mining*.

Launer, R. and G. Wilkinson (eds.). 1979. *Robustness in Statistics*, New York: Academic Press.

Lawley, D. N. and A. E. Maxwell. 1971. *Factor Analysis as a Statistical Method, 2nd Edition*, London: Butterworth.

Le, Q., T. Sarlos, and A. Smola. 2013. "Fastfood–approximating kernel expansions in loglinear time," in *Proceedings of the International Conference on Machine Learning*.

LeCun, Y., Y. Bengio, and G. Hinton. 2015. "Deep Learning," *Nature*, **521**:436–444.

Lee, D. and H. Seung. 2001. "Algorithms for non–negative matrix factorization," *Advances in Neural Information Processing Systems*, **13**:556–562.

Lee, J. A. and M. Verleysen. 2007. *Nonlinear Dimensionality Reduction*, New York, NY: Springer–Verlag.

Lee, J. A., A. Lendasse, N. Donckers, and M. Verleysen. 2000. "A robust nonlinear projection method," *Proceedings of ESANN'2002 European Symposium on Artificial Neural Networks*, Burges, Belgium, pp. 13–20.

Lee, J. A., A. Lendasse, and M. Verleysen. 2002. "Curvilinear distance analysis versus ISOMAP," *Proceedings of ESANN'2002 European Symposium on Artificial Neural Networks*, Burges, Belgium, pp. 185–192.

Lee, T. C. M. 2002. "On algorithms for ordinary least squares regression spline fitting: A comparative study," *Journal of Statistical Computation and Simulation*, **72**:647–663.

Levina, E. and P. Bickel. 2004. "Maximum likelihood estimation of intrinsic dimension," in *Advances in Neural Information Processing Systems*, Cambridge, MA: The MIT Press.

Lewis, N. D. 2016. *Deep Learning Made Easy with R: A Gentle Introduction for Data Science*, CreateSpace Independent Publishing Platform.

Li, C., J. Guo, and X. Nie. 2007. "Intrinsic dimensionality estimation with neighborhood convex hull," *Proceedings of the 2007 International Conference on Compuational Intelligence and Security*, p. 75–79.

Li, G. and Z. Chen. 1985. "Projection–pursuit approach to robust dispersion matrices and principal components: Primary theory and Monte Carlo," *Journal of the American Statistical Association*, **80**:759–766.

Lindsey, J. C., A. M. Herzberg, and D. G. Watts. 1987. "A method for cluster analysis based on projections and quantile–quantile plots," *Biometrics*, **43**:327–341.

Ling, R. F. 1973. "A computer generated aid for cluster analysis," *Communications of the ACM*, **16**:355–361.

Little, A. V., Y. Jung, and M. Maggioni. 2009. "Multiscale estimation of intrinsic dimensionality of data sets," *Proceedings Manifold Learning and its Applications: Papers from the AAAI Fall Symposium*, p. 26–33.

Loader, C. 1999. *Local Regression and Likelihood*, New York: Springer–Verlag.

MacKay, D. J. C. and Z. Ghahramani. 2005. "Comments on 'Maximum Likelihood Estimation of Intrinsic Dimension' by E. Levina and P. Bickel (2004)," **http://www.inference.phy.cam.ac.uk/mackay/dimension/**.

Manly, B. F. J. 1994. *Multivariate Statistical Methods – A Primer, Second Edition*, London: Chapman & Hall.

Manning, C. D. and H. Schütze. 2000. *Foundations of Statistical Natural Language Processing*, Cambridge, MA: The MIT Press.

Mao, J. and A. K. Jain. 1995. "Artificial neural networks for feature extraction and multivariate data projection," *IEEE Transaction on Neural Networks*, **6**:296–317.

Marchand, P. and O. T. Holland. 2002. *Graphics and GUIs with MATLAB, Third Edition*, Boca Raton: CRC Press.

Marchette, D. J. and J. L. Solka. 2003. "Using data images for outlier detection," *Computational Statistics and Data Analysis*, **43**:541–552.

Marsh, L. C. and D. R. Cormier. 2002. *Spline Regression Models*, Sage University Papers Series on Quantitative Applications in the Social Sciences, 07–137, Thousand Oaks, CA: Sage Publications, Inc.

Martinez, A. R. 2002. *A Framework for the Representation of Semantics*, Ph.D. Dissertation, Fairfax, VA: George Mason University.

Martinez, A. R. and E. J. Wegman. 2002a. "A text stream transformation for semantic–based clustering," *Computing Science and Statistics*, **34**:184–203.

Martinez, A. R. and E. J. Wegman. 2002b. "Encoding of text to preserve "meaning"," *Proceedings of the Eighth Annual U. S. Army Conference on Applied Statistics*, 27–39.

Martinez, W. L. and M. Cho. 2014. *Statistics in MATLAB: A Primer*, Boca Raton: CRC Press.

Martinez, W. L. and A. R. Martinez. 2015. *Computational Statistics Handbook with MATLAB*, Third Edition, Boca Raton: CRC Press.

McGill, R., J. Tukey, and W. Larsen. 1978. "Variations of box plots," *The American Statistician*, **32**:12–16.

McLachlan, G. J. and K. E. Basford. 1988. *Mixture Models: Inference and Applications to Clustering*, New York: Marcel Dekker.

McLachlan, G. J. and T. Krishnan. 1997. *The EM Algorithm and Extensions*, New York: John Wiley & Sons.

McLachlan, G. J. and D. Peel. 2000. *Finite Mixture Models*, New York: John Wiley & Sons.

McLachlan, G. J., D. Peel, K. E. Basford, and P. Adams. 1999. "The EMMIX software for the fitting of mixtures of normal and t–components," *Journal of Statistical Software*, **4**, `http://www.jstatsoft.org/index.php?vol=4`.

McLeod, A. I. and S. B. Provost. 2001. "Multivariate Data Visualization," `www.stats.uwo.ca/faculty/aim/mviz`.

Mead, A. 1992. "Review of the development of multidimensional scaling methods," *The Statistician*, **41**:27–39.

Meyer, D., A. Zeileis, and K. Hornik. 2008. "Visualizing contingency tables" in C. Chen, W. Härdle, and A. Unwin (editors), *Handbook of Data Visualization*, Springer Handbooks of Computational Statistics, pp. 589–616, New York: Springer–Verlag.

Milligan, G. W. and M. C. Cooper. 1985. "An examination of procedures for determining the number of clusters in a data set," *Psychometrika*, **50**:159–179.

Milligan, G. W. and M. C. Cooper. 1988. "A study of standardization of variables in cluster analysis," *Journal of Classification*, **5**:181–204.

Minh, V., K. Kavukcuoglu, D. Silver, A. Graves, I. Antonoglou, D. Wiersgtra, and M. Riedmiller. 2013. "Playing Atari with deep reinforcement learning," *NIPS Deep Learning Workshop*, `http://arxiv.org/pdf/1312.5602.pdf`

Minnotte, M. and R. West. 1998. "The data image: A tool for exploring high dimensional data sets," *Proceedings of the ASA Section on Statistical Graphics*, Dallas, Texas, 25–33.

Mojena, R. 1977. "Hierarchical grouping methods and stopping rules: An evaluation," *Computer Journal*, **20**:359–363.

Moler, C. 2004. *Numerical Computing with MATLAB*, New York: SIAM.

Montanari, A. and L. Lizzani. 2001. "A projection pursuit approach to variable selection," *Computational Statistics and Data Analysis*, **35**:463–473.

Morgan, B. J. T. and A. P. G. Ray. 1995. "Non–uniqueness and inversions in cluster analysis," *Applied Statistics*, **44**:114–134.

Mosteller, F. and J. W. Tukey. 1977. *Data Analysis and Regression: A Second Course in Statistics*, New York: Addison–Wesley.

Mosteller, F. and D. L. Wallace. 1964. *Inference and Disputed Authorship: The Federalist Papers*, New York: Addison–Wesley.

Mukherjee, S., E. Feigelson, G. Babu, F. Murtagh, C. Fraley, and A. E. Raftery. 1998. "Three types of gamma ray bursts," *Astrophysical Journal*, **508**:314–327.

Murtagh, F. and A. E. Raftery. 1984. "Fitting straight lines to point patterns," *Pattern Recognition*, **17**:479–483.

Nair, V. and G. E. Hinton. 2010. "Rectified linear units improve restricted Bolzmann machines," in *Proceedings of the 27th International Conference on Machine Learning*, p. 807–814.

Nason, G. 1995. "Three–dimensional projection pursuit," *Applied Statistics*, **44**:411–430.

Ng, A. Y., M. I. Jordan, and Y. Weiss. 2002. "On spectral clustering: Analysis and an algorithm," *Advances in Neural Information Processing Systems (NIPS)*, **14**:849–856.

Olbricht, W. 1982. "Modern statistical analysis of ancient sand," MSc Thesis, University of Sheffield, Sheffield, UK.

Ord, J. K. 1967. "Graphical methods for a class of discrete distributions," *Journal of the Royal Statistical Society*, Series A, **130**:232–238.

Pal, N. R. and J. Biswas. 1997. "Cluster validation using graph theoretic concepts," *Pattern Recognition*, **30**:847–857.

Panel on Discriminant Analysis, Classification, and Clustering. 1989. "Discriminant analysis and clustering," *Statistical Science*, **4**:34–69.

Parzen, E. 1962. "On estimation of a probability density function and mode." *Annals of Mathematical Statistics*, **33**:1065–1076.

Parzen, E. 1979. "Nonparametric statistical data modeling," *Journal of the American Statistical Association*, **74**:105–121.

Pearson, K. and A. Lee. 1903. "On the laws of inheritance in man. I. Inheritance of physical characters," *Biometrika*, **2**:357–462.

Pettis, K. W., T. A. Bailey, A. K. Jain, and R. C. Dubes. 1979. "An intrinsic dimensionality estimator from near–neighbor information," *IEEE Transactions on Pattern Analysis and Machine Intelligence*, **1**:25–37.

Porter, M. F. 1980. "An algorithm for suffix stripping," *Program*, **14**:130 – 137.

Posse, C. 1995a. "Projection pursuit exploratory data analysis," *Computational Statistics and Data Analysis*, **29**:669–687.

Posse, C. 1995b. "Tools for two–dimensional exploratory projection pursuit," *Journal of Computational and Graphical Statistics*, **4**:83–100.

Posse, C. 2001. "Hierarchical model–based clustering for large data sets," *Journal of Computational and Graphical Statistics*, **10**:464–486.

Prim, R. C. 1957. "Shortest connection networks and some generalizations," *Bell System Technical Journal*, **36**:1389–1401,

Raginsky, M. and S. Lazebnik. 2006. "Estimation of intrinsic dimensionality using high–rate vector quantization," in *Advances in Neural Information Processing*, **18**:1105–1112.

Rahimi, A. and B. Recht. 2007. "Random features for large–scale kernel machines," in *Advances in Neural Information Processing Systems*, pp. 1177–1184.

Rahimi, A. and B. Recht. 2008. "Uniform approximation of functions with random bases," in *IEEE Communication, Control, and Computing Conference*, pp. 555–561.

Rahimi, A. and B. Recht. 2009. "Weighted sums of random kitchen sinks: Replacing minimization with randomization in learning," in *Advances in Neural Information Processing Systems*, pp. 1313–1320.

Rand, W. M. 1971. "Objective criteria for the evaluation of clustering methods," *Journal of the American Statistical Association*, **66**:846–850.

Rao, C. R. 1993. *Computational Statistics*, The Netherlands: Elsevier Science Publishers.

Redner, A. R. and H. F. Walker. 1984. "Mixture densities, maximum likelihood, and the EM algorithm," *SIAM Review*, **26**:195–239.

Reinsch, C. 1967. "Smoothing by spline functions," *Numerical Mathematics*, **10**: 177–183.

Riedwyl, H. and M. Schüpbach. 1994. "Parquet diagram to plot contingency tables," in F. Faulbaum (editor), *Softstat '93: Advances in Statistical Software*, pp. 293–299, New York: Gustav Fischer.

Ripley, B. D. 1996. *Pattern Recognition and Neural Networks*, Cambridge: Cambridge University Press.

Robbins, N. B. 2013. *Creating More Effective Graphs*, Houston, TX: Chart House Press.

Roeder, K. 1994. "A graphical technique for determining the number of components in a mixture of normals," *Journal of the American Statistical Association*, **89**:487–495.

Ross, S. M. 2014. *Introduction to Probability Models, Eleventh Edition*, New York: Academic Press.

Rousseeuw, P. J. and A. M. Leroy. 1987. *Robust Regression and Outlier Detection*, New York: John Wiley & Sons.

Rousseeuw, P. J. and I. Ruts. 1996. "Algorithm AS 307: Bivariate location depth," *Applied Statistics (JRSS–C)*, **45**:516–526.

Rousseeuw, P. J. and I. Ruts. 1998. "Constructing the bivariate Tukey median," *Statistica Sinica*, **8**:827–839.

Rousseeuw, P. J., I. Ruts, and J. W. Tukey. 1999. "The bagplot: A bivariate boxplot," *The American Statistician*, **53**:382–387.

Roweis, S. T. and L. K. Saul. 2000. "Nonlinear dimensionality reduction by locally linear embedding," *Science*, **290**:2323–2326.

Rumelhart, D. E., G. E. Hinton, and R. J. Williams. 1986. "Learning internal representations by error propagation," in D. E. Rumelhart, J. L. McClelland, and the PDP Research Group (editors), *Parallel Distributed Processing: Explorations in the Microstructure of Cognition, Volume 1*: Foundations, MIT Press.

Rutherford, E. and H. Geiger. 1910. "The probability variations in the distribution of alpha particles," *Philosophical Magazine*, **20**:698–704.

Salton, G., C. Buckley, and M. Smith. 1990. "On the application of syntactic methodologies," *Automatic Text Analysis, Information Processing & Management*, **26**:73–92.

Sammon, J. W. 1969. "A nonlinear mapping for data structure analysis," *IEEE Transactions on Computers*, **C–18**:401–409

Saul, L. K. and S. T. Roweis. 2002. "Think globally, fit locally: Unsupervised learning of nonlinear manifolds," Technical Report MS CIS–02–18, University of Pennsylvania.

Schena, M., D. Shalon, R. W. Davis, and P. O. Brown. 1995. "Quantitative monitoring of gene expression patterns with a complementary DNA microarray," *Science*, **270**:497–470.

Schimek, M. G. (ed.) 2000. *Smoothing and Regression: Approaches, Computation, and Application*, New York: John Wiley & Sons.

Schwarz, G. 1978. "Estimating the dimension of a model," *The Annals of Statistics*, **6**:461–464.

Scott, A. J. and M. J. Symons. 1971. "Clustering methods based on likelihood ratio criteria," *Biometrics*, **27**:387–397.

Scott, D. W. 1979. "On optimal and data–based histograms," *Biometrika*, **66**:605–610.

Scott, D. W. 2015. *Multivariate Density Estimation: Theory, Practice, and Visualization*, Second Edition, New York: John Wiley & Sons.

Sebastani, P., E. Gussoni, I. S. Kohane, and M. F. Ramoni. 2003. "Statistical Challenges in Functional Genomics," *Statistical Science*, **18**:33–70.

Seber, G. A. F. 1984. *Multivariate Observations*, New York: John Wiley & Sons.

Shepard, R. N. 1962a. "The analysis of proximities: Multidimensional scaling with an unknown distance function I," *Psychometrika*, **27**:125–140.

Shepard, R. N. 1962b. "The analysis of proximities: Multidimensional scaling with an unknown distance function II," *Psychometrika*, **27**:219–246.

Shneiderman, B. 1992. "Tree visualization with tree–maps: 2–D space–filling approach," *ACM Transactions on Graphics*, **11**:92–99.

Siedlecki, W., K. Siedlecka, and J. Sklansky. 1988. "An overview of mapping techniques for exploratory pattern analysis," *Pattern Recognition*, **21**:411–429.

Silverman, B. W. 1986. *Density Estimation for Statistics and Data Analysis*, London: Chapman and Hall.

Simonoff, J. S. 1996. *Smoothing Methods in Statistics*, New York: Springer–Verlag.

Sivic, J. 2010. *Visual Geometry Group*
http://www.robots.ox.ac.uk/~vgg/software/

Slomka, M. 1986. "The analysis of a synthetic data set," *Proceedings of the Section on Statistical Graphics*, 113–116.

Smaragdis, P. and J. C. Brown. 2003. "Nonnegative matrix factorization for polyphonic music transcription," in *IEEE Workshop Applications of Signal Processing to Audio and Acoustics*, New York, USA, pp. 177–180.

Smolensky, P. 1986. "Chapter 6: Information processing in dynamical systems: Foundations of Harmony Theory," in D. E. Rumelhart, J. L. McClelland, and the PDP Research Group (editors), *Parallel Distributed Processing: Explorations in the Microstructure of Cognition, Volume 1*: Foundations, MIT Pres.

Sneath, P. H. A. and R. R. Sokal. 1973. *Numerical Taxonomy*, San Francisco: W. H. Freeman.

Snee, R. D. 1974. "Graphical display of two–way contingency tables," *The American Statistician*," **28**:9–12.

Solka, J. and W. L. Martinez. 2004. "Model–based clustering with an adaptive mixtures smart start," in *Proceedings of the Interface*.

Späth, H. 1980. *Cluster Analysis Algorithms for Data Reduction and Classification of Objects*, New York: Halsted Press.

Steyvers, M. 2002. "Multidimensional scaling," in *Encyclopedia of Cognitive Science*.

Stone, C. J. 1977. "Consistent nonparametric regression," *The Annals of Statistics*, **5**:595–645.

Stone, J. V. 2002. "Independent component analysis: An introduction," *TRENDS in Cognitive Sciences*, **6**(2):59–64.

Stone, J. V. 2004. *Independent Component Analysis: A Tutorial Introduction*, Cambridge, MA: The MIT Press.

Strang, G. 1988. *Linear Algebra and its Applications*, Third Edition, San Diego: Harcourt Brace Jovanovich.

Strang, G. 1993. *Introduction to Linear Algebra*, Wellesley, MA: Wellesley–Cambridge Press.

Strang, G. 1999. "The discrete cosine transform," *SIAM Review*, **41**:135–147.

Stuetzle, W. 1987. "Plot windows," *Journal of the American Statistical Association*, **82**:466–475.

Sturges, H. A. 1926. "The choice of a class interval," *Journal of the American Statistical Association*, **21**:65–66.

Swayne, D. F., D. Cook, and A. Buja. 1991. "XGobi: Interactive dynamic graphics in the X window system with a link to S," *ASA Proceedings of the Section on Statistical Graphics*. 1–8.

Tamayho, P., D. Slonim, J. Mesirov, Q. Zhu, S. Kitareewan, E. Dmitrovsky, E. S. Lander, and T. R. Golub. 1999. "Interpreting patterns of gene expression with self–organizing maps: Methods and application to hematopoietic differentiation," *Proceedings of the National Academy of Science*, **96**:2907–2912.

Tenenbaum, J. B., V. de Silva, and J. C. Langford. 2000. "A global geometric framework for nonlinear dimensionality reduction," *Science*, **290**:2319–2323.

Theus, M. and S. Urbanek. 2009. *Interactive Graphics for Data Analysis: Principles and Examples*, Boca Raton: CRC Press.

Tibshirani, R., G. Walther, D. Botstein, and P. Brown. 2001. "Cluster validation by prediction strength," Technical Report, Stanford University.

Tibshirani, R., G. Walther, and T. Hastie. 2001. "Estimating the number of clusters in a data set via the gap statistic," *Journal of the Royal Statistical Society, B*, **63**:411–423.

Tiede, J. J. and M. Pagano. 1979. "The application of robust calibration to radioimmunoassay," *Biometrics*, **35**:567–574.

Timmins, D. A. Y. 1981. "Study of sediment in mesolithic middens on Oronsay," MA Thesis, University of Sheffield, Sheffield, UK.

Titterington, D. B. 1985. "Common structure of smoothing techniques in statistics," *International Statistical Review*, **53**:141–170.

Titterington, D. M., A. F. M. Smith, and U. E. Makov. 1985. *Statistical Analysis of Finite Mixture Distributions*, New York: John Wiley & Sons.

Torgerson, W. S. 1952. "Multidimensional scaling: 1. Theory and method," *Psychometrika*, **17**:401–419.

Toussaint, G. T. 1980. "The relative neighbourhood graph of a finite planar set," *Pattern Recognition*, **12**:261–268.

Trunk, G. 1968. "Statistical estimation of the intrinsic dimensionality of data collections," *Information and Control*, **12**:508–525.

Trunk, G. 1976. "Statistical estimation of the intrinsic dimensionality of data," *IEEE Transactions on Computers*, **25**:165–171.

Tufte, E. 1983. *The Visual Display of Quantitative Information*, Cheshire, CT: Graphics Press.

Tufte, E. 1990. *Envisioning Information*, Cheshire, CT: Graphics Press.

Tufte, E. 1997. *Visual Explanations*, Cheshire, CT: Graphics Press.

Tufte, E. 2006. *Beautiful Evidence*, Cheshire, CT: Graphics Press.

Tukey, J. W. 1973. "Some thoughts on alternagraphic displays," Technical Report 45, Series 2, Department of Statistics, Princeton University.

Tukey, J. W. 1975. "Mathematics and the picturing of data," *Proceedings of the International Congress of Mathematicians*, **2**:523–531.

Tukey, J. W. 1977. *Exploratory Data Analysis*, New York: Addison–Wesley.

Tukey, J. W. 1980. "We need both exploratory and confirmatory," *The American Statistician*, **34**:23–25.

Udina, F. 2005. "Interactive biplot construction," *Journal of Statistical Software*, **13**, `http://www.jstatsoft.org/`.

Ultsch, A. and H. P. Siemon. 1990. "Kohonen's self–organizing feature maps for exploratory data analysis," *Proceedings of the International Neural Network Conference (INNC'90)*, Dordrecht, Netherlands, 305–308.

Unwin, A., M. Theus, and H. Hofmann. 2006. *Graphics of Large Data Sets: Visualizing a Million*, New York: Springer–Verlag.

Vempala, S. S. 2004. *The Random Projection Method*, Series in Discrete Mathematics and Theoretical Computer Science, vol 65, American Mathematical Society.

van der Maaten, L. J. P. 2007. *An Introduction to Dimensionality Reduction Using MATLAB*, Technical Report MICC 07–07, Maastricht University, Maastricht, The Netherlands (`http://homepage.tudelft.nl/19j49/Publications.html`).

van der Maaten, L. J. P. 2014. "Accelerating t–SNE using tree–based algorithms," *The Journal of Machine Learning Research*, **15**:3221–3245.

van der Maaten, L. J. P. and G. E. Hinton. 2008. "Visualizing data using *t*–SNE," *Journal of Machine Learning Research*, **9**:2579–2605.

van der Maaten, L. J. P. and G. E. Hinton. 2012. "Visualizing non–metric similarities in multiple maps," *Machine Learning*, **87**:33–55.

van der Maaten, L. J. P., E. O. Postma, and H. J. van den Herik. 2009. *Dimensionality Reduction: A Comparative Review*, Tilburg University Technical Report, TiCC–TR 2009–005.

Vandervieren, E. and M. Hubert. 2004. "An adjusted boxplot for skewed distributions," *COMPSTAT 2004*, 1933–1940.

Velicer, W. F. and D. N. Jackson. 1990. "Component analysis versus common factor analysis: Some issues on selecting an appropriate procedure (with discussion)," *Journal of Multivariate Behavioral Research*, **25**:1–114.

Venables, W. N. and B. D. Ripley. 1994. *Modern Applied Statistics with S–Plus*, New York: Springer–Verlag.

Verboven, S. and M. Hubert. 2005. "LIBRA: A MATLAB library for robust analysis," *Chemometrics and Intelligent Laboratory Systems*, **75**:128–136.

Verma, D. and M. Meila. 2003. "A comparison of spectral methods," Technical Report UW–CSE–03–05–01, Department of Computer Science and Engineering, University of Washington.

Verveer, P. J. and R. P. W. Duin. 1995. "An evaluation of intrinsic dimensionality estimators," *IEEE Transactions on Pattern Analysis and Machine Intelligence*, **17**:81–86.

Vesanto, J. 1997. "Data mining techniques based on the self–organizing map," Master's Thesis, Helsinki University of Technology.

Vesanto, J. 1999. "SOM–based data visualization methods," *Intelligent Data Analysis*, **3**:111–126.

Vesanto, J. and E. Alhoniemi. 2000. "Clustering of the self–organizing map," *IEEE Transactions on Neural Networks*, **11**:586–6000.

von Luxburg, U. 2007. "A tutorial on spectral clustering," Technical Report Number TR–149, Max Planck Institute for Biological Cybernetics.

Wahba, G. 1990. *Spline Functions for Observational Data*, CBMS–NSF Regional Conference Series, Philadelphia: SIAM.

Wainer, H. 1997. *Visual Revelations: Graphical Tales of Fate and Deception from Napoleon Bonaparte to Ross Perot*, New York: Copernicus/Springer–Verlag.

Wainer, H. 2005. *Graphic Discovery: A Trout in the Milk and other Visual Adventures*, Princeton, NJ: Princeton University Press.

Wall, M. E., P. A. Dyck, and T. S. Brettin. 2001. "SVDMAN – singular value decomposition analysis of microarray data," *Bioinformatics*, **17**:566–568.

Wall, M. E., A. Rechtsteiner, and L. M. Rocha. 2003. "Chapter 5: Singular value decomposition and principal component analysis," in *A Practical Approach to Microarray Data Analysis*, D. P. Berar, W. Dubitsky, M. Granzow, eds., Kluwer: Norwell, MA.

Wand, M. P. and M. C. Jones. 1995. *Kernel Smoothing*, London: Chapman and Hall.

Wang, S. 2015. A Practical Guide to Randomized Matrix computationsl with MATLAB Implementations, **http://arxiv.org/abs/1505.07570**.

Ward, J. H. 1963. "Hierarchical groupings to optimize an objective function," *Journal of the American Statistical Association*, **58**:236–244.

Webb, A. 2002. *Statistical Pattern Recognition, 2nd Edition*, Oxford: Oxford University Press.

Wegman, E. J. 1986. *Hyperdimensional Data Analysis Using Parallel Coordinates*, Technical Report No. 1, George Mason University Center for Computational Statistics.

Wegman, E. 1988. "Computational statistics: A new agenda for statistical theory and practice," *Journal of the Washington Academy of Sciences*, **78**:310–322.

Wegman, E. J. 1990. "Hyperdimensional data analysis using parallel coordinates," *Journal of the American Statistical Association*, **85**:664–675.

Wegman, E. J. 1991. "The grand tour in *k*–dimensions," *Computing Science and Statistics: Proceedings of the 22nd Symposium on the Interface*, 127–136.

Wegman, E. J. 2003. "Visual data mining," *Statistics in Medicine*, **22**:1383–1397.

Wegman, E. J. and D. Carr. 1993. "Statistical graphics and visualization," in *Handbook of Statistics, Vol 9*, C. R. Rao, ed., The Netherlands: Elsevier Science Publishers, 857–958.

Wegman, E. J. and Q. Luo. 1997. "High dimensional clustering using parallel coordinates and the grand tour," *Computing Science and Statistics*, **28**:361–368.

Wegman, E. J. and J. Shen. 1993. "Three–dimensional Andrews plots and the grand tour," *Proceedings of the 25th Symposium on the Interface*, 284–288.

Wegman, E. J. and J. Solka. 2002. "On some mathematics for visualizing high dimensional data," *Sankkya: The Indian Journal of Statistics*, **64**:429–452.

Wegman, E. J. and I. W. Wright. 1983. "Splines in statistics," *Journal of the American Statistical Association*, **78**:351–365.

Wegman, E. J., D. Carr, and Q. Luo. 1993. "Visualizing multivariate data," in *Multivariate Analysis: Future Directions*, C. R. Rao, ed., The Netherlands: Elsevier Science Publishers, 423–466.

Wehrens, R., L. M. C. Buydens, C. Fraley, and A. E. Raftery. 2004. "Model–based clustering for image segmentation and large datasets via sampling," *Journal of Classification*, **21**:231–253.

Weihs, C. 1993. "Multivariate exploratory data analysis and graphics: A tutorial," *Journal of Chemometrics*, **7**:305–340.

Weihs, C. and H. Schmidli. 1990. "OMEGA (Online multivariate exploratory graphical analysis): Routine searching for structure," *Statistical Science*, **5**:175–226.

Weisstein, E. W. 2016."Sigmoid Function," from *MathWorld–A Wolfram Web Resource*. `http://mathworld.wolfram.com/SigmoidFunction.html`.

West, M., C. Blanchette, H. Dressman, E. Huang, S. Ishida, R. Spang, H. Zuzan, J. A. Olson, Jr., J. R. Marks, and J. R. Nevins. 2001. "Predicting the clinical status of human breast cancer by using gene expression profiles," *Proceedings of the National Academy of Science*, **98**:11462–11467.

Wijk, J. J. van and H. van de Wetering. 1999. "Cushion treemaps: Visualization of hierarchical information," in: G. Wills, De. Keim (eds.), *Proceedings IEEE Symposium on Information Visualization (InfoVis'99)*, 73–78.

Wilhelm, A. F. X., E. J. Wegman, and J. Symanzik. 1999. "Visual clustering and classification: The Oronsay particle size data set revisited," *Computational Statistics*, **14**:109–146.

Wilk, M. B. and R. Gnanadesikan. 1968. "Probability plotting methods for the analysis of data," *Biometrika*, **55**:1–17.

Wilkinson, L. 1999. *The Grammar of Graphics*, New York: Springer–Verlag.

Wills, G. J. 1998. "An interactive view for hierarchical clustering," *Proceedings IEEE Symposium on Information Visualization*, 26–31.

Wing, J. K. 1962. "Institutionalism in mental hospitals," *British Journal of Social Clinical Psychology*, **1**:38–51.

Witten, I. H., A. Moffat, and T. C. Bell. 1994. *Managing Gigabytes: Compressing and Indexing Documents and Images*, New York, NY: Van Nostrand Reinhold.

Wolfe, J. H. 1970. "Pattern clustering by multivariate mixture analysis," *Multivariate Behavioral Research*, **5**:329–350.

Wood, S. 2006. *Generalized Additive Models: An Introduction with R*, Boca Raton: CRC Press.

Xu, W., G. Liu, and Y. Gong. 2003. "Document clustering based on non–negative matrix factorization," *Proceedings of SIGIR'03*, 267–273.

Yan, W. and M. S. Kang. 2002. *GGE Biplot Analysis: A Graphical Tool for Breeders, Geneticists, and Agronomists*, Boca Raton: CRC Press.

Yeung, K. Y. and W. L. Ruzzo. 2001."Principal component analysis for clustering gene expression data," *Bioinformatics*, **17**:363–774

Yeung, K. Y., C. Fraley, A. Murua, A. E. Raftery, and W. L. Ruzzo. 2001. "Model–based clustering and data transformation for gene expression data," Technical Report 396, Department of Statistics, University of Washington.

Young, F. W. 1985. "Multidimensional scaling," in *Encyclopedia of Statistical Sciences*, Kotz, S. and Johnson, N. K. (eds.), 649–659.

Young, F. W., and P. Rheingans. 1991. "Visualizing structure in high–dimensional multivariate data," *IBM Journal of Research and Development*, **35**:97–107.

Young, F. W., R. A. Faldowski, and M. M. McFarlane. 1993. "Multivariate statistical visualization," in *Handbook of Statistics, Vol 9*, C. R. Rao, ed., The Netherlands: Elsevier Science Publishers, 959–998.

Young, G. and A. S. Householder. 1938. "Discussion of a set of points in terms of their mutual distances," *Psychometrika*, **3**:19–22.

Young, F. W., P. M. Valero–Mora, and M. Friendly. 2006. *Visual Statistics: Seeing Data with Dynamic Interactive Graphics*, Hoboken, NJ: John Wiley & Sons.

Yueh, W. C. 2005. "Eigenvalues of several tridiagonal matrices," *Applied Mathematics E–Notes*, **5**:66–74.

Zahn, C. T. 1971. "Graph–theoretical methods for detecting and describing gestalt clusters," *IEEE Transactions on Computers*, **100**:68–86.

Zastrow, M. 2016. "Machine outsmarts man in battle of the decade," *New Scientist*, **229**:21.

Author Index

Subject Index

A

acyclic graph, *197*
adjacency matrix, *199*
adjacent values, *377*, *403*
agglomerative model-based
   clustering, *254*
alternagraphics, *420*
AMISE, *370*
Andrews' curves, *439*
Andrews' function, *439*
ANN, *114*
arithmetic operators, *537*
arrays, *529*
artificial neural network, *114*
at-random method, *143*
autoencoders, *127*
average linkage, *173*
axes, *542*

B

bag, *399*
bagplot, *399*
band matrix, *306*
bandwidth, *370*
bar plot, *483*
Batch Map, SOM, *116*
batch training method, SOM, *116*
Bayes decision rule, *266*
Bayes decision theory, *264*
Bayes factors, *258*
Bayes theorem, *265*
Bayesian Information Criterion, *259*
beanplot, *388*, *389*
beeswarm plot, *385*
Bernoulli random variable, *462*
best-matching unit, SOM, *115*

BIC, *258*, *259*
bigram proximity matrix, *9*, *10*
bin smoother, *327*
binary tree, *333*
binomial coefficient, *206*
binomial distribution, *462*
binomialness plot, *469*
biplot, *452*
bisquare, *291*
bivariate boxplot, *400*
blind source separation, *161*
BMU, *115*
Boltzmann machines, *129*
box-and-whisker, *374*
box-percentile plot, *380*
boxplot, variable width, *380*
boxplots, *374*
brushing, *426*
   modes, *426*
   operations, *426*
bubble plot, *545*

C

Calinski-Harabasz index, *224*
canonical basis vector, *141*
capacity, *74*
capacity dimension, *74*
Cartesian coordinates, *437*
case data, *461*
categorical variable, *461*
CCA, *122*
cell array, *531*
cell frequencies, *482*
centroid linkage, *174*
chaining, *173*
Chernoff faces, *410*
chi-square, *483*
chi-square index, *150*
city block metric, *505*
class, *535*
class-conditional probability, *265*
classical scaling, *88*
classification likelihood, *254*

Printed and bound by CPI Group (UK) Ltd, Croydon, CR0 4YY

17/10/2024

01775667-0006